Road User Charging and Electronic Toll Collection

For a listing of recent titles in the *Artech House ITS Series,* turn to the back of this book.

Road User Charging and Electronic Toll Collection

Andrew T. W. Pickford
Philip T. Blythe

ARTECH
HOUSE

BOSTON | LONDON
artechhouse.com

Library of Congress Cataloging-in-Publication Data
Pickford, Andrew T. W.
Road user charging and electronic toll collection / Andrew T. W. Pickford, Philip T. Blythe.
p. cm.—(Artech House ITS series)
ISBN 1-58053-858-4 (alk. paper)
1. Electronics in transportation. 2. Motor vehicles—Automatic location systems. 3. Tolls. I. Blythe, Philip T. II. Title.

TA1235.P53 2006
388.1'14—dc22 2006049867

British Library Cataloguing in Publication Data
Pickford, Andrew T. W.
Road user charging and electronic toll collection.—(Artech House intelligent transportation systems series)
1. Toll roads—Automation 2. Motor vehicles—Automatic location systems
3. Electronics in transportation I. Title II. Blythe, Philip T.
388.1'14

ISBN-10: 1-58053-858-4
ISBN-13: 978-1-58053-858-9

Cover design by Yekaterina Ratner

685 Canton Street
Norwood, MA 02062

10 9 8 7 6 5 4 3 2 1

To the memory of the late Professor Peter Hills, Newcastle University, who provided mentoring and thought leadership in the field of road user charging for more than three decades

Contents

CHAPTER 3

CHAPTER 4

Preface

It was more than 30 years ago that the possibility of using vehicle identification to help automate toll collection was first publicly acknowledged by officials of the Golden Gate Bridge Highway and Transportation District. However, it was not until October 1987 that the commercial use of this innovation, known as electronic toll collection, was first shown to the international press, as part of a small project to connect the small island community of Ålesund to the mainland of Norway. The business objective in this case was not to increase the efficiency of toll collection operations, but instead to enable its commercial viability.

Since that time, traffic in many developed nations has increased by as much as 40%, and capacity has struggled to keep pace. This long-term growth was masking another trend, not in technology, but in policy—the increasing desire to manage traffic demand through charging. The technology that had developed for electronic toll collection was being readied to support policies that sought to transfer the marginal cost of road use directly onto the road users themselves. Traffic in Singapore has been electronically charged to enter the central business district since September 1998, with the aim of benefiting all road users, enabling higher quality public transport, and providing all road users with more consistent journey times. In February 2003, London showed that road pricing could no longer be regarded as a curiosity but as a potentially mainstream traffic management tool for the urban environment. Here, as in many projects, technology was a trusted enabler to meet a policy end.

Delays cost money and time, and reduce economic efficiency. It is perhaps not so well known that the free flow of goods and labor are the essential lifeblood of an economy. However, these benefits are often readily observed, due to cheap energy, cheap cars, convenience, and the political expedience of nurturing each of these with more capacity and more freedom to roam. New routes that provide access to communities to enhance their economic well-being can be commercially operated with sophisticated but proven multilane free-flow technologies. However, new roads can also act as a short-term remedy for an underlying problem—unmanaged demand. Existing routes are either already creaking under the pressure of unrelenting growth in demand, or on a trajectory to the same unacceptable future.

This book is therefore addressed to governments, public authorities, technology developers, system integrators, students, and road users in developed and developing countries. In an era where the value of tangible goods is being overtaken by the value of intellectual property and the growth in services, the underlying rationale for road usage is changing. It is hoped that this text puts the technologies that enable charging for the use of roads into context with the policies that they serve, at a time when far-reaching questions are being answered: the continued relevance of 80-year-old fuel taxation policies, the automatic need to provide new capacity to solve congestion, the need to better use existing capacity, and, indeed, the need for travel itself.

Acknowledgments

There are so many individuals and organizations who have provided input, advice, images, figures, and intellectual capital to help us build up our knowledge of the fields, past, present and future. We wish to thank them, and without whose generosity, this book would not have been possible.

We are pleased to have the support of the following contributors to this book:

- Ian Catling, Ian Catling Consultancy, for his pioneering work in the field, going back to the 1983 Hong Kong Trials, and for specific inputs on both the 1983 and 1998 Hong Kong Trials.
- David Clark, John Givens, Caroline Shield, Mike Burdon, Tony Rourke, and Neil Thorpe, Transport Operations Research Group, Newcastle University, The ADEPT Team, leading the pioneering research through the early 1990s.
- Gino Dompietro and Ken Daley, both of Transurban—an operator that has competently "assembled the pieces" several times, and has provided pragmatic input to Chapter 7 with the same name.
- Chris Fowler, Transport Operations Research Group, Newcastle University, for drawing various images used in Chapters 2 and 9.
- Inger Gustafsson, BMT Transportation Solution, Germany, an ITS pioneer and colleague, for input on Stockholm and European HGV charging schemes.
- Professor Margaret O'Mahoney, Trinity College Dublin, for numerous contributions to the field, and for specific input on the U.S. HOT and HOV lanes.
- Jack Opiola, Booz Allen Hamilton, for discussions of VII (Chapter 9), and the 1998 Hong Kong Trials.
- Duncan Matheson, PA Consulting, and Don Mackinnon, DfT, for helpful input and discussions on the DIRECTS project and the U.K. National Road Charging Plans.
- Professor David Parker, head of the School of CEGS, Newcastle University, for continued support and encouragement throughout this project.
- Eva Schelin, SWECO, Sweden, an ITS pioneer and colleague, for input on Stockholm and European HGV charging plans.
- Arild Skadsheim, colleague, who showed that technology is only an enabler, when he managed a project in Ålesund, Norway, that commenced services in October 1987, which is universally regarded as the world's first commercial application of ETC (Section 8.3.1).

- Nick Patchett of Consulting Stream and Trevor Ellis for insight into many aspects of road user charging, including enforcement and central services.
- Doug Valgren, Norwich Union Insurance, for input on pay-as-you-drive insurance and its relevance to road charging schemes.
- Dr. John Walker, Thales and Artech House Series Editor, for his long-term sustained commitment to getting this book off the ground, and, more recently, his moral support, editing skills, and tolerance.
- Nigel Wall, Shadow Creek Consulting, for numerous discussions and inputs on Galileo, CALM, and infrared communications.
- Bob Williams, Convenor ISO TC204 WG16.1, for generous input and information on CALM and several U.S. toll schemes.
- Patrick Wappenhans, Tecsidel, for his valuable input to vehicle detection and classification techniques (Chapter 5).
- Dr. Miles Yarrington, Andrew Jackson, Christine McDougal, and Gordon Baker, OST-Foresight Intelligent Infrastructure Project Team, for assembling a team that gave us a glimpse of the future, an extract of which is included in Chapter 9.

We also acknowledge many other friends, contributors, and colleagues, including Marit Hammer, Stefan Hoepfel, and Vera Zimerman, without whom this text would not have been possible.

It is also difficult to fully convey our thanks and gratitude to those who have contributed to the art and the science of road user charging, with expertise that includes core technology development, standards, and policy development:

- Jesper Engdahl, RappTrans AG, for his continued contribution and as a valued colleague to the development of road user charging and its underlying standards in Europe.
- Jeremy Evans of Transport for London, and Paul Mellon of Integrate, for helping to push the frontiers of technical and operations knowledge on charging in the urban environment.
- Professor Bengt Henoch, for his technological leadership in remote identification that ultimately led to the technology being used as a mission-critical component in the Ålesund project in 1987.
- Dr. Stephen Ladyman, U.K. Minister of State for Transport, for political leadership in road user charging.
- Ken Livingstone, Mayor of London, for having the courage of his convictions to introduce the London Congestion Charging Scheme, and essentially letting the road pricing genie out of the bottle.
- Jenny Martin, Secretary General, ITS (United Kingdom), for continuing to allow us to stir up debate on road user charging in the United Kingdom.
- Gopinath Menon, formerly of the Land Transport Authority, for leading the way with the Singapore ALS and the Singapore Electronic Road Pricing schemes.
- Per Risberg, formerly of Saab Combitech AB, for encouraging innovation and entrepreneurism in creating new supply chain business models in road

user charging, and Lars Olsson, Jan Svedevall, Göran Andersson, Geir Engelsen, and Anders Dahlbëck of Kapsch, for providing a supply side view of this topic.

- Professor Eric Sampson, U.K. DfT head of the Vehicle Standards and Technology Division, for leadership in the road user charging domain, and for being so generous with his time in his capacity as a visiting professor at Newcastle University.
- All the colleagues from government, industry, and academia, who have kindly taken time to peer-review and comment on earlier drafts of the chapters of this book.
- Last, but no means least, our partners, Jane and Fiona (and the kids, Hope and Leo), for their support of this venture.

CHAPTER 1

Introduction to Road User Charging

1.1 Introduction

Charging for the use of a road has become a significant political issue, and, if not already being implemented, it is on the agenda of many governments, city authorities, and road operators across the world. There are essentially two reasons why road operators and city authorities would consider introducing a charge for the use of roads: to manage congestion or to finance the infrastructure. It is worth pointing out at this early stage that these two, often competing, objectives of charging for road use are quite distinct.

Tolling or *toll collection* are terms attributed to the collection of a road use fee on certain roads, bridges, or tunnels, where the toll is levied to recover all or part of the capital, operating, and maintenance costs for that infrastructure.

Road user charging, also known as *road use pricing* or *congestion charging*, is the levying of some fee or charge for road use that aims to use "price" as a means of influencing a proportion of the road users to change their driving and/or travel behavior to manage the demand for the use of the road space to within some predetermined limits.

The two objectives are quite separate, in the sense that toll facility operators wish to meet financial targets and recover their costs by setting a fee-level that will not discourage too many drivers from selecting an alternative route so that the necessary revenue is raised. On the other hand, in a road user charging scheme used for demand management, the objective is to set a fee-level that will encourage a proportion of users not to travel in a vehicle on those roads at a particular time of day, to relieve congestion, and to mitigate environmental or other negative impacts of road use and congestion.

The primary aim of the book is to examine electronic tolling and road user charging technologies, which are also known as *e-tolling, electronic fee collection* (EFC), *electronic toll collection* (ETC), *road user charging* (RUC), and by a range of other names and acronyms. One cannot look at the underlying technologies in isolation from the political, financial, and operational objectives of the scheme. These objectives and the interrelationship between policy and technology have led to a wide range of names and acronyms being coined to describe the charging for road use. The following terms are commonly used by the industry, in the literature, and in the chapters of this book:

- Tolling or toll collection (generic terms);
- Road pricing (recently in the United Kingdom by DfT);

- Electronic road pricing (ERP, in Singapore);
- Electronic fee collection (EFC, by the European Commission);
- Automatic debiting system (ADS, by the European Commission);
- Road user charging (in the United Kingdom by DfT);
- Road use charging (in the United States and Europe);
- Road pricing (the economists' term);
- Open road tolling (in the United States, especially recently);
- Value pricing (in the United States);
- Congestion charging (in London, United Kingdom).

The two objectives of tolling and road user charging are distinct, yet are beginning to blur. The management of demand on toll roads is rapidly becoming an issue, due to the continued increase in car ownership and use, particularly the 5% to 10% annual growth rate in car ownership that many of the developing economies, most noticeably in Asia, are experiencing. The distinctions between objectives of the fee collection schemes are, in some cases, beginning to converge, as will be discussed in Chapter 8 on case studies. The use of new information and communications have a role to play, as part of managing the demand for roadspace, be it a congested urban area, arterial, or toll road; and as part of the requirement to collect road charges with a minimum of fuss and delay, but with a high level of reliability and accuracy. This book will explore how these new technologies, in the operational and political frameworks and practical constraints of a particular system, can ensure that the vehicle owners or drivers who pay for the use of the road network, can do so in a convenient, reliable, and efficient manner that does not require the vehicle to slow down or stop, nor require the driver to perform any action other than normal driving where the charge is levied. Some form of automated electronic charging system is desirable to achieve this goal. The development, evolution, operation, and relative functionality of these charging systems, along with the multifaceted, diverse, and wide-ranging context in which they are used, are the topics of this book. The charging technology must also take into account the vast differences in the road environment where charges are to be levied, from the relative calm and order of a toll plaza, to congested urban roads, with different mixes of vehicles. Figure 1.1 illustrates urban road congestion from India, as a reminder of this reality.

1.2 Scope of This Book

This book aims to bring together a wealth of knowledge regarding the technical options, technologies, and systems for road user charging systems, whether these systems are used for conventional toll collection purposes, or for wider demand management measures, such as road use pricing and congestion pricing. Case studies and examples of schemes will be cited and described where appropriate. Some technical solutions for road user charging are too sophisticated and over-designed for the purpose, due to their experimental nature, or the evolution of the technology from some earlier version of the system, or specific operational

Figure 1.1 The challenge of charging for road use in a congested urban environment. (Courtesy of Findlay Kember.)

constraints and requirements for a particular scheme. Later chapters of the book will also attempt to put some of these design decisions into an historical context, to explain why a particular class of system or technology was adopted and has evolved into the present-day systems. Some of the "heroic failures" of road user charging systems will also be cited to give a view on the dead ends and unacceptable (by either the end user or operating or regulatory authorities) approaches to charging that have been experimented with and demonstrated over the years.

The book is divided into nine chapters, this being the first.

Chapter 2 defines charging, outlines the evolution of tolling and road user charging for demand management, describes the processes necessary to achieve a tolling or road use transaction, and suggests enforcement action for any vehicle passages that were unable to be associated with the correct charge for the use of the road. The chapter also provides some historical context for how such systems and schemes have evolved and the options that now exist for tolling and road user charging. It also illustrates the differences between tolling and road user charging implemented for demand management and congestion-mitigation purposes.

Chapters 3 to 6 consider in detail the main elements of electronic charging systems, namely:

- The feasible charging technologies and their technical requirements for operation in the different charging environments, from dedicated monolane toll collection facilities to free-flow multilane environments;
- The associated enforcement systems and processes;
- The options for the vehicle detection and classification;

- Alternative architectures for the so-called central services, which provide a myriad of online and off-line functions for the operation of any electronic tolling or road user charging system, whether it is an isolated scheme or a wider area scheme between different service providers and road operators.

Chapter 3 focuses on the options and classes of systems used for the electronic collection of tolls and road user charges. The pros and cons of different design solutions using different technologies are presented and discussed. These solutions include: (1) on-board units (OBUs) with dedicated short-range communications (DSRC) and systems that utilize virtual charging based upon the vehicles' measured position using Global Navigation Satellite Systems (GNSS); and (2) electronic systems that use the automatic reading of a vehicle license plate as the primary means of levying a charge (or as a secondary means for occasional and nonequipped users, and for users attempting to evade payment).

Chapter 4 deals with the options available for the enforcement of charging schemes, which requires the recording of the passage of a vehicle for evidential purposes for the enforcement of noncompliant drivers. This generally requires recording the license plate of the vehicle, since the plate represents an internationally recognized, independent, and unique identification mark for vehicles. For enforcement, however, it is critical that the processes used are credible and do not undermine the public's confidence in the system with delivery of false evidence or the misidentification of noncompliant users. The chapter discusses the issues of the handling of the evidence, the evidence basis itself, and the sensitive issues of privacy and data protection.

Chapter 5 considers the technical options for detecting a vehicle's presence independent of the charging equipment to mitigate situations where the vehicle's equipment may not be working, or the vehicle may not be carrying equipment at all. The equipment must also automatically classify the vehicle in some way that relates to the tariff of charges and thus confirm the classes of the charges to be levied. The classification parameters and technology are quite diverse, and vary considerably from country to country and scheme to scheme, depending on the parameters to be measured and the operating environment (e.g., monolane toll collection, or dedicated multilane free-flow operations).

Chapter 6 considers the central services that are essential to any electronic tolling or road-use pricing scheme. This includes the need for customer interaction, registration, billing, clearing, issuing, and other customer-focused and financial/auditing functions of the collection scheme. The central services must also be regarded as the crucial element in the delivery of the desired level of interoperability between schemes. This may be between toll schemes operated by different service providers within a state (as is the case in the United States), and cross-border interoperability, which, for example, is the goal of the Association Européenne des Concessionnaires d'Autoroutes et d'Ouvrages à Péage (ASECAP) countries in Southern, Eastern, and Central Europe [1]. The challenge of integrating the operations of citywide, regional, and national road user charging schemes, where users may have some form of payment account and sharing of central system functions to leverage the benefits of economies of scale, places significant technical and operational demands on any central system.

Chapter 7 attempts to bring together the knowledge gained from the previous chapters. The chapter discusses options available to a scheme designer, with the view to making design and commercial decisions on implementation of a road user charging scheme for a hypothetical interurban road network with an extension into an urban environment, using a demand management element for some of the tolls. The processes presented give a comprehensive insight into the complexity and multifaceted choices that must be considered and weighed in the design and implementation of a scheme.

Chapter 8 provides case studies of significant tolling and road user charging implementations and pilot schemes. The case studies provide a wide range of systems that meet differing policy objectives, such as the automation of toll lanes, area-wide congestion management schemes, heavy goods vehicle charging schemes, and large-scale interoperable and regional toll road networks. Descriptions are included of some of the pilots and demonstration activities that have provided milestones in either the implementation of charging schemes, or in the experimentation with new technologies that may have future significance or have otherwise led to iconic breakthroughs and new knowledge in the field.

Chapter 9 considers what impact new and emerging technologies may have on future charging scheme designs. The chapter also provides some thoughts on the future evolution of transportation, and the increasing role of road user charging. It is always risky to predict the future of road user charging and tolling; as the past 20 years have shown, the technologies and the way we introduce and operate road user charging schemes have rapidly evolved. Could we really have envisaged the improvements in information and communications technologies (ICT), such as miniaturization and cost reduction, that we have today? Indeed, 20 years ago did most of us even know about the Internet, or expect that 85% of the population would possess technically advanced telephones that communicate in a mobile environment, access computer services, take photographs and videos, and enable us to watch television from a handset the size of a TV remote control? If all this is possible and we still are in the bounds of Moore's Law [2], then how will tolling and road user charging technology evolve over the next 20 years? We can expect to see more sophisticated forms of road user charging, due to potential technology developments, and wider public acceptance of road user charging, with the recognition that "something needs to be done about congestion" and particularly its effect on the environment. The success of the Singapore ERP scheme [3], and the overwhelming accomplishment of Transport for London (TfL) in delivering an effective and largely accepted road user charging scheme in London [4], suggest that innovations in charging will continue to arrive. The fact that TfL increased the charge to enter the congestion charging area from £5 ($8.50) to £8 ($13) within 2 years of introducing the scheme suggests that continued innovations in the technology and fine-tuning of the pricing regime may be necessary to maintain and develop the necessary demand-restraint targets, as the public becomes familiar with the scheme and accepts the choice to pay the charge. Finally, the chapter looks into the future and examines the role road user charging may play in the future of transport, considering future challenges of finance, new intelligent transport systems (ITS) technologies, resource availability, energy, environmental considerations, and the effect of climate change on future transport networks.

1.3 Brief Overview of Road Charging Developments

1.3.1 The Social and Economic Rationale for Charging

The economic case for replacing the current fixed vehicle excise duty (VED) by a variable tax that is related to the use made of the road system was established in 1964 by the Smeed Report [5], with the economics being developed further by Vickrey [6] and Walters [7]. The reports suggest that the variable tax should be adjusted according to traffic levels prevailing at different times and at different nodes within the road system. The Smeed Report was essentially concerned with the economic and technical feasibility of road-use pricing, but it also warned of some of the social and political objections to such a strategy. These problems center on the equity of the various methods of taxation, the perceived threat to individuals' freedom, and the potential for fraud and evasion.

The reason why variable pricing tends to be more equitable than fixed taxation relates to the principle of vehicle users' responsibility for the costs that arise from their use of the roads. These include the private costs of buying and operating a vehicle (in this sense, the fuel tax is a form of variable pricing, where annual license fees/registration charges or car tax is not), the public costs of providing and maintaining the infrastructure, and the social costs of accidents, congestion, and pollution. These costs arise even under free-flow conditions and arguably are already covered, on average, by the existing fixed taxes. The element of cost that arises with congestion (disproportionately, as congestion increases) is the delay to all other road users and the negative effects of additional air pollutants, noise emissions, and other harmful effects. In an equitable economic system, each vehicle user should pay a "rent" equal to the marginal social costs of his or her road use. In this sense, the fuel tax is inadequate as a form of variable pricing, since delay costs on society accumulate far more rapidly with congestion than does overall fuel consumption [1]. The Transportation Research Board (TRB) is now questioning, in a report on the alternatives to the fuel tax, the future of the fuel tax as a dependable source of revenue in the United States. The TRB's report concludes that fuel taxes can remain the primary funding source for the nation's highways for at least another decade, but that replacing this tax with a system for metering road use and charging accordingly could benefit travelers and the public. The report also suggests that, while the current funding system maintains existing highways, builds new ones, and ensures that users pay most of these costs, it does not help transportation agencies to alleviate congestion or target investment in the most valuable projects [8].

1.3.2 Current Examples of Toll Facilities

The introduction of a fixed toll for the use of a road, crossing a bridge, or entering a charged area appears to meet the requirements of a road-pricing system. However, the fixed toll does not offer the desired flexibility to alter the charge based on prevailing traffic conditions or other relevant parameters. Before looking at the requirements and technological options that are now emerging for road pricing, we will examine what is currently the only means of pricing for road use—the simple road toll.

Tolling has been a means of raising revenue from travelers for thousands of years. Most ancient civilizations with written records mention tolls in one form or another, and turnpikes and toll houses were common features of the seventeenth century in the present-day United Kingdom. The first large-scale modern toll road networks were established in France and Italy in the early 1950s on arterial highways, on some interstate highways (turnpikes) in the United States, and for tunnels and bridges across river estuaries and other natural geographical obstacles in many other countries. These toll revenues have been used to maintain the quality of the highways and to repay the financial cost of constructing and operating these new facilities. Revenue from tunnels and bridges is generally able to repay only part of the capital costs of the construction. The revenue must first finance the necessary maintenance and operational costs of the facilities. These toll facilities are generally large and require a significant number of toll lanes to efficiently process the traffic passing through the plaza. Figure 1.2 is a photograph of the toll collection facility at the Dartford Thurrock Crossing in England, with a mixture of manual, coin machine, and automatic vehicle identification (AVI) collection lanes, with AVI being the passing of an identification code from an in-vehicle tag to the roadside system by some electronic means.

Until the 1980s road-revenue collection had been almost exclusively a manual operation, which is a slow and laborious process, and can be relatively expensive on a per-transaction basis. Toll facilities have benefited in recent years from the introduction of automatic coin-validating machines and magnetic cards whose credit units are deducted by a reading device located in the toll lane.

Figure 1.2 Toll collection facility at the Dartford Thurrock Crossing, England.

Although these machines have been introduced, the underlying requirement of the toll systems in countries where such installations are widespread, such as the United States, France, and Italy, is that the driver must stop and pay. Until recently, the lanes at the toll plazas in Bergen, Norway, have been divided into "stop-and-pay" lanes, and "nonstop" lanes for exempt vehicles and those possessing a pass that is prominently displayed in the windshield. The vehicles in the nonstop lane are under surveillance both by booth attendants and by surveillance cameras. Even so, the detection of noncomplying vehicles is by no means foolproof.

The majority of existing road toll facilities use a means of revenue collection that requires drivers to stop their cars, and to find the correct coins or a valid card, before the barrier is opened or a green light shown. The toll charge levied rarely takes into account the type of trip, time of day, prevailing traffic conditions, and other relevant factors. Usually, only the type of vehicle is differentiated. A versatile tolling system that does not require the vehicle to stop, and that may vary the charges according to any of these factors, requires a rapid means of communicating data between the vehicle and the roadside infrastructure.

1.3.3 New Technology Applied to Road-Revenue Collection

The drawbacks of conventional toll collection methods will be accentuated as the use of tolls becomes more widespread. The disruption of traffic flow from the need to stop at toll sites will become acute as the predicted increases in road traffic materialize. Traffic demand in the European Union (EU) has risen by 40% in the past decade, while road capacity has increased by only 5%. It is generally accepted that three conventional toll lanes are necessary to process each lane of highway traffic. A nonstop toll collection system would increase the vehicle processing capability of a single toll lane by a factor of three. The reduced congestion at the site would shorten travel times and reduce harmful impacts, such as localized environmental emissions and unacceptable levels of noise. Financial benefits for the road operator will include a reduction in labor costs and a reduction in the physical area needed for each toll site. The potential for debt, fraud, and evasion should be substantially reduced, since less cash is handled.

The Port Authority of New York and New Jersey in the late 1970s performed the first notable experiments in nonstop tolling [9]. The use of an automatic vehicle identification (AVI) system permitted toll payments to be charged to the user's credit account by the toll company. The first large-scale demonstration scheme of AVI was the Hong Kong electronic road-pricing (ERP) experiment in 1983, which used inductive loop technology to facilitate communications between a vehicle with a transponder on the underside of the vehicle and a roadside automatic tolling station. The technology was shown to be highly successful, but the ERP system was severely limited in its scope, due to the low rate of data transfer that could be achieved with inductive loop (in-ground) communications [10]. The low rate of data transfer, the size of the vehicle transponder, and the cost of installation and maintenance of the buried loops make the use of inductive loop communications unattractive for future AVI and tolling applications. However, the trial did show future possibilities in the use of vehicle-to-roadside communications technologies; more details on the trial are provided in Chapter 8.

Despite the growing number of experimental systems, the world's first automatic tolling system was installed for commercial use at the Ålesund tunnel in Norway, in October 1987 [11]. The Programmable REMote IDentification (PREMID) system had originally been developed for industrial automation applications, tracking products around a production line and recording the processes that had been completed. The system used a tag about the size of a cigarette pack, mounted in the side window of each vehicle. This tag contained coded information relating to the identity of the vehicle, and, when passing the toll site, the tag reflected the incident microwave signal from the roadside interrogator [12]. The successful operation of this installation demonstrated how charging technologies could improve the business case for tolling. These AVI systems that were introduced almost two decades ago were only capable of conveying a limited amount of information between the vehicle and the roadside computer at the toll site, and only at slow vehicle speeds (less than 30 km/hr). It was already clear that if any large-scale road-use pricing scheme were to be successfully introduced, it would need a more advanced automatic system for revenue collection. This was the challenge set by the operators and road owners, whether for nonstop tolling purposes, or for demand-management charging applications. We have seen these early AVI systems superseded in the past decade by more intelligent on-board unit (OBU) designs, which include the ability to process data held by the on-board unit and deliver an array of secure charging and transaction services in both monolane and multilane free-flow operations [13, 14]. These systems utilize a range of technologies, including short-range radio, microwave and infrared communications, cellular phone, GNSS, and video technologies. The operation of systems in Singapore for electronic road user charging, in Melbourne [15], and Highway 407 [16] for free-flow tolling has proven that these new systems have sufficient functionality, robustness, and accuracy to meet the requirements of high-speed operation, reliability, and scalability.

The following chapters introduce the broad aspects of charging for road use. The chapters discuss the charging options, the impact of policies on the technological solutions, and the new generations of systems. These architectures are sufficiently advanced and of a modular design, and are able to meet the requirements and challenges of any modern road revenue collection system or future road-pricing scheme.

References

[1] Hills, P. J., and P. T. Blythe, "Road-Pricing Solving the Technical Issues," *Journal of Economic Affairs*, Vol. 10, No. 5, June/July 1990, pp. 8–10.

[2] Moore, G. E., "Cramming More Components onto Integrated Circuits," *Electronics*, Vol. 38, No. 8, April 1965, ftp://download.intel.com/museum/Moores_Law/Articles-Press_Releases/Gordon_Moore_1965_Article.pdf.

[3] Olszwski, P., and L. Xie, "Modelling the Effects of Road Pricing on Traffic in Singapore," *Transportation Research Part A: Policy and Practice*, Vol. 39, No. 7–9, August–November 2005, pp. 755–772.

[4] Evans, J., "The London Congestion Charging Scheme," *Proc. IEE Seminar on Road User Charging Technologies*, London, U.K., December 2005.

[5] Smeed Committee Report, "Road Pricing: The Economic and Technical Possibilities," Ministry of Transport, HMSO, London, U.K., 1964.

[6] Vickrey, W. S., "Congestion Theory and Transport Investment," *American Economic Review*, Vol. 59 (Papers and Proceedings), 1969, pp. 251–260.

[7] Walters, A. A., "The Theory and Measurement of Private and Social Cost of Highway Congestion," *Econometrica*, Vol. 29, No. 4, 1961, pp. 676–697.

[8] TRB, "The Fuel Tax and Alternatives for Transportation Funding," *The Transportation Research Board*, Special Report 285, Washington, D.C., January 2006.

[9] Foote, R. S., "Prospects for Non-Stop Toll Collection Using Automatic Vehicle Identification," *Traffic Quarterly*, Vol. 35, 1981.

[10] Dawson, J. A. L., and I. Catling, "Electronic Road Pricing in Hong Kong," *Transportation Research A*, Vol. 20A, 1986, pp. 129–134.

[11] Waersted, K., and K. Bogen, "No Stop Electronic Toll Payment Systems," *Proc. 2nd Intl. Conference on Road Traffic Monitoring*, London, U.K.: Computing and Control Division of the Institution of Electrical Engineers, February 7–9, 1989, pp. 128–132.

[12] Hills, P. J., and P. T. Blythe, "The Automation of Toll Collection and Road Use Pricing Systems," *Proc. 2nd Intl. Conference on Road Traffic Monitoring*, London, U.K.: Computing and Control Division of the Institution of Electrical Engineers, February 7–9, 1989, pp. 118–127.

[13] Stoelhorst, H. J., and A. J. Zandbergen, "The Development of a Road-Pricing System in the Netherlands," *Traffic Engineering and Control*, Vol. 31, No. 2, February 1990, pp. 66–71.

[14] Guerout, F., "VITA: Vehicle Information and Transaction Aid," Reference Document, ASECAP and the European Commission, March 1990.

[15] Olsson, L. J., "The Melbourne City Link Multilane Toll Collection System," *Proc. IBC Conference, Electronic Payment Systems in Transport*, London, U.K., March 1998.

[16] Horton, J., "Overview of the Highway 407 ETCS," *5th ITS World Congress*, Seoul, Korea, 1998.

CHAPTER 2

Road User Charging and Toll Collection

2.1 Historical Context

Charging for road use is by no means a new concept. Toll roads can be traced back to at least Roman times, when travelers paid a fee for using a road/track maintained (and in many cases protected) by the authorities of the day. Across the world today toll roads make up a significant proportion of the arterial road networks, and in many countries the tolling of estuarial crossings is commonplace. Tolling is essentially the recovery of a fee from users of a facility to cover the capital building, operation, and maintenance costs of the road [1]. In many cases the responsibility for toll roads have been given over to private operators to design, build, finance, and operate (DBFO), or to operate as a concession for a particular period of time [2]. Other schemes may have a more demand management–led set of objectives, such as managing travel demand and the consequential congestion when demand (for travel by car) outstrips the supply (of roadspace) [3, 4].

A variety of electronic technologies in the 1970s and in the mid- to late 1980s [5] were developed and tested with the aim of speeding up the collection of tolls. Subsequently, microwave tags and radio frequency identification (RFID) devices were developed, so that queuing at manual tollbooths could be reduced or completely eradicated, allowing drivers to pass through toll plaza facilities without stopping, their transactions being made automatically, using appropriate road charging equipment, across the roadside-to-vehicle communications link [6–11].

The first commercial use of e-tolling technology was in 1987 when the Ålesund Tunnel in Norway was equipped with a simple identification (ID) tag using microwave technology. A profusion of similar tag-based schemes was introduced in the United States, Southern Europe, and Japan over the next few years [12]. Chapter 8 describes a number of these schemes in more detail. These schemes were largely limited to single-lane, drive-though tolling arrangements, since the technology could not yet meet the challenge of free-flow, multilane charging that would be required for urban road user charging without the need to build conventional toll plaza infrastructure [13]. The introduction of new technology allowed Trondheim in Norway [14–16] in the mid-1990s, and Singapore [17–21] in the late 1990s, to introduce electronic toll charging rings that were used for revenue raising, and had the ability to influence travel demand and reduce peak-hour congestion. However, these technologies were quite limited in what they could deliver.

In the United Kingdom, innovative road pricing trials were undertaken in Cambridge from 1992 to 1994, which used a set of microwave beacons, delivered by the Automatic Debiting and Electronic Payment for Transport (ADEPT) project.

Microwave beacons were placed in a cordon around the city to trigger a congestion meter in the vehicle, which then charged users based upon either the distance their vehicle traveled within the cordon or on the level of congestion measured by the in-vehicle meter, which had a sensor connected to the vehicle's odometer [22–24]. It took another 10 years to see further developments in innovative road pricing in the United Kingdom: first, with the launch of the Durham access control system in October 2002 [25], and second, with the launch of the London Congestion Charging Scheme in February 2003 [26–28]. The success of these schemes, and the potential for developing significant "intelligence" in the transport infrastructure and within vehicles themselves, encouraged the U.K. government to consider the introduction of a national distance-based road pricing system in the future. It is expected that future developments will enable innovative forms of road pricing that could have a significant demand-restraining effect, providing an additional tool to deal with traffic congestion [29]. The two currently preferred charging technologies are DSRC (microwave in-vehicle tags communicating with roadside antennas), and satellite-based location systems that locate the position of the vehicle on an on-board digital map (the vehicle is then appropriately charged, based upon cordon-, point-, or distance-based charging). Mobile wireless networks, RFID, mobile phone technology, or camera-based automatic number plate recognition (ANPR) solutions may also offer options that are appropriate to support future nationwide road pricing solutions [30].

This chapter will attempt to provide an overview of tolling and road user charging technologies: how they have evolved, what they can do, what we can learn from the developments and schemes of the past, and where the future will take this important tool for the traffic management and ITS business sector.

2.2 Charging for Road Use

2.2.1 Context

The trend in transport policy in many parts of the world, particularly in Europe and the developing economies, is increasingly towards the recovery of construction, operation, and maintenance costs of new roads by the use of tolls or road use charges. These charges have been also extended to the existing "free" road stock.[1] There has been a reemergence on the political agenda of many governments and city authorities for some form of road use pricing to address the management of traffic demand [31].

It is desirable to introduce an efficient charging mechanism that is able to automatically levy the tolls and road use charges from the drivers, that is, without the need for the drivers to perform any action, other than those associated with normal driving activities. The system should also enable the collection of these charges at normal highway speeds outside of the specific toll plaza environment, and without the need for the physical separation of lanes, as is the constraining requirement with conventional toll collection facilities.

1. As we know, nothing in life is for free. By free road stock, one means roads that are not directly charged for at the point of use through tolling or road user charges, but rather financed through general taxation, vehicle and fuel tax, shadow tolling, or other economic mechanisms.

It is infeasible and unworkable, in many locations, to implement manual means of fee collection, in which traffic must be segregated into lanes to allow drivers to stop their vehicles and pay a fee, either manually to an operator, or by inserting coins, cash, or a card into a collecting machine. Manual collection would require the building of plazas (such as across North America, in Europe, and increasingly in Asia), which are costly both to build and operate, and require a substantial land area. Such manual collection plazas may only be built when a new road is planned and sufficient land is purchased. It is generally not practical to retrofit a toll plaza to an existing road. This is especially true in urban areas, due to restrictions on land use; the likely creation of additional congestion due to queueing at toll lanes; the increase in noise and air pollution; and the inflexibility of the charging system that could be employed. Newly designed toll roads generally have a limited number of entry and exit points, while existing "free roads" usually are not so restricted, which creates an additional difficulty when introducing urban road charging. It is now mainstream for traffic management theory to consider the potential for introducing some form of road use fee that directly relates to the amount of use of the road. The introduction of these charges may have a restraining effect on the traffic demand, as well as having the obvious attraction of raising relatively large amounts of capital that may be put back into improving the transport infrastructure, supporting public transport, and generally offering alternatives to travel by private car. In the United Kingdom, this policy was enshrined in the Transport Act 2000, which specifically requires local authorities that implement local road user charging or private nonresidential (PNR) parking schemes to reinvest any revenue raised in local transport schemes. It is, however, likely that any national charging scheme will be a tax rather than a locally hypothecated charge. The use of conventional stop-and-pay plazas is unattractive to implement such a policy of efficiently charging motorists; thus, some form of nonstop automatic charging of road users must be considered.

First, we need to clarify what we mean by tolling and road user charging.

2.2.1.1 Toll Collection

The collection of a toll for the use of road infrastructure is the most common form of pay-as-you-drive fees. A private concessionaire or a government agency levies a fee to recover the costs of the building, operating, and maintenance of the infrastructure. This became a significant instrument for road building after World War II in Southern Europe, the United States, Japan, and Southeast Asia [32–34]. Other countries, including the United Kingdom, Australia, and the Scandinavian countries, until recently had limited the use of tolls to estuarial crossings and other major bridge and tunnel infrastructures. This division has now been blurred, as we shall see later.

Motorway schemes using electronic devices to automate existing toll collection facilities are quite widespread and include numerous examples in the United States, in the ASECAP countries in Europe, in new multilane tolling schemes on Toronto's Highway 407, and in the Melbourne City Link [35].

2.2.1.2 Road User Charging

The concept of direct road user charging is not new. Road user charging has been considered for many decades as a tool for managing congestion and raising revenue, although few trials and implementations have actually taken place, until the recent success of the Singapore and London schemes, among others. Pigou (the father of welfare economics) first proposed the economic theory on which the principle of road use pricing is based in 1920 [36]. Vickrey [37] and Walters [38] further developed the theory, relating it specifically to road traffic. The Smeed Report [39] in 1964 first officially acknowledged the technical possibilities of direct pricing at the point of use. A great deal of research has been subsequently undertaken, and a number of attempts to introduce urban road user charging have been made, most notably the Hong Kong trials (1983–1985 and 1998) [40–42]; the Singapore Area Licensing Scheme (ALS) (1975–1998), which is now replaced by an automatic electronic scheme; and the toll rings around Bergen, Trondheim, and Oslo (however, these latter three schemes in Norway [43–45] are primarily revenue-raising schemes).

The difference between road user charging and tolling is that the fee is calculated to meet some demand management objective, rather than just recovering a fee for using the infrastructure. In this sense, road operators attempt to internalize some of the external costs associated with transport, including those related to congestion, delay, and environmental impact.

2.2.2 Early Operating Models

Nonautomatic and nonelectronic forms of fee collection have been used at toll facilities since their inception. It is worth reviewing the manual forms of collection that are implemented [46] before proposing automated fee collection systems.

Manual collection methods vary in many ways, depending upon the characteristics of the road. However, the overriding requirement for manual collection is that the vehicle driver must stop the car, open a car window (or door), and either hand over cash or a card, or insert either of these into a machine. These plazas are common across Europe for the collection of road tolls. No actual road pricing scheme employs such methods, although arguably the Oslo and Trondheim toll rings in Norway could be regarded as road pricing installations [47].

Manual toll collection usually requires the building of a toll plaza that divides the free-flowing multilane road into a number of single lanes. Each lane is serviced by a tollbooth, which either houses an operator who manually collects toll payments, or has the equipment (e.g., card reader or coin-accepting basket) that the driver must use to pay the toll. The general rule for the design of toll plazas is that there should be at least three tollbooths to service each one lane of traffic leading into the toll plaza. A four-lane road will typically require 12 tollbooths to efficiently service the traffic [1]. This is clearly a nonviable option for road use pricing in urban areas, due to the size of the toll plaza required and the high volumes of traffic that could be expected in morning and evening peaks. Figure 2.1 shows a four-lane toll plaza servicing a two-lane low-flow road in Normandy, France. The number of service lanes in a toll plaza may be reduced on roads with low flows. However, it is necessary to compare the benefits of reducing the number of toll lanes (thus the

Figure 2.1 Typical toll plaza layout. (Courtesy of Blythe/CSEE.)

land required and the number of operators employed) against the costs associated with queuing traffic and their noise and air pollution. The physical security of storing and moving a large amount of coins and paper money can also cause some logistical problems. At the Mersey Tunnels in Liverpool, United Kingdom, in the mid-1980s, approximately 15% of revenue was stolen systematically by operators. This problem was addressed and solved by the tunnel management, once it was detected; however, it is cited here to illustrate some of the issues that may occur when cash is handled. Approximately one-half ton of coins was being moved daily, which was a time-consuming and costly process [48].

The enforcement of manual toll systems generally relies on the use of a barrier that is not opened until confirmation by the operator or the collecting machine that the correct toll has been paid. These systems are often augmented by vehicle detectors, to count the vehicles passing through the lane, and by some form of vehicle classification, to distinguish different classes of vehicles that pay different tolls; obviously operators can classify vehicles manually, provided the definition of classes is not too complex [49]. Classification is usually based upon axle counters and/or vehicle height-measuring equipment. A video camera may be employed when a barrier is not used. However, this practice is not very common, due to the extra cost with little benefit over the barrier, since the vehicles are expected to stop anyway. Another option is to use a flashing light and alarm on the tollbooth, which attracts the attention of supervisory staff and enforcement vehicles at the toll plaza when a vehicle has violated the system. This approach is used extensively on the U.S. Turnpike network. This is not a particularly workable deterrent for congestion pricing, where very high vehicle flows can be expected and lanes are not normally segregated. Thus, it would be difficult to identify the offending vehicle without sophisticated enforcement systems [50]. Figure 2.2 shows a diagram of a typical toll plaza arrangement with deceleration and acceleration areas and a mixture of payment lanes.

Let us consider the different types of manual system that exist.

2.2.2.1 Manned Tollbooths

Manual toll collection using an operator to collect money is probably still the most widely used method of collecting tolls. An operator is situated in a tollbooth servicing one lane of traffic. These booths must be air conditioned and heated for the comfort of the operator. It is generally also necessary to employ some simple

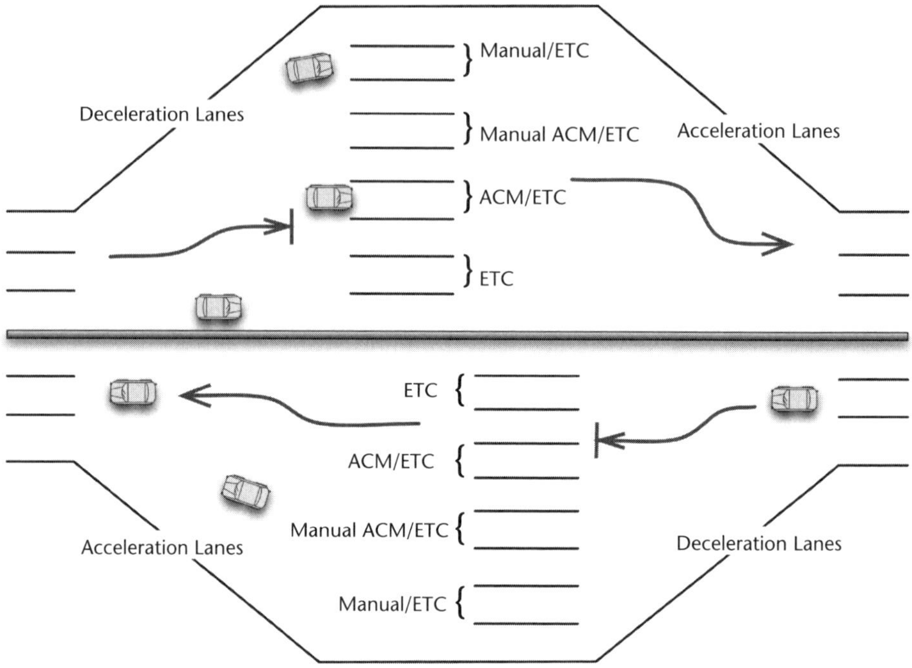

Figure 2.2 Typical mixed payment toll plaza arrangement.

auditing systems, such as counting the vehicles passing through the lanes and more commonly now providing a paper receipt on request for each transaction.

The collector takes coins, cash, tokens, or paper tickets from all the drivers passing through the lane. Where only the correct toll may be paid (i.e., no change given), or where prepaid tokens or paper tickets (vouchers) are used, the transaction itself takes only a few seconds. If the transaction requires that change is given or a paper receipt is provided, then this process takes longer. An experienced operator generally can achieve 300 or more transactions per hour, although this depends on the number of coins required to pay the toll. A $1 toll can generally be paid quicker than a $1.30 toll, for example.

2.2.2.2 Automatic Coin Machines

Automatic coin machines (ACMs) are widely used at many toll plazas to replace the need for a manned tollbooth. The coin machines are generally able to accept prepaid tokens (if used) and coins. Most of these machines use a basket or hopper, into which the drivers throw coins or tokens. These are generally read and validated within 2 to 3 seconds, and the barrier is raised (or some other indication of a correct toll charge given to the driver). The driver can press a button requesting a receipt to be printed. These basket/hopper arrangements are regarded as an efficient way to pay tolls, and are now quite reliable and environmentally robust (usually, they contain a heater/cooler to ensure operation in all conditions). The sophistication of the coin validation unit enables the machine to reject foreign currency and other objects thrown into the hopper. Figure 2.3 shows a combined stop-and-pay coin hopper, card reader, and read-only tag reader on a highway near Lyon, France.

Payment may be relatively quick for regular users of a toll plaza who are familiar with the operation of the basket and the coins it requires. Where barriers are not used, many regular drivers do not completely stop at the baskets, but throw their coins in the basket from their slowly moving vehicle. Inexperienced users of the system can considerably hamper the proceedings, particularly if they do not possess the correct coins, or if they miss the basket.

A single lane of a toll system may service up to 400 vehicles per hour, based upon the results of studies in the United States, where these hopper arrangements are widespread. These figures are exceptionally high, compared with throughput figures on most toll roads in France and Italy. Figures 2.4 and 2.5 illustrate the reduction in transaction and stopping time that can be achieved by a drive-through system, when compared to a stop-and-pay system.

2.2.2.3 In-Lane Card Readers

Prepaid cards (magnetic or paper-based), credit cards, and smart cards are all now used for toll payment purposes. All of these methods of payment require that the driver inserts an appropriate card into the card reader, waits for that card to be debited (or validated), and then collects the returned card (together with a receipt, if requested) before continuing the journey. Contactless "proximity" smart cards, which communicate using a radio frequency interface and comply with the International Organization for Standardization (ISO)/International Electrotechnical Commission (IEC) 14443 standard, are increasingly being used for tolling. These cards only need to be presented to the reader (usually at a range of less than 10 cm), rather than being inserted into a reader, which speeds up the overall transaction process [51]. An example of a contactless smart card–based tolling system was introduced in Turkey in 2005 [52]. New generations of smart cards that use the "vicinity" standard (ISO/IEC 15693) may be read from a range in excess of 1m, but as of yet, these cards have not been deployed in toll applications.

Prepaid cards (purchased in advance from the toll operator) are the most common cards to be used. These usually hold the "rights" for a given number of journeys, or the right to use the toll road at will for a particular period of time.

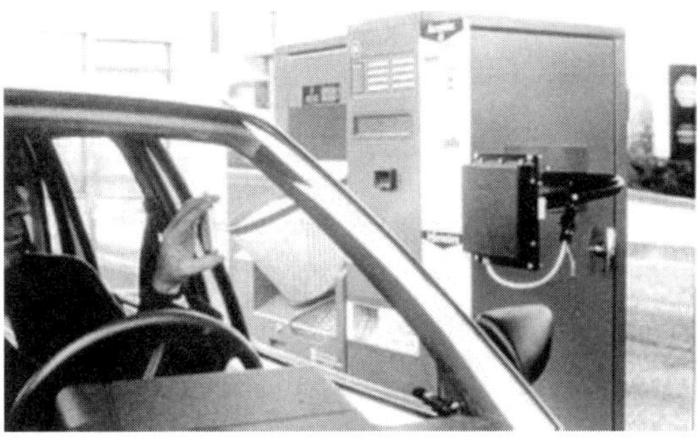

Figure 2.3 Coin, card, and tag payment booth. (Courtesy of Blythe/CSEE.)

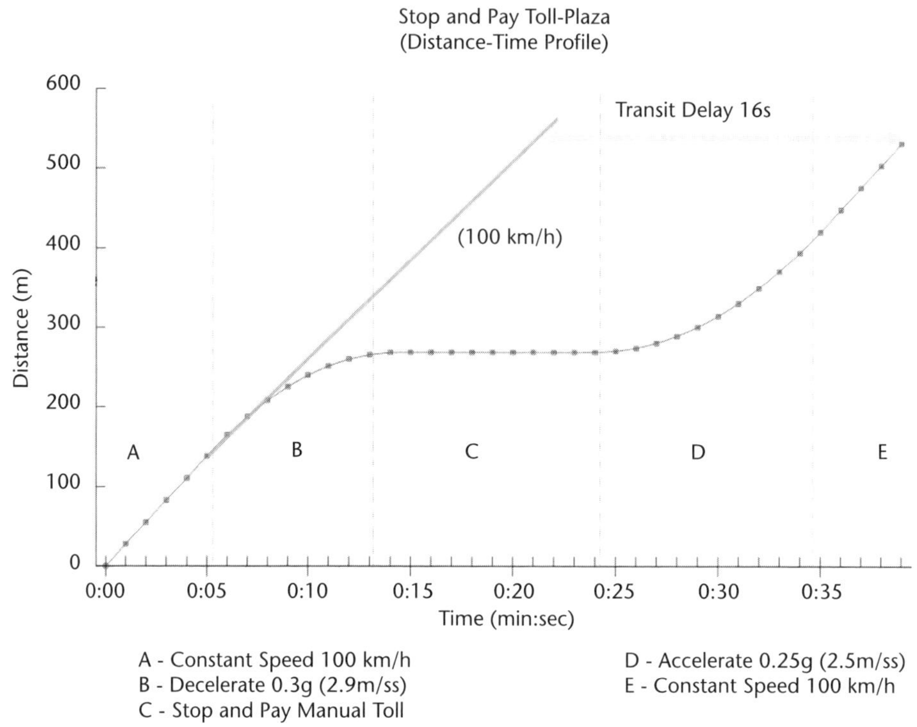

Figure 2.4 Distance-speed profile for stop-and-pay toll collection.

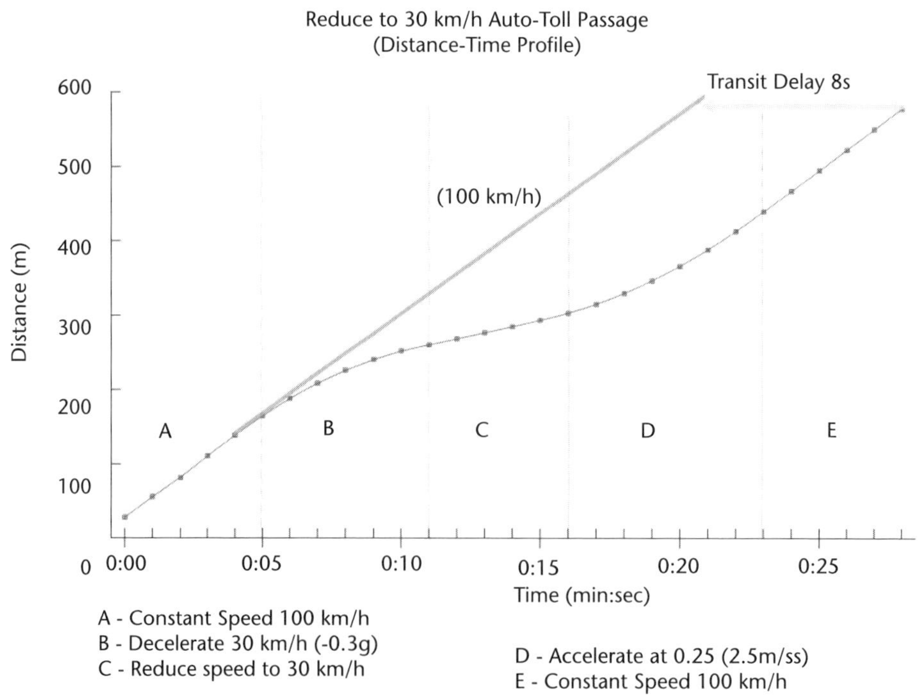

Figure 2.5 Distance-speed profile for vehicle passing through toll site at 30 km/hr.

Smart cards may also hold the same information, and they may be used to hold electronic cash or credit, which is deducted from a card's balance for each toll transaction. Smart cards also can be recharged with credit or subscription rights. Such usage could spread now that numerous banks have adopted electronic accounts held on smart cards.

The use of credit cards is not so widespread, for two good reasons:

- The value of the toll transaction is generally low, and credit card operators see no commercial viability in allowing credit card payments for such small amounts. One exception is on long-distance closed toll highway networks in Italy, Spain, and France, where drivers pay a charge related to the journey distance on the network, which may amount to several tens of dollars, make credit card payments viable.
- The time it may take the credit card reader to validate a transaction (typically from 10 to 15 seconds, if dial-up lines to a card validation computer are used) make this form of payment less than attractive at tollbooths, where long lines may develop if this form of payment were employed. The recent introduction of compulsory chip and personal identification number (PIN) payment in many countries may further slow down this transaction process. However, some PIN reader credit card machines speed up the transaction by only validating the PIN locally and not connecting to the card's central system.

Based upon an ergonomic study in France, the total time required for payment using journey tickets is 15 seconds (a rate of 240 vehicles per hour), while the total time for a credit card payment is 22 seconds (a rate of less than 170 vehicles per hour).

2.2.2.4 Paper Stickers, Area Licenses, and Vignettes

Systems that use paper permits or vignettes are an additional nonelectronic system. A driver purchases an additional license to use a particular toll road or road network on a specific day or time of day. Many toll road operators introduced such an option for regular travelers, prior to the introduction of electronic systems. Manual reading of the sticker or vignette, often supplemented by ANPR, enforces this system. The most significant examples of their use are found in citywide access control schemes.

A paper sticker–based system is a nonautomatic means of identification, which conveys to a manual observer (or camera) visual information regarding the rights of that vehicle user to drive on a specific road network for a specific period of time, or during certain times of day.

A paper sticker or license can only convey a small amount of fixed information, and, depending on the sophistication of the sticker, the information may only be read from a short distance in slow-moving or stationary traffic. This is usually achieved using brightly colored stickers, prominently displayed in the vehicle's windshield. The stickers are practically impossible to read with any degree of accuracy in fast-moving or multilane traffic, although Singapore did employ such a method.

The paper sticker has the advantage of being easy to implement and easy for drivers to understand. The difficulty with the approach lies in the enforcement of the system. The licenses must be read at a distance, either by a manual operator at the toll site or by a random inspection by police or another agency. It is also necessary to make the permits fraud-resistant and flexible enough for the different subscriptions and licenses that may be offered in a scheme. However, the potential for the counterfeiting of these printed permits is increasingly a risk, due to modern desktop publishing systems and high-quality color printers/copiers.

Manual reading of paper stickers was used effectively at toll sites in Bergen, Norway, for more than a decade. Special drive-though toll lanes were dedicated to those drivers possessing a paper sticker. This system was effective for enforcement, but it required that the road be divided into single lanes and that a manned tollbooth be used. A video camera was used to take digital photographs of all vehicles that did not possess a valid license. It was estimated that up to 600 vehicles per hour in Bergen could be checked manually. However, the system relied on the vigilance and integrity of an operator to perform a repetitive and less-than-fulfilling job. The Bergen scheme was upgraded to use "Autopass" (the Norwegian National charging technology) in 2001.

In Singapore, the Area Licensing Scheme (ALS) was successfully employed from 1975 to 1998. This scheme used paper licenses of different colors to depict different access rights. The entrance roads into the central zone, where the licenses applied, were clearly marked by gantries that used lights to indicate when the zone was "active," as shown in Figure 2.6. These roads were generally two- or three-lane roads, and there was no restriction in traffic flow. Enforcement was manually

Figure 2.6 Gantry indicating the boundary of the restricted zone, Singapore, 1994.

performed by inspectors in booths at the side of the road, although it is not known how effective they were at detecting violators across three crowded lanes of traffic. Police patrol cars were also used to check licenses through random inspections. Violators faced a hefty fine, and the official figures in Singapore suggested less than 1% for violations.

2.3 From Policy to Technology

2.3.1 Background

A degree of technology sophistication is needed to ensure that road user charges are collected in an efficient and effective way. Technology is the enabler in every system that ensures that road user charging policies can be delivered. Technology can be the means of introducing demand management policies for cities gripped by gridlock, or as the means of enabling a cost-effective toll collection scheme for a privately operated (concession) highway that provides access between areas of employment and the residential areas of labor. Technology can also enable the efficient collection of taxes from road users who are paying according to other parameters of travel, such as distance traveled or a fee reflecting the environmental impact of the journey. Technology is a means to an end and not the end itself; without technology, many of the opportunities opened by road charging would not be feasible [53].

Technology offers a range of options for a user to pay for road use. The charging technology may also include a means of measuring the road usage in parameters that are defined by local needs and charging policy, such as the distance traveled by the vehicle on a road segment that is charged at a higher tariff than an alternative parallel link. This tariff may depend on the vehicle classification. Heavier commercial vehicles may be required to pay more than light goods vehicles, and highly polluting vehicles more than environmentally friendly vehicles, for example. The charging technology may also provide a means of instant communication with a road user. It may confirm that the means of payment was accepted, or allow the user to modify the information on which the charges are based (e.g., declaring that a truck has a trailer attached).

Historically, toll collection operators have employed RUC as part of a pay-per-use service. The evolution of new charging, communication, and enforcement technologies also enable the principles of RUC to be implemented at a local and national level for selected road users.

Technology can deliver services that depend on several factors: the local charging policy, user preferences for information relating to the fees accrued, the means of payment, and the classification of the vehicle. This list is not static; the charging policy may vary, depending on the location of the vehicle, and the user preferences may vary over time and by journey. The vehicle classification may vary according to the local classification scheme; a commercial tractor unit may be able to lift one of its axles if it is not carrying any substantial load, and this lower axle count may enable the road user to claim a discount [54].

2.3.2 Policy Options

Wherever there is a need to differentiate categories of road users for charging or enforcement, or to define a boundary between areas of different charging levels, such as entering a charged area or passing to a lower tariff zone, there is a need for technology. The required technology may be situated at the roadside or in the vehicle, and should be capable of detecting and recording that the vehicle has, or is about to, cross a tariff boundary on a charged road or network of roads [55].

Charging technologies are most likely to be found where the charging policy requires an action, as the examples in Table 2.1 show.

Some or all of the functional requirements are also needed to enable a charge for road use to be calculated and applied. If the road use is measured by equipment located within the vehicle, or if a roadside system is triggered by equipment in the vehicle, then a means of connecting the in-vehicle equipment to the roadside system is also needed.

2.3.3 Basis of Charging

If RUC is based on a network of separate chargeable road segments, then the subsystems that perform the tasks listed above will need to be integrated at some point, to enable full-service "roaming" between geographically disparate operators, otherwise known as interoperability.

As mentioned previously in this chapter, there are a number of different ways of implementing a charging scheme based upon the charging objectives and the type of road network to be charged [12, 56]. The following section briefly considers a selection of these options, although many of the examples given are described in more detail in the case studies of Chapter 8.

2.3.3.1 Open Toll Road

An open toll is the term given to a tolling scheme that implements a charge at a specific point on a road, as illustrated in Figure 2.7. This usually applies to a particular piece of managed infrastructure, such as a bridge or tunnel at an estuarial crossing, or some significant geographic barrier, such as passing through a mountain range. A toll is levied on vehicles passing through the toll plaza. The toll generally is a fixed charge and does not relate to the distance the vehicle travels on the road network, but instead is purely a charge for the use of the infrastructure. The charge may vary by time of day as an attempt to spread peak-hour traffic. The first example of peak-hour charging on a toll road was implemented on the Paris-Lille toll road in the mid-1990s, as a means of controlling the high traffic demand generated on a Sunday evening for Parisians returning after a weekend in the countryside.

It is important to note the distinction between the term *open toll* defined here as a fixed fee for use of a single facility and the "open road toll" (ORT), which is now a commonly used term, particularly in North America to describe a toll road with two or more express tolling lanes using electronic tolling equipment, such as Highway 407 near Toronto and some of the EZ-Pass installations in Illinois [57]. Toll plazas can also be converted to ORT.

Table 2.1 Policy Requirements

Policy Requirement	*Policy Requirement (Subset)*	*Examples*
Detect entry to chargeable area or across boundaries between different tariffs	Detect when a vehicle crosses a tariff boundary or measurement of vehicle position relative to tariff boundary	Entry to toll road, exit from toll road (e.g., closed toll road); entry to or travel within a charged area; entry to different tariff road (e.g., highway)
	Area	Transition from one charged area to another at a different tariff
	Time of day	Proxy for measured congestion—charges depend on time of day as a simple charge/no-charge scheme, or graduated charges applied over whole day
Measure road usage	Distance traveled	Measuring distance traveled on chargeable road segments by identification of road segment, or by incremental distance traveled
	Congestion	Measure vehicle's contribution to congestion, or measure overall congestion with external fixed sensors
	Class of road	Identify road type on which the vehicle travels (e.g., motorway, public versus privately-owned roads)
Declare vehicle attributes	Emissions class	Manufacturer-declared emissions class
	Weight	Manufacturer-declared gross carrying capacity, dynamically measured axle weight on the vehicle, in-ground dynamic measurement of weight
	Quantity of axles	Total or separated into tractor and trailer
Correlate charging and enforcement records	Off-line payment	Off-line payment linked to declared vehicle registration
	Online payment	Spatially and/or temporally correlate means of payment with vehicle at point of payment
Communicate with roadside infrastructure	Interface to roadside system	Temporary or off-line connection to deliver payment-related information and to permit declarations to be transferred for charging and enforcement purposes
	Interface to road user	To communicate result of payment transaction, allow declarations to be changed, or add other value-added services.
	Interface to other in-vehicle system	To capture incremental distance traveled information from odometer
Payment	On-board account	Account specific to toll operator or electronic account containing authenticated value

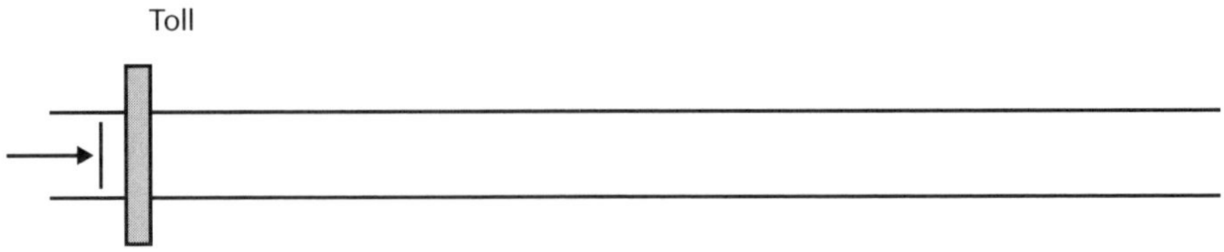

Figure 2.7 Open toll road.

Such toll schemes usually charge on pay-per-use methods, whether the user pays in cash or pays electronically by using a tag to identify a centrally held user account. This is often the simplest approach, as used by the M6 Toll or Dartford Thurrock Crossing in the United Kingdom, and by various electronic toll roads elsewhere. The trend in the United States is moving towards interoperable toll tags at such sites.

2.3.3.2 Closed Toll Road

The most common form of interurban highway tolling is closed tolling, in which the toll is related to the distance the vehicle travels on the toll road. The toll charge is measured by registering when and where the vehicle enters the toll road network, and when and where it leaves the network. Thus, there is a need for a series of entry and exit points on the toll road network, as illustrated in Figure 2.8.

The system can generally be configured in two ways when using automatic tolling technology. In the first configuration, the in-vehicle tag identifies itself to the toll system upon entry to the network, and again upon exit from the network, where the appropriate toll is calculated. In the second configuration, the entry data is recorded onto the tag itself and then presented back to the toll system on exit from the network, so the appropriate toll can be calculated. Closed toll systems increasingly are migrating towards open road free-flow tolling systems, as seen on Highway 407 [58] around Toronto, the Melbourne City Link, and the recently opened toll facilities in Chile [59].

Wide area systems that calculate the distance traveled using on-board equipment could be used for closed tolling. They are not necessary if a dedicated toll plaza has been built to service entry and exit points. Where distance-based charging is introduced to previously free road stock, as may happen in the United Kingdom, then wide area systems utilizing GNSS and Groupe Spéciale Mobile (GSM) may be viable. This is also the case for national schemes that have been introduced for

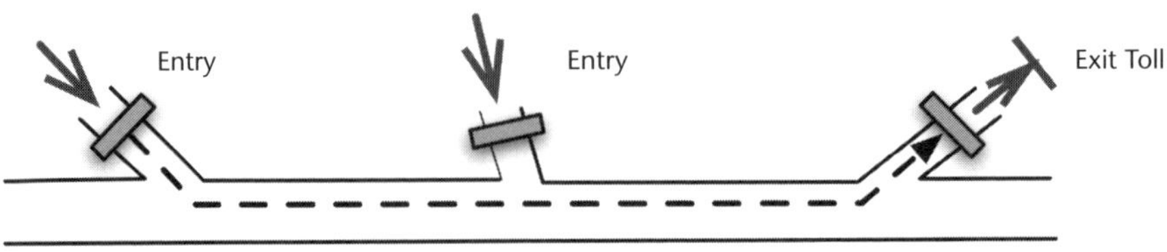

Figure 2.8 Closed toll road.

heavy goods vehicles, such as in Germany, Austria, Switzerland, and parts of the United States.

2.3.3.3 Cordon and Area Charging

The charging of a fee for crossing a cordon is by far the most common configuration for urban demand management. There is a boundary into a central business district or environmentally sensitive area that will incur a charge if crossed, as illustrated in Figure 2.9. The charging rings around Trondheim, Norway, and the Singapore ERP are often cited as examples of such an approach [60]. The cordon need not necessarily be operated on a charging basis, and may be configured to allow certain users to cross the cordon without penalty [61]. Another early example was the access control scheme established around some of the residential areas of Barcelona, to restrict access only to residents and business owners during the 1992 Olympics [62]. This scheme, partly funded under the EU's DRIVE II Programme GAUDI project, used first generation Q-Free (Køfri) AVI tags.[2]

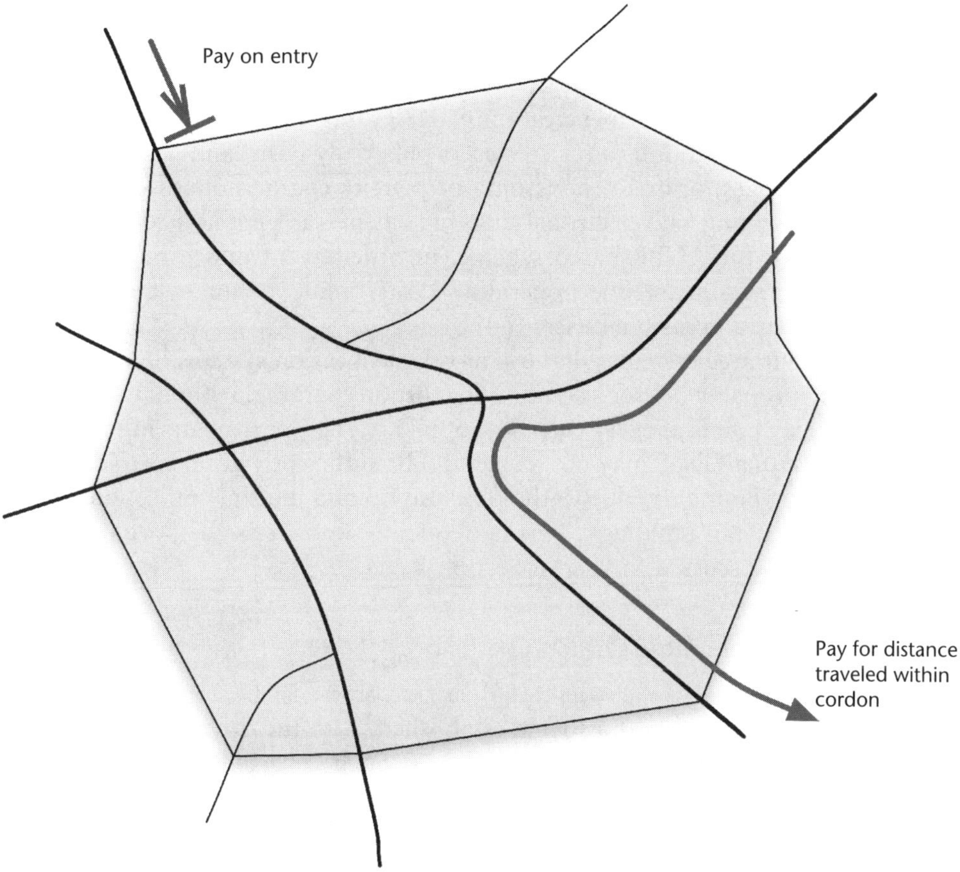

Figure 2.9 Cordon charging.

2. Refer to Chapter 8 for further details on the Norwegian systems.

One problem with a cordon is that it is a relatively blunt instrument—if you travel into the cordon area, then you pay a fee, regardless of the time spent and road space used by the vehicle. Experiments have been undertaken to reflect more specific charging once the vehicle enters a cordon. This essentially changes cordon charging into an area charge. In 1992, Cambridge tried a cordon-based scheme, in which an in-vehicle meter was activated using microwave beacons as vehicles entered the city. Once inside the cordon, the vehicle only accrued charges when the vehicle was deemed to be in a congested situation [22, 24]. The same system demonstrated the accrual of charges based upon the measured distance that the vehicle traveled inside the cordon, as illustrated in Figure 2.9. In the same year, GEC ESAMS[3] demonstrated a variant of this scheme, which charged for time spent in a cordon in Richmond, United Kingdom. Another possibility could be to charge a fee related to the levels of environmental pollution generated by vehicles in a particular area.

Most cordon-based systems currently use microwave tags to initiate payment. However, London introduced their congestion charging scheme based upon the preregistration of vehicle license plates, which are then checked, and violators recorded, using ANPR [28, 63]. The advantage of the London scheme is that no vehicle is required to have electronic equipment installed, so regular users and occasional users pay in the same way. The scheme is fairly inflexible, because it is difficult to vary the charge, and relatively costly to operate in comparison to schemes with a high penetration of DSRC tag usage. This is due to the need for manual intervention to register users on a daily basis, and to check unclear license plate images, prior to the issuing of penalty charge notices. London is currently experimenting with electronic charging schemes as a possible replacement or supplement for the ANPR-based scheme, in order to introduce more flexibility in the charging regime, reduce operating costs, while retaining charging options for unequipped occasional users [64].

Wide area systems[4] that use in-vehicle location systems linked to a digital map could probably deliver a solution for cordon charging, without the need for physical charging points at every entry location [12, 31]. Experiments in several major cities suggest that GNSS may not (currently) be sufficiently accurate to define the cordon charging boundary, due to the obscuration and multiple reflections of the satellite signals by tall buildings. This is frequently known as the "urban canyon" effect, which is discussed further in Section 3.5.3.

2.3.3.4 Concentric Cordon Charging

A variation on the conventional single cordon is the concentric cordon scheme. Outer and inner cordons were established, with the driver required to pay at both boundaries, as illustrated in Figure 2.10. Such arrangements may be used to reflect the additional demand management measures required to deal with the congestion in the center of a city. The inner cordon could also be used to encourage park-

3. Refer to the section on the Cambridge trial and in-vehicle metering systems in Chapter 8 for more details on this system.
4. Wide area systems are also often referred to as mobile positioning systems (MPS), and virtual positioning systems (VPS). The technology options available for such systems are presented in more detail in Chapter 3.

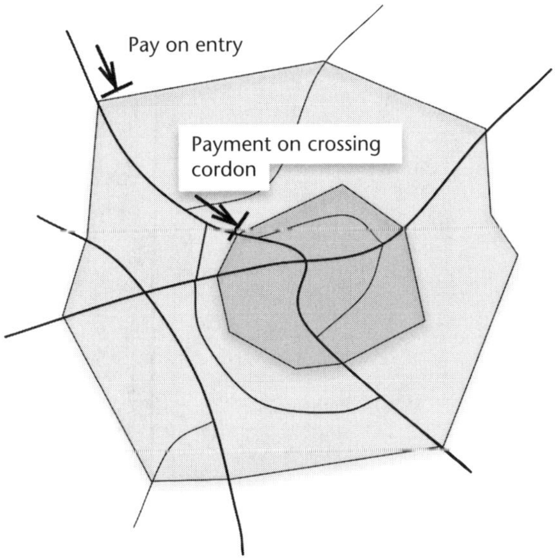

Figure 2.10 Concentric cordon charging.

and-ride and modal shift before reaching the inner cordon. Charge levels can be different at each cordon, and be operated on an area-pricing arrangement, as discussed in Sections 2.3.3.3, 2.3.3.5, and 2.3.3.6. In all the cases of the cordon and zonal configurations, it would be possible to implement charges in both directions of travel, which could be used to tackle problems associated with the evening rush hour.

The concentric cordon approach has not yet been implemented in an urban charging scheme. It was the basis for the proposed Edinburgh, Scotland, road pricing scheme, but this scheme was rejected by the residents of Edinburgh in a referendum in February 2005 [65, 66]. An inner cordon was initially proposed for the Stockholm congestion charging solution, although rejected in favor of a single-cordon scheme, which began a 9-month experimental period in January 2006.

2.3.3.5 Area Charging with Through Route

It may be necessary to allow some through traffic where a cordon scheme has been implemented, to avoid generating a large number of trips by circular routes around the cordon. Figure 2.11 shows a dedicated "free" corridor that could be established to enable these transits. The extension to the Central London Congestion Charging Scheme to the Royal Borough of Kensington and Chelsea allows for such a transit route.

2.3.3.6 Quasidistance/Zonal Charging

Another arrangement is to introduce a series of interlocking minizones, where a charge is levied at the interface of each zone, as illustrated in Figure 2.12. Such an arrangement would assume charging using tags or ANPR, as if using a wide area scheme. The charge could be fine-tuned in a different manner, such as using distance-

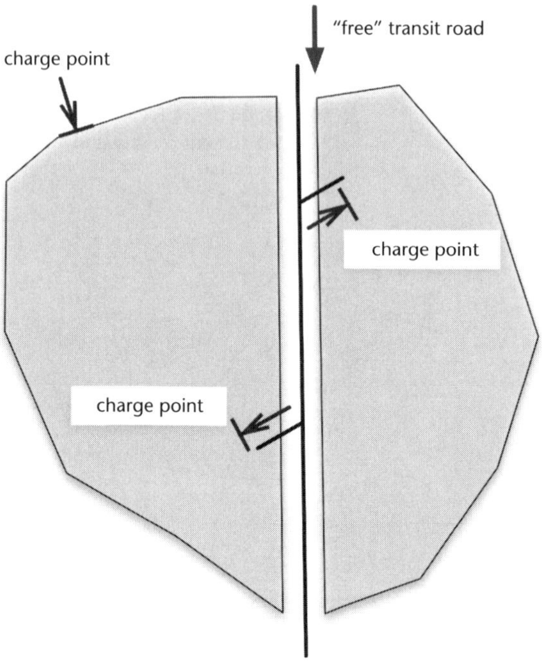

Figure 2.11 Area charging with through route (liability to be charged on entry and travel within area).

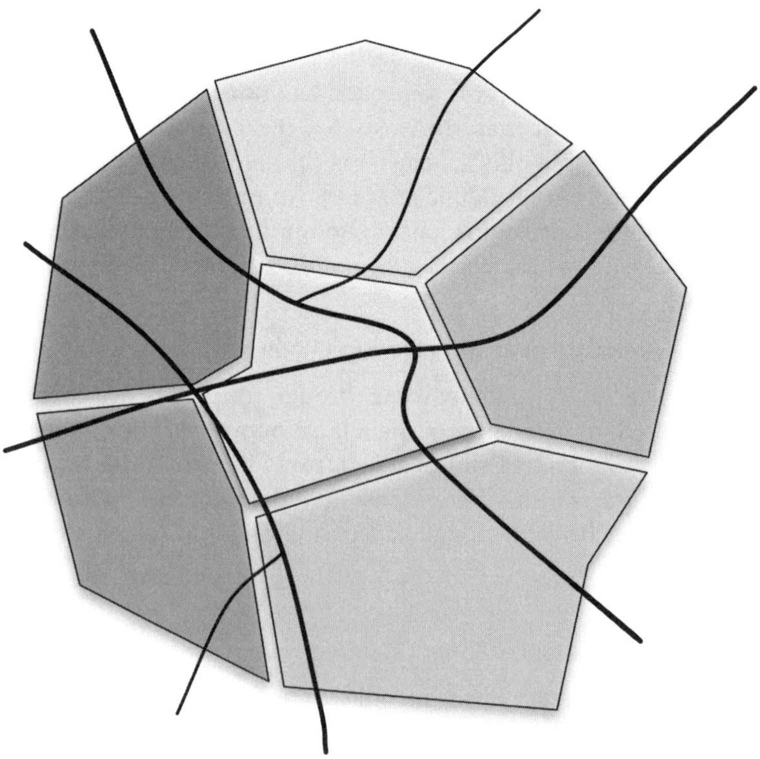

Figure 2.12 Quasidistance/zonal charging.

based charging [67]. The Hong Kong ERP trial in 1983–1985 came the closest to such a configuration, since the trial scheme had four charging zones [2]. The scheme also varied the charge for crossing the cordon by time of day, with a peak, shoulder[5] peak, off-peak, and no-charge fee bands [42].

2.3.3.7 Road Segment Charging

Road segment charging is a charging configuration specifically designed for wide area charging systems. A boundary is defined around the road that is to be charged, which usually extends to some distance beyond the boundary of the road section to account for errors in the location calculations made by the vehicle on-board unit [68, 69]. The road segment identification may be performed within the on-board unit or central system. Once it is recognized that the vehicle is within the boundary, charging is initiated. A network of such segments could be defined to cover a large network of roads, or the entire national network of mapped roads [70, 71]. See Figure 2.13.

2.3.4 Operational Requirements

There are several basic questions that must be considered when introducing an automatic road user charging system. These questions are presented in this chapter

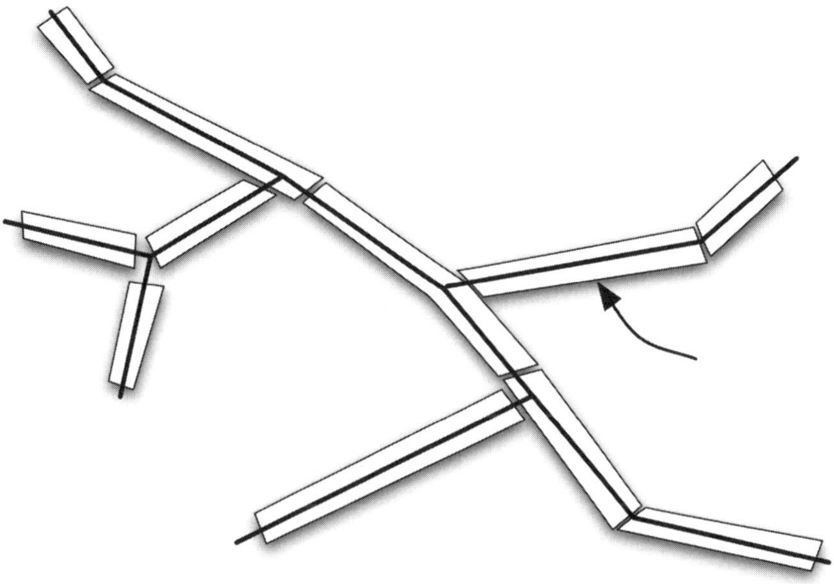

Figure 2.13 Road segments for use with wide area/VPS charging systems.

5. A shoulder charge is an intermediate level charge established to offset the step-change between a high-priced peak charge of, say, \$5, and off-peak low-cost charge of, say, \$1. The purpose is to try and discourage many users from waiting for the charges to switch from the high price to the low price, and thus cause unnecessary levels of congestion and queuing. An intermediate level charge of, say, \$3 may offset this effect. For more details of the Hong Kong trials, refer to Chapter 8.

as generic options. The more specific detailed implementations of each element are considered in Chapters 3 to 6, where the operator and customer requirements are considered in much more detail.

These basic considerations are as follows:

- What payment methods will be allowed, and what types of accounts will be offered by the operator? The sophistication of the methods may change, based upon whether it is a toll collection scheme or a road user charging system.
- What is the likely traffic flow through the charging locations, and what kind of toll collection facility will be used? The operation may incorporate a toll plaza, or may be a more free-flow operation, as we are beginning to see with new toll road implementations (e.g., Toronto 407 ETR and Melbourne City Link). It may be for urban road user charging (e.g., Trondheim, London, and Singapore).
- How will the fees be collected? Will it be manual, or by some form of automated system?
- Will the charging basis be a point charge on a toll road, a distance-based charge, or some other parameter? Is the road a single road, a network, a cordon, or some multizonal arrangement? What parameters will be measured to calculate the charge?
- What level of enforcement is required, and what complexity of vehicle detection and classification is necessary?

Many of the products and services required to successfully implement road user charging depend on technical innovation, technology development, and end-to-end systems deployment. The role of technology in enabling a charging scheme can be viewed from several perspectives, including national government, local government, road operator, technology vendor, system integrator, and road user.

Looking beyond the front end of the system that actually facilitates the on-road charge, a complete charging scheme will require many or all of the following systems and services:

- Service provider and clearing operator system development to manage high volume payment collection, clearing, and funds transfer;
- Customer relationship management (CRM), billing, and general support, particularly immediately following the start-up of a scheme;
- A system to distinguish between vehicles that are equipped with technology [often known as "tags" or "on-board units (OBUs)"] to facilitate charging from those that are not equipped;
- Identification of suspected violators and management of the evidence of the violation;
- IT infrastructure development, deployment, and maintenance [e.g., wide area network (WAN) backbones], for distribution of tariff information to vehicle-based equipment;

- Road use information collection, dissemination, and display [e.g., to signal charge levels and alternative means of travel, using roadside variable message signs (VMS) or in-cab displays];
- The manufacture, personalization, distribution, delivery, and installation of OBUs (if used);
- Evidential enforcement record management, registered user identification, and penalty collection;
- Service quality level auditing, security risk assessments, and environmental impact reduction for all on-road infrastructures.

Project management, financing, risk absorption, integration, maintenance, and operations are the elements that would also be needed for a complete scheme. While a scheme for 400,000 heavy trucks may be appropriate for a single service provider, it is likely that a national, mass-market scheme serving 30 million vehicles, for example, would need a multitiered national and local service and maintenance operation. This may be further complicated by the possibility that the objectives, and operation, of a local scheme may be very different than a national scheme if both are operated simultaneously.

2.3.5 Functional Requirements

Several functions are generally needed for all road user charging schemes that require the use of a tag or OBU installed in participating vehicles. The following functions need to be supported, in order to meet the operational objectives listed in Section 2.3.4.

- User registration and access to in-vehicle equipment;
- Declaration of user and vehicle-related information, to allow the correct charge to be determined;
- Enforcement if the correct charge cannot be applied (e.g., missing or incorrect user declarations);
- Collection and management of records relating to user and vehicle charging and enforcement events;
- Collection and settlement of charges and penalties.

Every scheme that depends on electronic means of payment, from the simplest to the most complex, needs to employ a selection of these service elements. The business model of a local scheme may suggest that several charging products must be offered by the scheme operator, depending on the frequency of user access to the charged road segments, the vehicle type, the payment options, and the level of privacy and anonymity required by the driver or allowed by the scheme operator.

2.3.5.1 User Registration and Access to In-Vehicle Equipment

The in-vehicle equipment needs to uniquely and unambiguously point to the means of payment, so at the time of issue, the equipment must be linked to the charge payer and optionally to the vehicle. This linkage may be physical, such as a simple adhesive fixing or a permanent tamper-resistant installation, and/or logical, by relating the in-vehicle equipment to the vehicle in a central system database, which is discussed in Chapter 6.

In-vehicle equipment may not always be required (e.g., as in the London Congestion Charging scheme). Some scheme operators offer a product for occasional users that requires the registration of the license plate of a vehicle against a means of payment. A road user would be encouraged to register for an occasional user scheme before traveling on the chargeable road network, but grace periods could range to as much as 5 days later. Trondheim, Norway, offered occasional users the option of paying for entry into the city using coin machines on the cordon entry roads. Such an option would only be feasible in small-scale schemes. The Stockholm, Sweden, pilot only allowed postpayment within 5 days of the vehicle passage. These and other approaches to dealing with occasional users are described in Chapter 3. The strategy for dealing with occasional users is very important. A number of potential urban charging schemes in the 1990s were shelved because no credible occasional user scheme could be established at an acceptable cost and level of complexity. This is the beauty and pragmatism of the present-day London Scheme; all users of the congestion charging zone, whether occasional or regular, use the same method of registration and payment, through license plate registration and enforcement with ANPR [27, 28]. If and when the TfL migrates to some form of electronic on-board unit for regular users, the occasional users would still be able to utilize the license plate registration scheme as an alternative form of payment, as well as for enforcement, since the infrastructure already exists.[6]

2.3.5.2 Declaration of User- and Vehicle-Related Information to Allow the Correct Charge to Be Determined at the Point of Provision of Road Use

The in-vehicle equipment needs to provide the means for a road user (or the entity responsible for the vehicle) to make declarations of the vehicle type and other attributes to enable the correct charge to be calculated. The user's ability to influence the content of this declaration is likely to be very limited (e.g., informing of the existence of a trailer or caravan). Other attributes are either static (e.g., a vehicle's emissions class) or dynamic (e.g., entry point to closed toll road, quantity of road segments traveled, and time of day), but in most cases cannot be modified by the road user. As will be discussed in Sections 5.5 and 9.3.3, new sensing and monitoring technology may provide options for more dynamic declarations, such as using real-time environmental measurements as a basis for calculating a component of the change.

Declarations that have a direct relationship with the calculation of road usage, such as a vehicle's classification, may be subject to independent external checking. These declarations and the results of any other external checks or measurements are related to the enforcement process (see Chapter 4), rather than to the charging process.

6. The London scheme is discussed in more detail as a case study in Chapter 8.

2.3.5.3 Enforcement If the Correct Charge Cannot Be Applied

Ensuring compliance with the locally enacted charging policy is crucial to an effective, credible charging regime. Charging cannot exist without enforcement [54].

A vehicle's license plate must be used to enable a penalty to be issued if a user drives through a toll lane and the vehicle is not properly equipped to interact with the electronic payment system, or if the charging process fails for any other reason. If the toll lane has a barrier, the responsible person is the driver who would be required to pay by another means.

The choice between automatically triggering an enforcement process and attempting to apply a charge based on a vehicle's license plate will depend on the enforcement policy of the operator. For example, if the electronic payment system requires the vehicle's license plate to be registered, and if the license plate number is captured correctly, this would be sufficient to apply the charge. This process would cost more to the operator than would a transaction generated by in-vehicle equipment. This approach to the enforcement process forms part of the Stockholm Congestion Charging pilot scheme [72]. The alternative is to treat the lack of in-vehicle equipment as an offense. This may incur a higher cost to the operator, which is offset by revenues from penalties or fines, depending on the policies of the particular jurisdiction. However, cross-border enforcement is difficult and costly, so the revenue recovered may not be as high as anticipated. A business case analysis allows enforcement policy options to be compared, although the choice will invariably depend on other factors, such as the intended purpose of the scheme (e.g., demand management or tolling). The enforcement strategy also must consider the cost of enforcement and the probability of the violation being detected and the user identified. There may be issues with the availability and accuracy of the vehicle license plate database with cross-border or cross-state operation.

Permitting a vehicle to register for a payment scheme *after* traveling on the chargeable road network could result in the deferral of enforcement processes until the registration (and payment) deadline has passed. However, a mismatch between declarations and independently measured vehicle attributes (where they directly relate to the amount of the charge) would immediately trigger the enforcement process.

Finally, the charging technology itself may support the enforcement process by providing the physical location of any in-vehicle equipment to enable it to be matched with the relevant vehicle at the point of enforcement.

2.3.5.4 Collection and Management of Records Relating to User and Vehicle Charging and Enforcement Events

There are different modes of charging, including cordon, area, distance-based, and time-based; see Section 2.3.3. Road user charging also includes annual registration fees, fuel duty, and other charges and taxes. Some means of recording road usage is required either by means of the roadside equipment (e.g., identification of road usage on every road segment), or by the in-vehicle equipment (e.g., recording whenever a new chargeable road segment is being used).

The location of the measuring process will depend on the charging policy, for example. The economics of a scheme based on a single toll plaza and 100,000 vehicles suggests that the charge will probably be assessed by the toll plaza equipment, and that vehicles will be expected to carry a simple tag. A scheme based on 500,000 vehicles and 5,200 interconnected road segments (e.g., similar to the German truck tolling scheme) suggests that the most economically favorable solution may be that the in-vehicle equipment should play a greater role in measuring the road usage. In practice, roadside infrastructure is always required, particularly for enforcement. The decision remains to be made on the level of complexity of the in-vehicle equipment and communication channel requirements to a data collection center [73]. Obviously the cost of "the many" OBUs against the cost of "the few" roadside charging points needs to be balanced.

2.3.5.5 Collection and Settlement of Charges and Penalties

Paying the charge means transferring funds from a road user's account to the account of the road operator or some agent acting on behalf of the road operator, whether this be postpayment, immediate payment, or, in many cases, prepayment. The transfer of funds can be triggered, for an isolated scheme, simply by the collection of a record of a vehicle passage that can be related to an account. In a network of operators linked contractually, the transfer of funds may require a higher standard of proof, such as a certificate generated by a transaction with in-vehicle equipment that is authenticated during the passage on the charged road network.

2.3.6 Payment Methods

A number of payment means have been formally defined in international standards, some of which apply to a particular scheme or objective.

2.3.6.1 Automatic Account Identification: Postpayment

From the 1970s up to the mid-1980s, automatic account identification (AAI) was the most widespread system, since it generally required only the use of simple read-only tags and a relatively low level of sophistication in computing capability at the roadside. Such systems required communications to be established in only one direction (i.e., vehicle-to-roadside), and, in most schemes, little data is required to be transferred. This method was also widely (but incorrectly) known as AVI.

Upon interrogation, the roadside equipment records the unique account identity of the vehicle owner's tag and the time of day that the vehicle passed through the charging site. The validation of the identity code is generally performed as an online process, but the collection and accounting of the actual revenue are off-line processes.

Threats to privacy problems may occur, due to the necessity of having a central computer record of the information regarding each vehicle's movement and identity. However, some relationship between the user and the central system needs to be defined for the purposes of an audit trail. The record must be maintained for as long as it takes for the recovery of the outstanding charges from the user, or until it meets the requirements of the audit trial. The information may only be recorded for a few hours or days if a direct debiting facility is used. However, if postpayment billing is used, then the information must be stored for at least the period between successive bills (e.g., monthly or quarterly).

Most operators have moved away from offering the postpayment option. The operators have a clear advantage in using prepayment options, since they receive users' money in advance of the transaction actually occurring. Prepayment also offers the operator the benefit of a simple and secure "audit trail." The additional costs of recovering money from a roadside postpayment operation may be considerable.

2.3.6.2 Automatic Account Identification (AAI): Prepayment

Prepaid AAI is the method of road use revenue charging and collection that is favored in most current automatic tolling and cordon-pricing schemes. The data acquired from the tag or OBU is usually validated in real time, which allows a check that the user's in-vehicle device is legitimate, and that the user's account has adequate credit and is not blacklisted for any reason.

The financial transaction takes place immediately after validation of the identification code, by deducting the appropriate charge from the vehicle owner's account that is held with the toll authority. The transaction may be performed by means of electronic funds transfer, which ensures the security of the information.

Once the transaction has been completed, the information gathered could be destroyed. The vehicle owner should have access to a record of recent transactions carried out with his or her in-vehicle device, in case it is necessary to contest the validity of the transaction charges. With read-write tags or automatic debiting transponders, only the user could actually request as a preference that this data be written into the device's memory for later access. The only record of the transaction in almost all current schemes is held by the operating authority, with access available to the user on demand. Few, if any, on-board units record the transactions as an independent record and audit trial. However, in past demonstration projects, such as the Cambridge congestion metering trial, up to 50 of the most recent transactions were recorded on the user-held smart card that was inserted into the OBU. This option may again be offered; an electronic or a printed receipt is almost universally provided as a record of credit card or Internet transactions [24, 51] and so the card or tag log would replace the need for paper records.

2.3.6.3 Subscription Account Based upon Identification

Subscription involves the advance purchase of a "service right." This may be either the right for the user to pass a specified number of times without incurring any further charges (a concept similar to the Paris and Brussels underground networks' CARNET), or the right to use the road network an unlimited number of times within a given time period, like a season ticket (a concept similar to a London Underground TravelCard).

Subscription with identification is usually (but not exclusively) associated with the fixed-number-of-journeys principle. The information regarding the number of journeys that remains on a user's tag is usually held by the scheme operator. This information is checked and adjusted, in real time, with each passage through the toll site by the user.[7]

2.3.6.4 Anonymous Subscription Account

Anonymous prepayment subscription is generally operated on the same basis as a travel card (i.e., permission to use the road network as often as desired for a predetermined period of time). The time of day may also be differentiated in terms of an "off-peak" and a more expensive "peak" or "all-day" road use subscription. The subscription may also be arranged in terms of access allowed into differently priced zones.

The Spanish Association of Toll Road Operators (ASETA) established this system in the early 1990s with a read-and-write tag system. Certain data is written onto the transponder, which, upon interrogation, indicates that the tag is programmed with a code that indicates a right to use the roadspace without incurring an additional charge during the specified period, while still maintaining tag and use anonymity. Similar schemes have been successfully tried using colored stickers and vignettes, such as in Bergen, Norway [16, 43].

2.3.6.5 Automatic Debiting—On-Board Electronic Credit (Anonymous)

Automatic debiting tags and OBUs are emerging as a new generation of devices for automatic tolling and road user charging. The device allows for flexible on-board processing of data, and the facility to store user-held credit that has been purchased in advance from an operating authority. This credit may be stored directly in a secure memory area of the tag or OBU, or, more conveniently, in a portable value-card or smart card connected to them.

The ability to electronically store credit in the in-vehicle device allows for great flexibility in the charging of a variable fee (e.g., dependent on time of day, vehicle class, traffic conditions) for the use of the road, and the ability to inform the driver of the charges he or she is incurring. A flexible charge could be made using a tag system, but it would be difficult for the user to keep track of incurred charges. This is important not just for point-charging, but also if a vehicle-metering system is to be used. The main benefit of holding the credit on-board is that the transaction with the roadside can be achieved without the need for the identity of the user to be conveyed to the roadside system (under correct operating conditions). This will overcome the most serious of the concerns associated with current road use revenue collection systems—the threat to privacy, which may not be an issue when users choose to "opt in" to an optional e-tolling scheme, but may be if a road user charging scheme is mandatory.

The price to pay for this anonymity is added complexity of the software (and to some extent the hardware) required at both the roadside and in the vehicle's transponder. Nevertheless, it can protect the system from fraud and other misuse, which is a particular concern where actual electronic credit is being passed over the communications link from the vehicle to the roadside charging station.

7. Further discussion on how such information is held and made available to the user is provided in Chapter 6.

2.4 New Methods of Charging

2.4.1 Business Considerations

Less than 20 years ago, there were no automated systems for the collection of road user charging fees and tolls. If a road authority wished to collect fees for road use and tolls, then it required a largely manual process, in which the vehicle stops and the driver hands cash to an operator. The most advanced systems of the day were automated coin machines or magnetic cards that were inserted into a card reading device. Microelectronics had not yet really entered the transport domain, and the main form of communications between a vehicle and a roadside system was probably inductive loop-initiated communications, or citizens band (CB) radio.[8] The first transponders were developed for the transport sector in the mid-1980s to track railway vehicles, buses, freight containers, and for other rudimentary vehicle tracking and identification applications [5, 46]. Some of these systems used bar codes that were either optically or magnetically read, while others used radio frequency, and were coined RFID systems. These operated at different frequencies in different parts of the world. In the United States, 400 MHz and 902–928 MHz were used (902–928 was not used in Europe); in Japan, 2.45 GHz and 13.5 MHz were used; and in Europe, a range of frequencies were used, including most of the above, as well as 5.8 GHz and millimeter-wave systems in the 60-GHz region. Most of these systems read a small amount of data from the vehicle-mounted transponder to identify the vehicle (or the load).

These early technologies did show the toll collection industry the possibilities that future technological developments could offer. The first two systems that demonstrated this tolling and road use charging were the Hong Kong ERP trial in 1983–1985 [74], which used inductive loop communications from a buried loop in the road to an in-vehicle transponder on the underside of the vehicle; and in 1987, when the Ålesund toll road in Norway was the first to be commercially implemented at a toll site, and illustrated to operators from around the world that some automation of the toll collection process was possible, and that the benefits were apparent and quantifiable.

Operators of toll roads saw great advantages in electronic means of payment. They noted that it speeds up the toll collection process and reduces some of the major disadvantages of toll collection facilities, such as the congestion generated at peak periods of use, the noise and air pollution, and the delays that drivers experience [75].

Here is where the technology requirements for tolling and road user charges begin to diversify. The toll plaza has a fairly controlled and in most cases monolane operation, while the urban or wide area road user charging scenario must move away from the toll plaza concept, since such structures are impediments to traffic flow and are unacceptable in most environments. We are left with the requirements for toll road facilities where automated lanes are fitted to existing toll plazas to offer options other than manual or subscription payment, and more free-flow systems for road user charging and congestion charging, in urban areas where

8. Twenty years hence, it is now difficult to convey the message of how low-tech the road to vehicle communication was when the concept of electronic tolling began to be considered.

building a toll plaza is not possible. The mechanism for charging is often similar, but the differences in operation between monolane systems and those that must support free-flow multilane operation can be significant [76].

2.4.2 Monolane Operation

Manual toll collection has always been regarded as inefficient, due to the need for vehicles to stop, causing congestion and creating unnecessary noise and air pollution. The area (and cost) of land needed for a conventional toll plaza is great, with at least three manual lanes of toll collection equipment required for each lane of highway feeding into the toll plaza. This land is not readily available when building new roads, and is not available when toll collection or road use pricing is to be introduced on existing road infrastructure.

The concept of collecting user fees from a vehicle's driver without the need for the driver to slow down, stop, or perform any actions (other than driving) at the point of collection, is not new. Until a few years ago most automatic tolling systems had one or more lanes of a toll plaza equipped with automatic reading equipment, enabling drivers to pass through the toll lane at a reduced speed, without stopping, and without the need for the driver to hand over coins, cash, or a card. There was a real need for some form of automation of the toll collection process, and where such systems have been installed they have generally been met with a high level of acceptance from both the driver and the toll site operator. Many of these systems now exist across Europe and the United States. Early systems used extremely short-range communications between the in-vehicle tag and the roadside reading device. Communications technologies included inductive, low-frequency radio, and optical or magnetic barcode systems. These early systems were limited to a very short communications range, which required the passage speed of vehicles to be very slow, or in some cases, even required the vehicle to stop [8, 11]. Many systems of this type are still in use with operators, generally using radio or microwave frequencies for communications, and allowing vehicle passage speeds of up to 60 km/h. Examples of such tags can be found at the Mersey Tunnels, Severn Bridges, and Tyne crossings in the United Kingdom. The limitation of these systems are largely due to the fact that the toll collection procedure still requires barriers (as with the other collection lanes) and must adhere to the traffic management scheme prevalent at the collection site.

The other shortcoming of these early toll systems was that they were limited to conveying only a fixed identification code to the roadside system. A generalized schematic of such a system is shown in Figure 2.14. This fixed code relates to an account that the vehicle's owner has set up with the collection agency. These systems are known as read-only or AVI systems. However, many monolane systems now also use read-and-write capable transponders, which widens the options for payment and functionality. Systems developed for the monolane market are generally not suitable for use in a free-flow, multilane road use charging context [77, 78].

2.4.3 Multilane Systems

2.4.3.1 General System Design

Toll roads that were not specifically designed for multilane toll collection create a number of difficulties. First, the physical area required for a conventional stop-

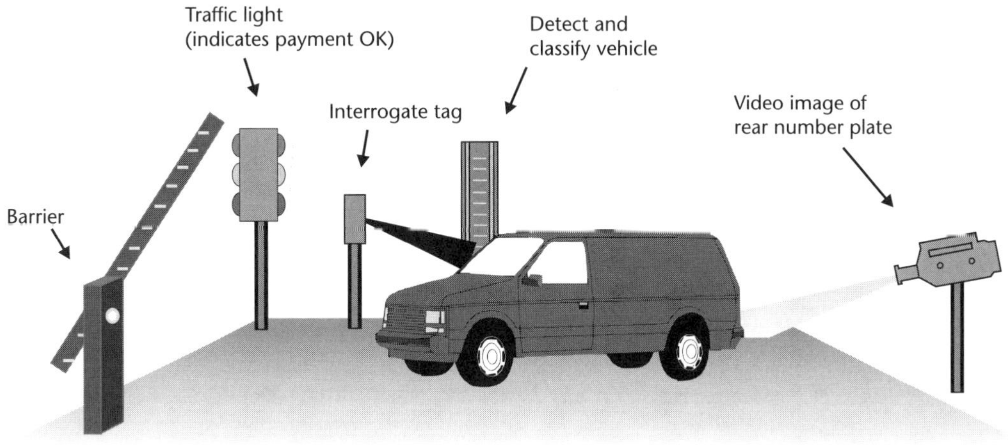

Figure 2.14 AVI system generalized architecture for monolane operation.

and-pay manual toll plaza or drive-through single lane AVI system is not available. Second, the number of entry and exit points on the road where tolling is applied retrospectively are generally much higher on a road specifically designed to permit the collection of tolls. Finally, the technical and procedural problems of how to electronically detect vehicles at the toll site, levy the correct toll electronically, and, where necessary, perform real-time enforcement of noncompliant vehicles without restricting the traffic flow, must be solved. This is the so-called multilane problem.

It should be reiterated that the multilane problem is a problem associated with certain technical systems. Systems that use short-range DSRC tags and transponders must address the multilane problem. The charging, classification and enforcement processes are all related to knowing with which vehicle the gantry system is in contact, and that any noncompliant vehicles can be identified and located.

Wide area systems that use GNSS, GSM, or some form of in-vehicle metering do not have the same requirement for charging. However, when the checking and the enforcement of these systems are performed, it is necessary to be able to identify and locate the position of the vehicle on the road for enforcement purposes; thus, free-flow multilane solutions are required for these processes [12, 28, 70]. A typical short-range tag multilane layout is illustrated in Figure 2.15.

This solution may seem cumbersome but it is necessary in many cases. The challenge is to design a reliable system that could use a single gantry. The challenge is also to achieve this in two distinct scenarios: on high-speed roads with high vehicle speeds, and on urban roads where there may be congestion and consequently many vehicle transponders within the range of a single roadside transceiver [79].

In between multilane and monolane operations, vehicles operate in a free-flow situation, but with the requirement that they stay in their lane when passing through an automated road charging point. This is often called quasimultilane operation. These three scenarios are illustrated in Figure 2.16.

2.4.3.2 Challenges

Vehicles are allowed to pass through the toll site or road user charging site in multilane free-flow situations, without any additional restrictions on speed or lane

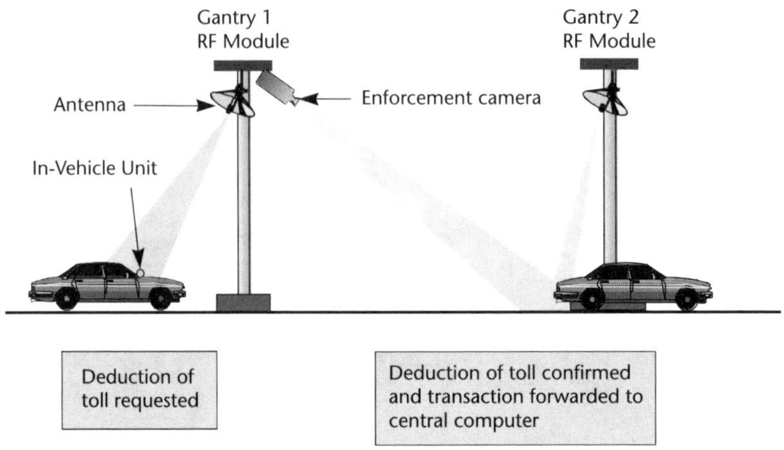

Figure 2.15 Typical arrangement for multilane road-user charging.

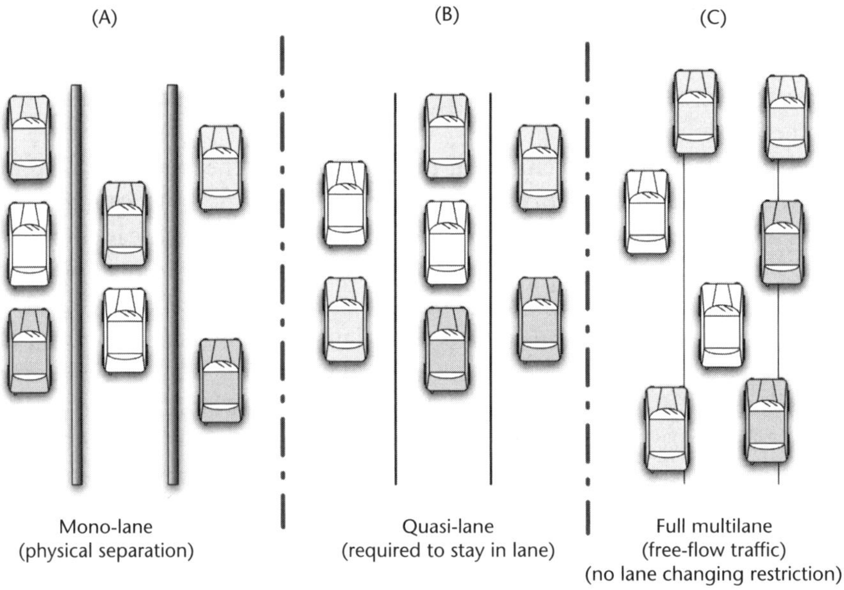

Figure 2.16 Multilane free-flow: operational scenarios.

discipline, other than those required for normal driving behavior. This means that vehicles are not restricted from passing or changing lanes at the toll site, but they are free to move as they would in normal traffic on a multilane highway.

This poses two problems for a multilane debiting and enforcement system. The first problem is communication between the vehicle's tag or OBU and the roadside tolling system, and the second is enforcement [80].

The communication problem arises because of the need to have an orderly dialogue with several vehicles transponders simultaneously, when more than one vehicle may be in the communication zone at any one time. This means that the system must maintain a secure logical communication link with each transponder for the period of time necessary for the debiting transaction to be completed.

The enforcement problem is determining which vehicle has not performed a valid payment and recording the details of the vehicle. There can be two reasons why a vehicle has not performed a valid transaction: (1) when a vehicle does not have a tag or OBU; and (2) when a vehicle does have a tag or OBU, but the payment transaction has not been performed correctly, either due to some system failure or an attempt to defraud the system. The spatial position of all the correctly paying tags or OBUs must be known to some reasonable degree of accuracy by the roadside system, in order for the system to perform a correlation (match) with the vehicle detection and classification (VDC) system, which must detect and classify vehicles passing through the toll site independently from the transaction system.

The problem of designing a system to operate correctly in a multilane environment is regarded as one of the most technically demanding challenges in ITS. These technical difficulties are due to the following distinct points:

1. For DSRC-based toll systems (or DSRC-based enforcement points in a GNSS or other wide area scheme), the time constraints imposed on the system due to the short communications window in which the debiting transaction may take place (typically 100 ms) at high vehicle speeds [81];
2. The need for the roadside system to communicate with, and perform a correct transaction with, all the equipped vehicles in the communications zone;
3. The requirement for the system to detect and classify all vehicles passing through the communications zone;
4. The need to determine which vehicles have correctly performed a transaction;
5. The identification of nonpayers;
6. The identification of unequipped vehicles;
7. Recording the identity of nonpaying vehicles for enforcement purposes.

For systems that use wide area communications, such as the global positioning system (GNSS)-based TollCollect System for lorry road user charging in Germany, the requirement to carry out the above steps (except step 1) is still valid. However, the point where these steps are performed may be distributed and not at the actual point where the transaction takes place [82]. There will still be some need for on-street multilane enforcement functionality to check that vehicles are correctly paying their charges, even if the charging is performed without the need for a multilane arrangement like that required for DSRC (transponder-based systems). The specific solutions will depend on the system configuration, charging policy, or enforcement regime [83, 84].

2.5 Complementary Systems

2.5.1 Vehicle Classification

As illustrated in the previous section, one of the ways to detect a violation is to check that the toll or fee being paid by a vehicle is the correct amount for that vehicle class. Where charges are differentiated by vehicle class, there must be an automatic vehicle classification scheme that can discriminate between vehicles of predetermined classes on a multilane highway in real time [27, 28].

Automatic vehicle classification is performed by measuring some parameters of the vehicle and comparing them to parameters stored in a database that defines the classes in use.

No common, clearly defined classes of vehicles exist for automatic toll collection systems. Some current implementations have as few as 6 classes, while some European operators document 32 different classes. The cost of a system that can accurately discriminate between 32 different classes of vehicles would put it beyond the means of most operators of multilane systems. Experienced operators suggest that a pragmatic approach to classification should be taken, since there is a great deal of trade-off between accuracy, cost, and complexity of the system.

The classification process is also complicated by the wide variety of vehicle designs, which gives rise to marked differences between vehicles within the same class and to similarities between vehicles of different classes. A wide range of vehicle dimensions exists, which must be collected and interpreted to correctly classify the vehicles. This is difficult to achieve with some vehicles, particularly two-wheel vehicles, due to their relatively small size.

Many new techniques for online vehicle detection and classification have been demonstrated and deployed in the past few years. Although remote classification measurements may be used, many toll operators rely on the vehicle's OBU declaring the class of the vehicle when it is in communication with the roadside system. The automated classification systems are used as a backup and a threat to operators and individuals who choose to defraud the system. Chapter 5 focuses on the details of vehicle detection and classification.

2.5.2 Enforcement

To enforce an automatic road use pricing scheme, noncompliant vehicles must be detected and their identities recorded to provide evidence valid for prosecution [53, 56]. All vehicles are currently required by law to clearly display a license plate on the front and rear of the vehicle, in most European countries, and in most of the United States, with the exception of motorcycles that are required to display only a rear license plate. Recording of the license plate number provides a means of uniquely identifying noncompliant vehicles. Other information, such as the make, model, and color of the vehicle, may support prosecution of noncompliant vehicles and prevent the unjust prosecution of compliant drivers, particularly if drivers have falsified their license plates to avoid identification.

Simply manually noting the license plate details is one means of enforcing a road pricing system, although it is impractical for a heavily used widespread system. Some form of photographic method is necessary to maintain an efficient and effective enforcement procedure recourse to automatic recording equipment [57].

Two variants of automatic recording equipment currently available, based on photographic and video cameras, are used to provide pictures of license plates, which are then read manually off-line. ANPR systems are also available that may be considered to automate online or off-line processing of license plates.

In the online case, the system is located at the roadside and the license plate numbers of noncompliant vehicles are recorded as the vehicles are detected. In the off-line case, the system is located in a central control station, and replaces manual reading of images captured by photographic or video cameras [63].

The reading of vehicle license plates by automatic methods involves image-processing techniques for vehicle-presence detection, the accurate location of the license plate in the image, the processing of the license plate image to isolate the characters from the background, and the identification of the characters. High-quality, high-contrast images are required for accurate reading [85].

Video and photographic techniques to detect and locate license plates, record the images, and read the characters online are among the most rapidly changing fields in ITS [86]. Many vendors are offering innovative and high-performance solutions, motivated by the success of the ANPR technologies utilized as the primary form of charging and enforcement in the London Congestion Charging Scheme. Enforcement technology options are the focus of Chapter 4.

2.6 Summary

This chapter has introduced the concept of road user charging and tolling, and the technical issues and options associated with the implementation of such policies. The choices presented in terms of technology and operational modes of the charging system are complicated by the policy-orientated goals that the systems must meet. The following four chapters consider each of the key elements of the system in more detail, and the design options, trade-offs, and choices are then discussed.

References

[1] Blythe, P. T., and P. J. Hills, "Automatic Toll Collection and the Pricing of Road-Space," Ch. 7, *Advanced Technology for Road Transport*, Ian Catling, (ed.), Norwood, MA: Artech House, 1994, pp. 119–143.

[2] Hau, T. D., "Congestion Charging Mechanisms for Roads," *World Bank Policy Research Working Papers*, Series WPS 107, 1992, pp. 1–99.

[3] Button, K., and E. Verhoef, *Road Pricing, Traffic Congestion and the Environment: Issues of Efficiency and Social Feasibility*, Cheltenham, U.K.: Edward Elgar, 1998.

[4] Paulley, N., "Recent Studies on Key Issues in Road Pricing," *Transport Policy*, Vol. 9, No. 3, 2002, pp. 175–177.

[5] Martin, B., and A. Scott, "Automatic Vehicle Identification: A Test of Theories of Technology," *Science, Technology, & Human Values*, Vol. 17, No. 4, Autumn 1992, pp. 485–505.

[6] Hills, P. J., and P. T. Blythe, "Paying Your Way," *IEE Review*, Vol. 35, No. 10, November 1989, pp. 377–381.

[7] Armstrong, J., "Breakthroughs in Vehicle Identification," *Railway Age*, June 1984, pp. 40–48.

[8] Catling, I., "Automatic Vehicle Identification," in *Information Technology Applications in Transport*, P. Bonsall and M. Bell, (eds.), Ch. 3, Utrecht, the Netherlands: VNU Science Press BV, 1987, pp. 41–64.

[9] Davies, P., "Testing and Appraisal of Automatic Vehicle Identification Systems," *Proc. 5th World Conference on Transport Research: Transport Policy, Management & Technology Towards 2001*, Yokohama, Japan, July 10–14, 1989, pp. B355–B363.

[10] De Lozier, E. J., "How and Why AVI?" *International Bridge, Tunnel and Turnpike Authority International Symposium on AVI Technology for Toll Collection*, New York, 1990.

[11] Foote, R. S., "Prospects for Non-Stop Toll Collection Using Automatic Vehicle Identification," *Transportation Quarterly*, Vol. 35, No. 3, 1981, pp. 445–460.

[12] Blythe, P. T., and B. Laurent, "Electronic Tolling in Europe," *Proc. EU-China Conference on ITS Applications*, Beijing, China, June 1997.

[13] Sommerville, F., "Applications of Automatic Vehicle Identification Technology," *Transport Reviews*, Vol. 11, No. 2, 1991, pp. 173–191.

[14] Hoven, T., "Experience with the Norway Toll Ring Project," *Proc. IBC Conference on Electronic Payment Systems in Transport*, Amsterdam, March 1996.

[15] Tretvik, T., "The Trondheim Toll Ring: Applied Technology and Public Opinion," *Joint OECD/ECMT/GVF/NFP Conference in Urban Travel Management*, Basel, Switzerland, 1992.

[16] Larsen, O. I., "The Toll Cordons in Norway—An Overview," *Journal of Transport Geography*, Vol. 3, No. 3, 1995, pp. 187–198.

[17] Menon, A. P., et al., "Singapore's Road Pricing System: Its Past, Present and Future," *ITE Journal*, Vol. 63, No. 12, December 1993, pp. 44–48.

[18] McCarthy, P., and R. Tay, "Economic Efficiency vs. Traffic Restraint: A Note on Singapore's Area License Scheme," *Journal of Urban Economics*, Vol. 34, No. 1, 1992, pp. 96–100.

[19] Toh, R., "Experimental Measures to Curb Road Congestion in Singapore: Pricing and Quotas," *Logistics and Transportation Review*, Vol. 28, No. 3, 1992, pp. 289–312.

[20] Toh, R., "Road Congestion Pricing: The Singapore Experience," *Malayan Economic Review*, Vol. 22, No. 2, 1977, pp. 52–61.

[21] Blythe, P. T., "Electronic Road Pricing in Singapore—Report on Demonstration Phase," *RTI Focus Report*, September 1994.

[22] Oldridge, B., "Congestion Metering in Cambridge City," *PTRC Conference on Practical Possibilities for a Comprehensive Transport Policy With or Without Road Pricing*, Cambridge: Cambridgeshire County Council, 1991.

[23] Ison, S., "A Concept in the Right Place at the Wrong Time: Congestion Metering in the City of Cambridge," *Transport Policy*, Vol. 5, No. 3, July 1998.

[24] Clark, D. J., et al., "The ADEPT Project: 3. Congestion Metering the Cambridge Trial," *Traffic Engineering and Control*, April 1994.

[25] Leromonachou, P., M. Enoch, and S. Potter, "All Charged Up—Early Lessons from the Durham Congestion-Charging Scheme," *Town and Country Planning*, Vol. 72, No. 2, February 2003, pp. 44–48.

[26] Livingstone, K., et al., "Driving Through Change—Introducing Congestion Charging in London," *Planning Theory and Practice*, Vol. 5, No. 4, December 2004.

[27] Banister, D., "Critical Pragmatism and Congestion Charging in London," *Int. Social Science J.*, Vol. 55, No. 2, 2003, pp. 249–264.

[28] Blythe, P. T., K. Walker, and P. K. Knight, "The Technical and Operational Feasibility of Automatic Number-Plate Recognition as the Primary Means for Road-User Charging," *Journal of the Royal Institute of Navigation*, Vol. 54, No. 3, 2001, pp. 345–355.

[29] Hensher, D. A., and S. M. Puckett, "Road User Charging: The Global Relevance of Recent Developments in the United Kingdom," *Transport Policy*, Vol. 12, No. 5, September 2005, pp. 377–383.

[30] Blythe, P. T., "RFID for Road Tolling, Road-Use Pricing and Vehicle Access Control," *Proc. Colloquium on RFID Technology, Institution of Electrical Engineers*, London, U.K., October 1999.

[31] Blythe, P. T, "Congestion Charging: Technical Options for the Delivery of Future U.K. Policy," *Transportation Research Part A*, Vol. 39, No. 7–9, August 2005, pp. 571–587.

[32] Kim, K. S., and K. Hwang, "An Application of Road Pricing Schemes to Urban Expressways in Seoul," *Cities*, Vol. 22, No. 1, February 2005, pp. 43–53.

[33] Nakamura, N., and K. M. Kockelman, "Congestion Pricing and Roadspace Rationing: An Application to the San Francisco Bay Bridge Corridor," *Transportation Research Part A: Policy and Practice*, Vol. 36, No. 5, 2002, pp. 403–417.

[34] Borins, S. F., "Electronic Road Pricing: An Idea Whose Time May Never Come," *Transportation Research A*, Vol. 22A, No. 1, 1998, pp. 37–44.

[35] Lay, M. G., and K. F. Daley, "The Melbourne City Link Project," *Transport Policy*, Vol. 9, No. 3, 2002, pp. 261–267.

[36] Pigou, A. C., *Wealth and Welfare*, London, U.K.: Macmillan, 1920.

[37] Vickrey, W. S., "Congestion Theory and Transport Investment," *American Economic Review* (Papers and Proceedings), Vol. 59, 1969, pp. 251–260.

[38] Walters, A. A., "The Theory and Measurement of Private and Social Cost of Highway Congestion," *Econometrica*, Vol. 29, No. 4, 1961, pp. 676–697.

[39] Smeed Report, "Road Pricing: The Economic and Technical Possibilities," *U.K. Ministry of Transport, HMSO*, London, 1964.

[40] Fong, P. K. W., "Issues of the Electronic Road Pricing System in Hong Kong," *Transportation Planning and Technology*, Vol. 30, 1985, pp. 29–41.

[41] Harrison, B., "Electronic Road Pricing in Hong Kong: 3. Estimating and Evaluating the Effects," *Traffic Engineering and Control*, Vol. 27, January 1986, pp. 13–18.

[42] Hau, T. D., "Electronic Road Pricing: Developments in Hong Kong 1983–1989," *Journal of Transport Economics and Policy*, Vol. 24, No. 2, May 1990, pp. 203–214.

[43] Lauridsen, H., "The Toll Rings in Bergen and Oslo: Evolution and Experience," *East Europe Transport Policy Seminar*, Paris, France, 1990.

[44] Langmyr, T., "Learning from Road Pricing Experience: Introducing a Second-Generation Road Pricing System (The Norwegian Experience of Road Pricing)," *Planning Theory and Practice*, Vol. 2, No. 1, April 2001, pp. 67–80.

[45] Odeck, J., and T. Skjeseth, "Assessing Norwegian Toll Roads," *Transportation Quarterly*, Vol. 49, 1995, pp. 89–98.

[46] Blythe, P. T., "State of the Art Review of Automatics Debiting and AVI Systems," DRIVE Project V1030 PAMELA, Commission of the European Communities, Brussels, Belgium, 1991.

[47] Larsen, O. I., "The Toll Ring in Bergen. Norway—The First Year of Operation," *Traffic Engineering & Control*, April 1988.

[48] Gillard, J., "Electronic Toll Collection at the Mersey Tunnel," *Proc. IBC International Conference Electronic Payment Systems in Transport*, Amsterdam, March 1996.

[49] Hong, S. P., "A Study of Real Time Vehicle Classification Using Image Sensors," *Proc. 5th World Congress on Intelligent Transport Systems*, Korea, October 1998.

[50] Burden, M. J. J., and P. T. Blythe, "The Enforcement of a Pan-European Multilane Debiting System," *Proc. Intelligent Transportation Systems 4th World Congress*, Berlin, Germany, October 1997.

[51] Blythe, P. T., "The Integration of Smart Cards with Microwave Transponders for Road-Use Pricing and Tolling Applications," *Proc. Intl. Symp. on RF and Microwave Communications*, London, U.K., September 1998.

[52] Kolay, R., B. T. Oranc, and M. D. Barat, "Automatic Tolling with Contactless Smart Cards," *Proc. 12th Intl. Congress on Intelligent Transport Systems*, San Francisco, CA, November 2005.

[53] Ison, S., and T. Rye, "Implementing Road User Charging: The Lessons Learnt from Hong Kong, Cambridge and Central London," *Transport Reviews*, Vol. 25, No. 4, 2005, p. 451.

[54] Perret, K., "Vehicle Classification and Enforcement Systems for Use with Electronic Fee Collection," *Proc. 5th World Congress on Intelligent Transport Systems*, Korea, October 1998.

[55] Grieco, M., and P. Jones, "A Change in the Policy Climate? Current European Perspectives on Road Pricing," *Urban Studies*, Vol. 31, No. 9, November 1994, pp. 1517–1532.

[56] Blythe, P. T., "Road Use Pricing Technology," *EU Conference on the Promotion of Road Pricing*, Commission of the European Communities, DGVII, Brussels, Belgium, September 1997.

[57] Pickford, A., "The Short Guide to EFC," *Traffic Technology Intl.*, April/May 2003.

[58] Horton, J., "Overview of the Highway 407 ETS," *Proc. 5th World Congress on Intelligent Transport Systems*, Korea, October 1998.

[59] Kauer, A. I., "Urban Freeflow in Santiago—Turning Up the Heat on Chile's Infrastructure Goals," *Tolltrans*, October/November 1997.

[60] Enoch, M., "Road-User Charging—Lessons from Scandinavia and the Far East," *Town and Country Planning*, Vol. 70, No. 11, November 2001, pp. 297–299.

[61] Opiola, J., "The Role of Private Motor Car Electronic Road Pricing in Hong Kong," *Proc. 5th World Congress on Intelligent Transport Systems*, Korea, October 1998.

[62] Vera, P. E., S. Hayes, and J. Burgell, "Findings from a GAUDI: Zone Access Control Field-Trial in Barcelona," *Traffic Engineering and Control*, Vol. 34, No. 3, 1993, pp. 114–121.

[63] Gaunt, G., and A. Stevens, "Toll Enforcement Using Number Plates," TRL (Transport Research Laboratory), Crowthorne, Berkshire, 1999.

[64] Evans, J., "Update on the London Congestion Charging Scheme," *IEE Seminar on Road User Charging*, London, U.K., March 2003.

[65] McQuaid, S., and M. Grieco, "Edinburgh and the Politics of Congestion Charging: Negotiating Road User Charging with Affected Publics," *Transport Policy*, Vol. 12, No. 5, September 2005, pp. 475–476.

[66] Rye, T., S. Ison, and M. Enoch, "Lessons from Edinburgh's No," *Town and Country Planning*, Vol. 74, No. 7/8, July/August 2005, pp. 228–239.

[67] O'Mahony, M., D. Geraghty, and I. Herbert, "Distance and Time Based Pricing in Dublin," *Traffic Engineering and Control*, Vol. 41, No. 1, January 2002, pp. 17–19.

[68] Birle, C., "Use of GSM and 3G Cellular Radio for Electronic Fee Collection," *IEE Seminar on Road User Charging*, London, U.K., June 9, 2004.

[69] Patchett, N., and D. Firth, "Overview of Results from Phase 1 Congestion Charging Technology Trials in London," *Intl. Workshop on Future Road User Charging Research Challenges*, Newcastle, DfT/Newcastle University, February 2005.

[70] Thorpe, N., and P. J. Hills, "Experiences from Designing and Implementing a GPS-based Road-User Charging System," *Proc. 5th Intl. Congress on ITS*, Korea, October 1998.

[71] Catling, I., "Road User Charging Based on Satellite Positioning Systems and Cellular Network Communication—Progress on Standardisation and Interoperability," *Proc. 10th ITS World Congress*, Madrid, Spain, November 2003.

[72] Schelin, E., "Current Status of Road User Charging in Sweden," *IEE Seminar on Road User Charging*, London, U.K., June 9, 2004.

[73] Mackinnon, D., "The DfT Road User Charging Research On-Road Programme," *IEE Seminar on Road User Charging*, London, U.K., March 2003.

[74] Dawson, J. A. L., and I. Catling, "Electronic Road Pricing in Hong Kong," *Transportation Research A*, Vol. 20A, 1986, pp. 129–134.

[75] Blythe, P. T., and M. J. J. Burden, "Electronic Toll and Traffic Management—New Developments in Technologies and Systems," *Proc. Asia Roads and Highways*, Hong Kong, September 1994.

[76] Delaney, T. D., and T. Davis, "Developing a Regional Payment System to Meet the Needs of Transit Tolls and Parking," *Proc. 11th Intl. Congress on Intelligent Transport Systems*, Nagoya, October 2004.

[77] Okamoto, T., "A Study of the Deployment of Electric Toll Collection System," *Proc. 10th Intl. Congress on Intelligent Transport Systems*, Madrid, Spain, November 2003.

[78] Savion, E., "Cross Israel Highway Toll Road," *Proc. 10th Intl. Congress on Intelligent Transport Systems*, Madrid, Spain, November 2003.

[79] Blythe, P. T., "Electronic Tolling in Europe: State of the Art and Future Trends," *Operation and Maintenance of Large Infrastructure Projects*, Balkema, 1998, pp. 85–102.

[80] Skadsheim, A., "Electronic Payment in Denmark's First Toll System," *IBC Conference, Electronic Payment Systems in Transport*, London, U.K., 1998.

[81] Stogis, Y., "Systems Management and Traffic Telematics Implementation on the Egnatia Motorway in Greece," *Proc. 10th Intl. Congress on Intelligent Transport Systems*, Madrid, Spain, November 2003.

[82] Kossak, A., "Tolling Heavy Goods Vehicles on Germany's Autobahns," *IEE Seminar on Road User Charging*, London, U.K., June 9, 2004, http://www.iee.org/oncomms/pn/auto.

[83] Pickford, A., "Pay Time (Lorry Road User Charging—Europe Wakes Up)," *Annual Review, Traffic Technology Int.*, 2004, pp. 82–86.

[84] Egeler, C., and M. Bibaritsch, "Enforcement of the Austrian Heavy Goods Vehicle Toll," *Proc. 10th World Congress on Intelligent Transport Systems and Services*, Madrid, Spain, November 2003.

[85] ROCOL, *Road Charging Options for London: A Technical Assessment*, Report, HMSO, London, U.K., 2000.

[86] Miles, J. C., and K. Chen, *ITS Handbook*, PIARC, 2004.

CHAPTER 3

Technology Options for Charging

3.1 Background

Historically, tolling via cash at discrete locations on the route had been the only direct means of paying for road use. The traditional policy of using tolls to help pay back the cost of construction and operations has since been supplemented by several new forms, including area pricing, cordon pricing, and distance-related charging, largely for demand management purpose. Technology availability and capability helps influence policies, and vice versa: Policy development guides future direction of technology evolution. This chapter focuses on the collection of charges for road usage based on measurement of road usage, and the capture of vehicle-related information to support the enforcement process when a charge cannot be properly levied. For charging to be effective, it cannot depend on every vehicle being equipped with technology. If the use of an OBU is not mandatory, then the occasional user that does not have an OBU needs to be included, and alternative payment methods need to be offered, including cash.

Perhaps the first notable study of charging technologies was the Smeed Report [1] published in 1964, which examined the economic and technical issues associated with road user charging as a restraining and demand management measure. In the context of congestion charging, the report made the following observations.

> Vehicles must carry identification units which enable their presence to be recorded by roadside apparatus. The recording must be in a suitable form to comprise the input data of the computing equipment. The system must be capable of distinguishing between, say, 30 million different vehicle identities [. . .] We have enquired about optical, electromagnetic, radar and sonic methods, and the only serious proposal put to us was the electromagnetic Link Tracer suggested by Professor William Vickrey for vehicle identification in Washington DC. The capital cost quoted for the vehicle, roadside and computing equipment was £12 10s 0d per vehicle [. . .] a good deal higher than the £5 that we allowed. [Note: £5 in 1964 is about £64 ($112) today.]

A suggested alternative scenario was based on time spent within a priced zone. Vehicles would be required to install an automatic meter.

> The automatic meter tries to eliminate much of [the responsibility of both driver and traffic authority] by placing control apparatus in the road [. . .]. The setting of the meter is performed for [the driver] by [a] switching circuit which operates in response to signals received for road-sited transmitters installed at the zone entry and exit points and intermediate points within the zone.

The technologies available when the report was written to implement charging systems were severely restricted to electromechanical devices, with almost no communications capabilities available the time. Nevertheless, the principles of vehicle identification, location-specific charging, and automatic metering within charged zones described over 40 years ago underpin today's policy approaches to charging. Building on Chapter 2, which translated policy options into functional requirements, the following sections map these onto feasible technologies, and present the pros and cons of the options available.

For the 10 years beginning with 1987, the majority of pay-per-use charging services were based on ETC plazas. Whenever the vehicle enters the toll lane, the vehicle's OBU is accessed to identify the means of payment and other account-related information, in a process known as AAI. AAI provided a simple solution for locally focused charging schemes that are based on the pay-per-use policy, although some of the earliest projects offered subscription accounts. Trondheim, one of Europe's first ETC installations, also applied a maximum fee payable in any month. After opening an account, the user installed a small OBU on the inside of the vehicle's windshield. An example of an OBU design is shown in Figure 3.1.

The use of the term *tolls* reflects the underlying rationale for funding of the infrastructure and its operation, in principle, although any automated process that enables the measurement and charging of road usage for the same purpose can also be described as an ETC. Chapter 2 distinguishes between the policy objectives of tolling and road user charging, and this distinction is continued here to show how charging policies influence the selection of charging technologies, and how these technologies, in turn, must be combined to meet policy requirements.

Chapter 2 also identified a range of possible charging policies, including tolling and other forms of pricing based on crossing cordons, traveling within a charged

Figure 3.1 Typical DSRC OBU.

area, and variations of these policies. Charging can also be applied to all road users in selected geographic areas, such as an interurban highway or a city. Furthermore, vehicles may be charged only if the entry to the charged area is within a specific time period. The technologies required in the vehicle and roadside infrastructures have become more complex as the charging policies have evolved. Conversely, in many cases, the technical possibilities have often led to the consideration of new policy options.

Section 3.2 defines the minimum operational requirements for charging for road use, and Section 3.3 highlights how precedence can influence scheme designs. The dilemma is whether or not to allow a progressive evolution to more advanced forms of charging, since this approach may encourage organizational and institutional inertia, limit policy innovation, and reduce the long-term benefits that tolling and road user charging could offer. The alternative is more rapid change as technology capability permits.

Automating the charging process means that payment is no longer linked to charging. Section 3.4 explains why this is the case and what this means for future charging schemes. Since the choice of technologies is guided by the underlying charging policy, Section 3.5 identifies technology building blocks (e.g., traditional plaza-based ETC schemes, and advanced city-wide, regional, or national pricing schemes), and shows how these technologies can be combined to deliver various charging policies. This section also shows how scheme operators can accommodate all road users, even those without any in-vehicle technology. Section 3.6 introduces standardization and the different levels of interoperability that enable road users to travel within a charged road network made up of different schemes, each with their own charging policy. The evolution to increasingly more complex charging policies places more diverse demands on the charging technologies themselves. Section 3.7 focuses on how the technology building blocks will evolve, and how closer integration with the vehicle may be required to improve the efficiency and effectiveness of the charging and enforcement processes. Finally, Section 3.8 summarizes this chapter.

3.2 Minimum Operational Requirements for Charging Technologies

The use of tolling and road user charging has increased as an efficient means of funding infrastructure development, operation, maintenance, and demand management, both in the urban environment and increasingly on strategic arterial routes. Today, a road user, whether in a developed or a developing country, is more likely than ever to come into contact with such a scheme. In regions where toll collection is already widespread, a typical journey may include traveling on two or more separately charged road segments or zones.

Each scheme operator is likely to be presented with a bewildering array of technology options for charging and enforcement. Although the imposition of tolls or charges is enabled by technology, the charging policies have been shaped by technologies themselves. Policymakers need to know that the policy can be delivered at an acceptable risk. In turn, the requirements on charging technologies are indirectly determined by the charging policies themselves.

The starting point to identify charging technologies is the set of minimum operational objectives that need to be met by a charging scheme:

- To uniquely identify the vehicle, since it is the vehicle's use of the road that is chargeable;
- To measure road usage, either as discrete events or on a more continuous basis, to determine the correct charge;
- To uniquely identify an authorized means of payment;
- To inform the driver or account holder that a charge has been levied, either at the point of charging or via a periodic statement;
- To support the enforcement process, ensuring payment if a vehicle cannot be linked to an authorized means of payment, or if other charging discrepancies exist.

Many of the products and services that are required to successfully implement a charging scheme depend on technical innovation, technology development, and deployment. The user requires that the service must be fair, understandable, easy to use, safe to use while driving, and convenient. Developing user confidence, accessibility, and a high level of compliance are all critical to the long-term economic success of a charging scheme.

In-vehicle equipment must communicate the vehicle's road usage and other declarations (e.g., exemptions, discounts, or user-related information) to external systems. For example, an AAI system only needs to know the account information at the point of vehicle detection, whereas a distance-related charging scheme needs to know the distance traveled on chargeable roads. If there is no in-vehicle equipment, then the enforcement process needs to be based on the only unique information that can be observed on the vehicle, namely, its license plate. Chapter 4 elaborates on the relationship between charging and enforcement.

3.3 The Dilemma of Precedence

Technology selection is not an automatic process. Existing technology is often used as an excuse to do more of the same in the future, without consideration of changes that are occurring in the fiscal, political, technical, and legislative processes that are often inextricably linked to charging. Historical precedence provides lessons on what could work, and offers reassurance that a specific technology will meet the requirements where substantial public or private investment is required (e.g., building a new road). This leads to a combination of past and present technologies coexisting in a single scheme, particularly for tolling, where toll plazas allow the simultaneous operation of both drive-through ETC lanes and less automatic forms of payment.

This simultaneous view on what has been shown to work and what will be required for the future often presents a dilemma. In the worst case, operators act independently, resulting in a fragmented approach to technology selection, based entirely on satisfying local needs and minimizing risk. Technology choice should instead reduce the cost and improve the efficiency or effectiveness of the charging process, while meeting policy objectives for tolling and road user charging, and, if possible, enabling new service offerings to road users. However, as road user charging is adopted at local, regional, and national levels, road users will typically travel on several chargeable road segments, each based on a different charging policy. Users should not have to understand the differences between the increasing number of charging schemes, even if the charging technologies are apparently identical. Instead, users should expect to experience seamless roaming between these policy areas, in the way that mobile phones roam between networks and across international boundaries. The complexity of an individual scheme and its relationship to other schemes should therefore be invisible to users.

If each policy area required a different charging technology (e.g., tariff structures, payment channels, and so forth), then the user would face functional and usability barriers that are unrelated to any other costs of paying for road use, which could undermine the user's understanding and support for the principles of charging. The technology choices should be limited, but may be more than one. Technology choice should therefore aim to make road user charging more accessible and understandable for road users. This aim must also consider the privacy and data protection expectations of road users, particularly when there are multiple scheme operators, as discussed in Section 4.4.4.

3.4 Charging Versus Payment

Cash payment of tolls highlights the simplicity of the charging process. Traditional cash-based toll collection systems combine charging and payment into one event, simply by the transfer of cash from the road user to the toll collection attendant at the point of payment.

As automated charging methods are introduced, we need to clearly differentiate between charging and payment. The charging process is strategically important for all scheme operators; it uses all the information relating to the vehicle's passage to establish the amount due. Conversely, payment is the obligation of road users (or accountholders) to transfer funds to the scheme operator, or to an intermediary established to accept fees relating to the road usage.

Road usage and payment for road usage are usually separated in time, at least for electronic payment methods. A driver may either prepay or postpay for road usage. For example, closed toll roads (see Section 2.3.2) depend on the issuance of a ticket (physical or electronic) on entry, which is then used to calculate the fee at exit. The toll road operator requires the user to provide a valid means of payment at the point of exit, which could be an electronic record provided by in-vehicle equipment that contains enough information to uniquely identify an authorized account.

The account itself may be prepaid or postpaid, but nevertheless, the scheme operator would need sufficient confidence (i.e., a financial risk assessment embodied in business rules) to allow the vehicle to leave the chargeable road segment without enforcement. For example, if a barrier-controlled ETC toll lane cannot identify the account information (or if none were provided), then the enforcement barrier would prevent the vehicle from leaving the lane. However, on an open highway, drive-through nonstop toll plaza, or in an urban charging scheme, enforcement would typically be based on digital imaging systems used to capture evidence of a vehicle's identification and presence. The charging and payment processes are inextricably linked to the enforcement process, regardless of the choice of charging technology. Chapter 4 further discusses the relationship between charging and enforcement, while Chapter 6 explains the matching of payments with charges.

The measurement of distance traveled would trigger a payment after the road usage has occurred. The collection of records that enables a charge to be computed may occur hours or days after the recorded road usage, simply to reduce the load on the record collection and billing system. Chapter 6 discusses central system operations and billing in detail.

3.5 Functional Requirements and Technology Choice

3.5.1 Technology Building Blocks

The first step in identifying charging technologies is to determine the functional requirements, and the second is to translate them into technology options.

The apparent choice between technologies is more likely to be a choice between a cluster of complementary technologies that, when coordinated, measure, report, and calculate road usage. The charging policy itself will determine whether it is necessary to measure the distance traveled by the vehicle, or whether it is sufficient to only detect and identify the vehicle once (e.g., on entry to an open toll road). The appropriate technology building blocks sometimes will be obvious due to local precedence. The introduction of ETC at a single isolated plaza requires no more than vehicle (account) identification and notification to the user that a charge has been made. If vehicles are charged for the use of all roads based on distance traveled, then the technology building blocks will need to include distance measurement, reporting, notification to the user, and integration with fixed and mobile enforcement. There are intermediate cases in which the technology options are not clear, but the steps remain the same; policy requirements must be translated into functional requirements, and then the functional requirements used to outline the technology building blocks.

Table 3.1 shows the relationship between functional requirements and technology building blocks. Since a charging policy cannot exist without the means to enforce it, Table 3.1 adds another function—the need to support the enforcement process. Additional technologies are needed to make the charging process secure, robust, accurate, and auditable. A short list of these essential elements is also provided.

There are three main approaches to charging, each comprising a cluster of the technology building blocks:

- DSRC;
- CN/GNSS/DSRC and augments;
- ANPR.

DSRC and GPS have evolved in parallel from very different origins, and both were conceived as tangible technologies in the mid-1970s. Both have passed through

Table 3.1 Functional Requirements and Technology Building Blocks

Function	*Technology Building Blocks*
Vehicle identification	ANPR RFID Dedicated short-range communication (DSRC)
Discrete location determination	ANPR + video image capture RFID DSRC Future methods, such as continuous air interface for long and medium range initiatives (CALM) (multiple communication methods), and ultrawideband (UWB) for localization within discrete zones
Continuous location determination	Satellite-based positioning: GNSS (including GPS, GLONASS, Galileo, and Loran-C) Terrestrial positioning systems, such as Enhanced Observed Time Difference (E-OTD), time of arrival (TOA), angle of arrival (AOA), and their variants/hybrids Proximity and vicinity detection In-vehicle positioning augments and assisted global positioning system (A-GPS) provided by the network
Measurement of distance traveled	Identification of individual segments and addition of their separate lengths Odometer/tachograph Integration of position estimates over time, matched to a map of the road network
Reporting from in-vehicle equipment to enable road usage to be charged	Vehicle-infrastructure communications: Localized discontinuous communications, such as DSRC Cellular networks (CN), such as GSM, code division multiple access (CDMA), wideband CDMA (WCDMA) Future options: Wi-Max Secure memory card or smart card Other methods of reporting, such as manual pay stations
Notification to road user or accountholder	Audible indicator or man-machine interface (MMI) (e.g., display or keypad) Off-line notification by e-mail, short message service (SMS), and so forth
Enforcement support	OBU localization Electronic vehicle identification (EVI), electronic registration identification (ERI) Localized vehicle-to-infrastructure communications, such as via DSRC
Additional essential functions	Integration with enforcement Data encryption and security key schemes to protect charging data from tampering or modification OBU authentication at charge points to protect accounts from fraudulent OBU Vehicle detection and classification to ensure that the correct charge relating to vehicle type is made Application support, such as on-off board map matching, and route reconstruction to help build the final bill for road usage

several generations, both are now available in mass-market products, and both are well supported by an internationally competitive industry. GPS and DSRC perform completely different functions (positioning and communications, respectively), but this has not stopped frequent, direct comparison and misleading claims of the relative split of cost between the vehicle and infrastructure by industry segments that have historical roots (and significant R&D investments) in either GPS-based or DSRC-based developments.

ANPR was initially used in closed user group access control schemes from about 1985. It then provided support to manual enforcement processes for toll plazas from about 1990. It has generally been accepted as an essential enforcement tool for tolling and road user charging applications.

A scheme designer making decisions on charging technology choice will also need to consider the degree of automation influenced by several factors, including the quantity of charging events, vehicles, and accounts. However, the potential quantity of unusual conditions can be the most significant operational cost driver. These exceptions include misread license plates, errors in measured distance, dependency on the user at the time of charging, process errors, and so forth. The main determinants of technology choice include the charging policy, type of road user (measured by frequency of use), capture accuracy (detected events), data capture accuracy (accuracy of reporting events), and the business case for the technology itself. Figure 3.2 shows the relationship between three technology forms differentiated by usage.

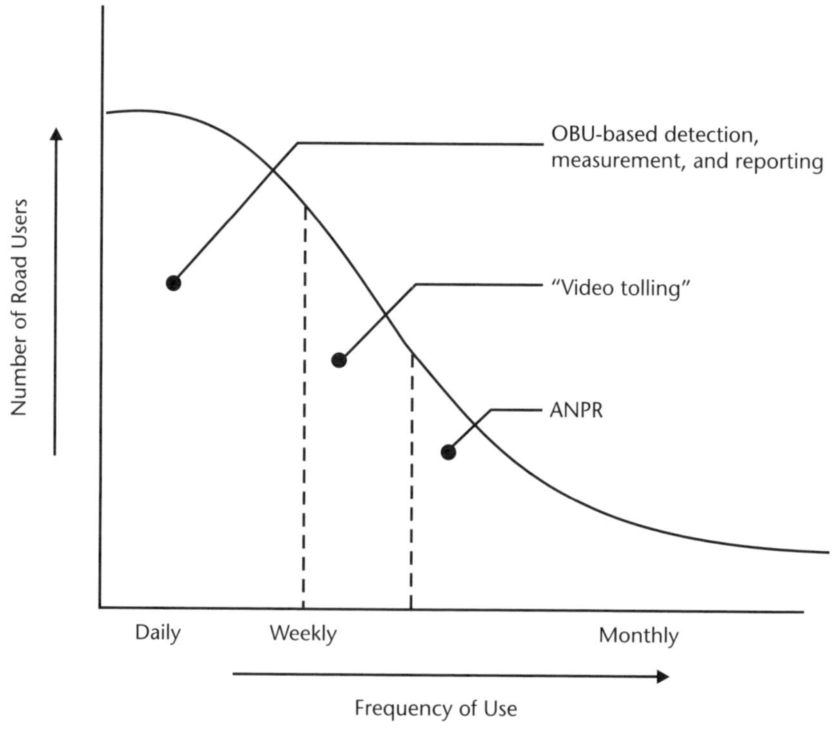

Figure 3.2 Technology choice and usage.

- OBU-measured usage or OBU-triggered charging events;
- Image-triggered charging events (video tolling);
- ANPR, enforcing a period licensing scheme (such as a day pass).

The importance of usage relates directly to the business case; higher usage is best satisfied with greater automation to capture the benefits of economies of scale and reduced transaction costs. This is analogous to capital-intensive mass production compared with handcrafted, low-volume production. The investment in OBUs (by the scheme operator and user) and related roadside infrastructure needs to be offset by the savings in transaction costs over the lifetime of the investment, as described below.

The boundary lines between the approaches shown in Figure 3.2 are not to scale, and will depend upon the transaction costs for each type of transaction, which in turn depend upon the investment in charging process capacity in each type and lifecycle costs for each subsystem. The relationship between data capture accuracy, the business case, and charging policy is also described in the following section.

3.5.1.1 Accuracy and Business Case

Frequent users of a road network generate more chargeable events, so it makes sense to use the most efficient, automated means of recording and reporting their road usage. This uses in-vehicle equipment where single-point capture accuracy is required, and video tolling or ANPR where multiple detection points are possible. The equipment costs are outweighed by the operational cost savings through more accurate and automatic recording of road usage. The cost to the road user (e.g., time, effort) is also reduced through this automation. The frequent road user and the ETC scheme operator both benefit from the use of OBUs (also called tags). The operational cost saving made by the operator can be shared with the road user in the form of a per-transaction discount, as offered to all EZ-Pass accountholders in the United States, for example. This can increase the adoption of OBUs, which further reduces the operational cost for each charging event. The data capture accuracy of an OBU (DSRC and CN/GNSS) is virtually 100%. With adequate security management this means that the data can be trusted, and used to levy a charge without any manual intervention. Overall encouraging regular users to adopt an OBU means that the highest possible volume of charging events can be automatically handled.

The cost/benefit ratio changes for infrequent road users. The cost of an OBU to an operator includes handling, personalization, packaging, distribution, replacement, and customer support. The adoption of tags by infrequent users would not make economic sense, unless the OBU could be made interoperable with other operators, or the toll charge is sufficiently high (e.g., the Storabælt Bridge in Denmark, where passenger vehicles pay €28 or $34 per crossing). ANPR offers the opportunity to identify the vehicle of an infrequent user by its license plate. ANPR can be used to enforce a charging scheme (e.g., London Congestion Charging), or can be configured for video tolling.

ANPR cameras typically have a low data capture accuracy, so video tolling relies on the capture of multiple images (e.g., front and rear license plates) at a single location to improve data capture accuracy for a single charging event. This requires manual validation to ensure that the charge is applied to the correct account (e.g., Melbourne City Link, 407 ETR, Cross Israel Highway).

Section 3.5.4 gives further information on the use of ANPR for charging.

3.5.1.2 Charging Policy

DSRC is typically used as the primary method of charging where a charge is to be applied at one of a discrete number of specific points, such as a toll plaza or a location on the open highway. Over 60 million DSRC OBUs are in use worldwide, mainly for ETC. The Austrian truck tolling scheme uses DSRC for segment-by-segment charging on motorways (see Figure 3.3).

Table 3.1 shows that enforceable, distance-based charging schemes from continuous measurements can be provided by a combination of GNSS (continuous measurement determination), CN (reporting), and DSRC (identification for enforcement). Accurate GPS-based position estimates can be compared with an on-board or off-board database of the road network to work out the most likely road segment on which the vehicle is traveling. Each road segment could have its own tariff (probably proportional to its length and time of day), which means that it is possible to determine the charge for the road segment. The OBU contains functions to filter out any noise in the measurements, the effect of reflections from nearby objects such as buildings, and distortions due to atmospheric disturbances.

Figure 3.3 DSRC charge point (LKW Austria). (Courtesy of Kapsch TrafficCom AG.)

The OBU may also be able to get external assistance data from the scheme operator that alerts the OBU to available satellites, and provides corrections for short-term distortions to improve the acquisition time of satellites. The acquisition time from an initial start is known as the time to first fix (TTFF). Section 3.5.3 discusses further variants to improve OBU positioning performance through augmentation. The alternative solution that uses only DSRC (i.e., discrete location determination and identification for enforcement) could be equally technically viable. The business case would reveal which is more economically appropriate, after considering the enforcement infrastructure for all methods of charging, the extent of the chargeable roads, quantity of vehicles, interoperability with other schemes, and the need for discrete DSRC infrastructure for charging compared with the operationally more complex GNSS OBU.

The distance traveled by a vehicle can also be based on direct measurement from the vehicle odometer, although this method alone does not identify the road type, so would not permit charges to be differentiated between road types. An in-vehicle OBU that incorporates a GPS module can be used to estimate the vehicle position, although positioning information by itself is not always accurate enough to determine distance traveled [2].

The Swiss heavy truck tolling scheme Leistungsabhängige Schwerverkehrs-abgabe (LSVA) has used a feasible hybrid solution since 2001, which relies on an odometer to measure the distance traveled by the vehicle, DSRC to turn on and off at international borders, and GPS to provide redundancy and to audit the odometer reading. Other variants are expected to emerge, depending on whether there are one or two tariff boundaries (e.g., motorways and other roads), or more than two boundaries (e.g., charges differentiated on all road types). The increased quantity of tariff boundaries generally increases the dependency on continuous positioning. The Austrian and U.S. schemes, including PrePass, Norpass, and Commercial Vehicle Information Systems and Networks (CVISN) [3, 4], depend on detection of the vehicle at discrete locations on strategic routes to enable the allocation of fees or gas taxes to the states in which trucks pass.

By comparison, the New Zealand truck tolling scheme [5, 6] is based exclusively on manually reading the distance traveled from a certified odometer fixed to the hub of trucks (all diesel engine vehicles), although this scheme is not able to identify the road type.

Overall, 6 million CN/GNSS OBUs are in use, with small-scale pilots for distance-related charging underway in Europe, the United States, Australia, Southeast Asia, and Japan, potentially for all vehicles. A sample OBU that incorporates CN, GNSS, and DSRC technologies is shown in Figure 3.4.

We can already see that simple requirements may need more than a single technology. These examples also show that technology choice is not a choice between charging technologies, but rather a selection of an appropriate bundle that meets local needs, and, if they exist, regional and national policies.

Sections 3.5.2 to 3.5.4 outline the three main technology groups. Section 3.5.5 deals with occasional users. The ability to roam between schemes that apply different charging policies depends on the regional interoperability strategy, as discussed in Section 3.6.

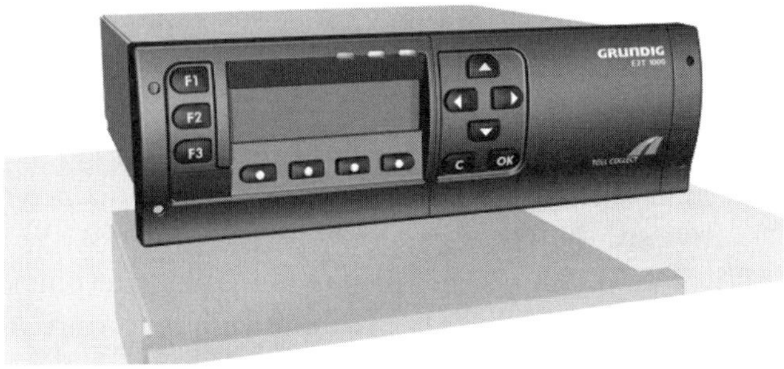

Figure 3.4 Hybrid GNSS/CN/DSRC Toll Collect OBU designed specifically for Toll Collect to be used in HGVS for in-dash mounting. (Courtesy of Efkon Mobility, Delphi Grundig, and Toll Collect.)

3.5.2 Dedicated Short-Range Communication

3.5.2.1 Background

DSRC is a localized, bidirectional, high-data-rate channel that is established between a fixed roadside system and a mobile device installed within a vehicle. The most widely used frequency bands for DSRC are 902 to 928 MHz (mainly North America); 5.8 GHz; or 5.9 GHz, depending on locally applicable standards; and infrared frequencies (mainly selected countries in Southeast Asia). See Table 3.2. Other frequencies have been used in the past, including 2.45 GHz (still used

Table 3.2 Variants of DSRC

Frequency Band (Primary and Secondary)	*Applicable Standards*	*Communication System*	*Dominant Regions of Use*	*Dominant Application Area*
5.850 to 5.925 GHz	IEEE P1609.1 to P1609.4 and ASTM- E2213-03 WAVE Platform	Active	United States	Road user charging and electronic toll collection
5.875 to 5.815 GHz	CEN DSRC Specifications	Modulated backscatter	Europe, South America, Australia, Southeast Asia	Safety, public services, road user charging, and electronic toll collection
850 nm (Wavelength)	CALM IR ISO CD 21214	Active	Malaysia, Taiwan (planned), and Germany	Road user charging and electronic toll collection
5.790 to 5.810 GHz and 5.83 to 5.85 GHz (primary); 5.770 to 5.790 GHz and 5.81 to 5.83 GHz (secondary)	ARIB STD-T75	Active	Japan	Electronic toll collection
902 to 928 MHz	Title 21	Modulated backscatter	United States, Canada, Mexico	Electronic toll collection

in Hong Kong and Singapore), and 850 MHz (SAW technology, initially used in Oslo, Norway). The standardization process saw the migration to 902 to 928 MHz (mostly the United States) and 5.8 GHz (Europe, South America, and Southeast Asia), using so-called modulated reflectance or backscatter techniques for communication. Since 1990, the Telepass-branded ETC system in Italy has been based on a single-vendor 5.9 GHz solution complying with a local standard [7]. The standardization of DSRC in Europe has been slow, although there are examples of national and cross-border schemes.

A modulated reflectance OBU is able to rapidly vary the reflective property of its antenna, which is known as a patch antenna, and is typically a single patch of copper less than 5 cm^2, to transfer incident RF energy generated by a roadside DSRC transceiver, back to the transceiver. The OBU does not generate any RF, but it merely modulates the reflected energy. When using RF or microwave frequencies, these systems work in a master-slave (S/M) mode. The roadside antenna transmits data to the OBU using a modulated carrier. When the OBU needs to transmit data, the roadside antenna transmits an unmodulated carrier signal, which is received by the OBU, modulated on the carrier, and then reflected back to the roadside antenna. The fact that the OBU reuses the signal from the roadside transmitter severely limits the range of the DSRC systems, since the attenuation of the reflected signal follows the R^4 power law (i.e., the received signal is attenuated by a power of four proportional to R, the range of the communications).

The use of modulated reflectance for communication allows the OBU to operate at very low power levels, requiring either a long-life battery (DSRC 5.8 GHz), or no battery at all (902 to 928 MHz), where regulations permit sufficient energy to be transferred to the OBU. The communication distance typically ranges from 10m to 20m. This is sufficient to enable localized, lane-specific communications at toll plazas and OBU localization for tracking and enforcement in open road charging schemes known as open road tolling (ORT), which is a combination of a toll plaza alongside open lanes, and multilane free-flow (MLFF), which is an open road without any plaza.

The most common applications of DSRC are electronic toll collection (ETC) at toll plazas and MLFF/ORT schemes, and as localized communication for enforcement as part of GNSS solutions (e.g., the German truck tolling scheme). Figure 3.5 shows a scheme that employs DSRC as the primary means of charging.

DSRC technologies have traditionally been considered as simply another payment option within the tolling application area. DSRC roadside systems (e.g., transceivers and lane controllers) have evolved to provide a simple (although proprietary) interface to existing toll lane equipment, along with magnetic and smart card readers, manual toll terminals (MTTs), and ACMs. The technology initially could only cope with very low vehicle speeds (less than 25 mph), and only limited amounts of application data could be exchanged between the OBU and DSRC roadside system (RSS). Following 20 years of development, speeds up to 100 mph/180 km/hr and integration with high-performance enforcement equipment is now considered routine, which is confirmed by the willingness of financiers to back MLFF schemes worldwide.

The main functions of a DSRC-based charging point, highlighted in Figure 3.5, are:

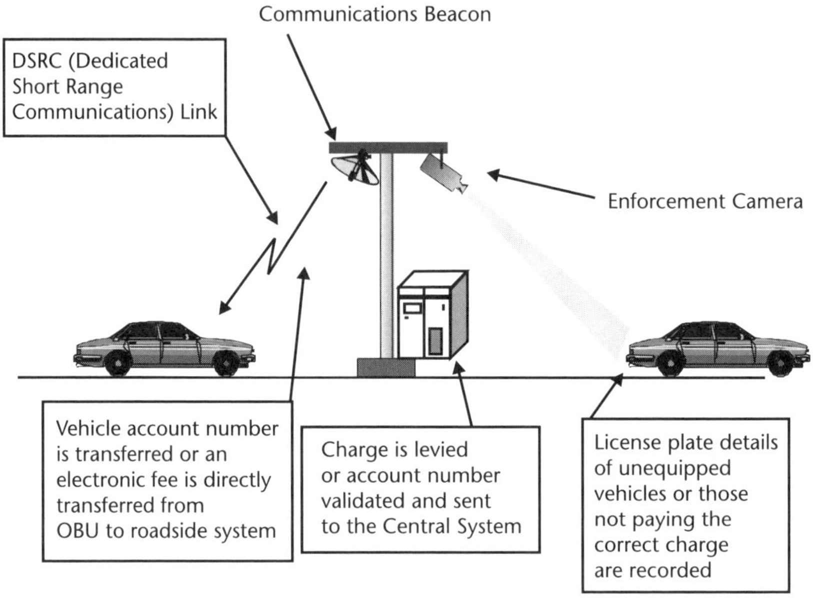

Figure 3.5 Schematic of DSRC scheme.

- Storage of account-specific and optionally vehicle-specific data within an OBU for declaration to a roadside system;
- Transfer of the OBU data to a roadside system over a directional DSRC interface;
- The ability to spatially localize the OBU in ORT/MLFF systems, or to limit communication to a single vehicle within the toll lane;
- Interpretation of the received information, packaging, and transmission to the central system;
- Detection and management of occasional (unequipped users);
- Capture of images, if any discrepancy is detected between the OBU's declarations, locally held account information, and direct measurements.

DSRC could be deployed at the boundary points between road types that are differentiated by charging rates, if the charging policy and functional requirements allow this. The number of boundary points (defined by the underlying charging policy) represents a significant cost factor for DSRC-based charging systems. For ETC, a significant cost factor is the number of toll lanes that offer ETC services. For all DSRC implementations, the number of tags issued is also a cost factor, although unit prices for at least 100,000 tags would be approximately €17 (approximately $21).

Triggered by the FCC's allocation of 75 MHz of spectrum to ITS applications, future U.S. development efforts [8] will include the 5.9-GHz band, with the active participation of the Institute of Electrical and Electronics Engineers (IEEE) and the American Society for Testing and Materials (ASTM) [9]. The most recent addition to DSRC is the IEEE P1609 family of standards [10–14] and ASTM E2213-03 [15], which comprise the 5.9 GHz Wireless Access for Vehicular Environments

(WAVE) platform. This platform uses active transceivers at both ends of the communication link to achieve operating ranges up to 1 km; although the focus is primarily on safety, it also enables a broad range of ITS applications, including ETC. The U.S.-led OmniAir consortium is developing certification specifications and related over-the-air transaction definitions to enable multivendor support for WAVE-compliant products. Prestandard WAVE products are being readied for application testing, ahead of the scheduled publication of IEEE 802.11p in June 2007 [16]. WAVE forms track 2 of the U.S. Department of Transportation (DOT)–led vehicle infrastructure integration (VII) initiative [17], which aims to incorporate communication technologies in all vehicles and on all major U.S. roadways. Consumer access to WAVE-related services will depend on collaboration with the automotive industry, and will be subject to the vehicle planning life cycles of these companies. Chapter 9 gives further information on WAVE and the VII.

3.5.2.2 Extended OBUs

Some OBUs have a modular design, facilitating add-on peripheral equipment (e.g., smart card readers, keyboards, displays, and connections to other in-vehicle equipment). Such OBUs were first developed in the early 1990s by the EU-funded ADEPT project [18, 19], led by the Transport Operations Research Group (TORG) in the United Kingdom, Sweden, Portugal, and Greece. The modularity in the design of these prototypes allows several different forms of payment (all of them cashless) with one device. Possession of this form of OBU offers users the possibility of holding a positive (or a limited negative) credit balance, either directly in the OBU's memory or on a separate smart card interfaced to the OBU. The smart card, being portable, can then be used for other payment purposes, and hold an audit record of incurred transactions.

The key limiting factor in on-board automatic debiting systems is the processing speed of the smart card. In Singapore, each charging point has two gantries: one to start communications with the vehicle and a second (further down the road) to complete the transaction and perform enforcement measures. Nevertheless, despite the speed limitations of mainstream products, smart card–based solutions are well proven in plaza-based ETC schemes in other countries, including Italy and Malaysia (see Figure 3.6). Turkey uses a smart card for in-lane use.

Schemes that use DSRC as the primary means of charging usually use ANPR as an enforcement system. The license plate is currently the only available unique identifier that can identify the vehicle if the charging equipment is not working properly, or is not installed. Chapter 4 discusses this further.

The Singapore ERP scheme, Melbourne City Link (Australia), Cross-Israel Highway (Israel), Costanera Norte (Chile), and Highway 407 (Canada) are the most familiar DSRC schemes, since they were the first in their respective regions.

The lowest cost OBUs are monolithic; that is, the only external interface is via an ultrahigh frequency (UHF), microwave, or infrared (IR) link. The payment transaction result traditionally was communicated to the driver via lights or variable message signs located in toll lanes. The evolution of multilane, open highway systems resulted in a simple interface being added to the OBU, typically a monophonic beep and light emitting diode (LED) indicators. Enhanced versions have a

Figure 3.6 OBU with integrated smart card reader. (Courtesy of Q-Free.)

direct external interface to the vehicle (as demonstrated by the ERTICO-led DELTA project [20]), a utility serial interface, multilane display, and an integrated smart card reader.

The current markets served by DSRC have the following typical characteristics and requirements:

- *Focused application:* The systems should support tolling in single lane environments, and tolling and road user charging in ORT/MLFF environments.
- *Inexpensive end-user equipment:* Low-cost, mass-produced OBUs should have an operational lifetime of at least 5 years (ideally 7 years).
- *User-installed:* OBUs are designed to be distributed through retail outlets, automated vending machines, or by post. This ensures high market penetration with limited (or no) installation support from the highway operator, although there is always a risk that a small percentage of the units will be incorrectly fitted.
- *Minimal interface capability:* Minimal interaction with the user is required.
- *High speed:* Performance should be predictable and reliable in constrained low-speed toll lanes and in high-speed (typically more than 100 mph/ 180 km/hr) lanes. Transaction error rates are claimed to be less than 1 in 10 million in MLFF environments.
- *Harsh environment:* They should be capable of operation between extremes of ambient temperatures, from parked vehicles sitting in direct sunlight to subzero temperatures.
- *Autonomous:* The OBU is simply fixed to the windshield using a proprietary holder, with no interface to the vehicle. Tamper detection is available.
- *Low lifetime cost:* Battery life should range from 3 to 10 years for a simple interface. The roadside system can notify the user at a DSRC charge point by a simple audio/visual indication to return the OBU to the issuer at a predetermined time interval for a replacement unit.

- *High volume:* An estimated 60 million units have been deployed worldwide, with typical project batch sizes between 50,000 and 100,000. Start-up volume batch sizes are sometimes greater, based on forecasts of initial adoption rates.
- *Limited support for other ITS applications:* The limited communication range of modulated reflectance devices (from 10m to 20m, depending on applicable standard) means limited support for other ITS applications. The WAVE platform promises a range up to 1 km.

Competition for large-scale projects between 1996 and 1999 in the United States led manufacturers to compete on OBU unit price rather than on roadside system price. This precedent impacted European vendors, leading to an early establishment of a unit cost (to a highway operator) of between €17 and €30 (approximately $20 to $36) for OBUs, which is estimated to fall to less than €15 (approximately $18) within 5 years.

Specialized OBUs are also available to meet local requirements, including:

- Taxi-Tag available from Melbourne City Link (Australia), which increments the taxi meter with total charges for the trip;
- Explosion-Proof OBU required by Dartford Thurrock Crossing (United Kingdom) for petrochemical fleet operators;
- Motorcycle OBU offered by the Singapore Land Transport Authority (LTA), comprising a weatherproof enclosure to protect the smart card and balance display;
- OBUs with mounting brackets for passenger cars and heavy goods vehicles with various types of windshields;
- External antenna OBUs, offered by Autobahnen und Schnellstrassen-Finanzierungs-Aktien Gesellschaft (ASFINAG) (Austria) to trucks that have metallized windshields;
- An OBU with an external connector to allow a manual lane operator to read the tag without a DSRC reader; for example, a simple serial interface and display used by some Télépeage Inter-Société (TIS) operators in France to access on-board data.

These variants do not modify the DSRC interface and therefore do not impact the communications interoperability with roadside equipment. However, the different mechanical configurations and display capabilities limit the direct exchange of one manufacturer's tag for another, although this is rarely an issue.

The impact of standards, the development of interoperability specifications, and the separation of procurement of roadside systems from OBUs have broken the interdependence between pricing strategies for OBUs and roadside systems. The legacy of this is a broad array of OBUs, differentiated by cost, brand name, user interface, and availability of an integrated reader smart card. The most important factors in a global market are unit cost, standards compliance, and ability to meet interoperability specifications, although isolated schemes may continue to benefit from proprietary solutions.

3.5.2.3 Failure Rates

The main cause of failure of a DSRC transaction at a single point of detection is incomplete or no communication with the roadside system.

Under a controlled environment, using systematic testing with trained drivers, the probability of incomplete or no communications is typically 0.005%. Under live conditions at several MLFF DSRC projects (i.e., optimal geometry), the long-term average is between 0.3% and 0.5% at a single point of detection. This error rate can be reduced in proportion to the number of detection points along a defined route by logically rebuilding the journey between locations where the OBU was detected.

The most common causes of incomplete or no communication failures are as follows:

- *Incorrectly mounted OBUs:* This can be mitigated by high levels of user compliance achieved by clear installation instructions, and by associating OBUs with specific vehicles.
- *Unmounted OBUs:* OBUs may be on a dashboard, on the seat, or held in the hand. This can be mitigated by making it more difficult to swap tags between vehicles (contractual restrictions), and suppression of the user's belief that the OBUs contain value.
- *Dead OBU (faulty):* This can be mitigated by encouraging road users to contact the operator if the OBU does not provide any audible notification at a charge point.
- *Dead OBU (battery):* This can be mitigated by battery management within the OBU. Examples include shutting the battery down automatically when the terminal voltage reaches a predetermined level, notification to the driver to return the OBU to the operator, battery voltage monitoring and reporting, low-battery fault monitoring, and activity timers for reactive OBU management methods. These policies permit the road operator to plan in advance when to replace an OBU to reduce the probability of in-service failure.

3.5.2.4 Integration with Enforcement

Figure 3.5 highlighted typical features of a DSRC charge point with enforcement capability. The geometric arrangement of communications, vehicle detection, classification, and enforcement permits vehicles to be detected, tracked, and spatially paired with OBUs at the point of charging. Depending on the charge point configuration, vehicles may be tightly constrained within a toll lane, which simplifies the enforcement function. The DSRC subsystem merely has to confirm that the OBU declares sufficient information to be consistent with an in-lane vehicle classification subsystem, and associate the OBU with a valid account. Vehicle detection and (optionally) classification subsystems with unconstrained toll lanes are required to provide spatial information, which enables a vehicle to be matched with an OBU localized with the DSRC subsystem. The precise methods of the matching process are dependent on the vendor and project. Figure 3.7 shows an example from Stockholm, and Figure 3.8 shows an example from London, both of which employ matching techniques.

Figure 3.7 Communication and enforcement (Stockholm Congestion Charging Pilot, Sweden). [Courtesy of ITS (UK).]

The Stockholm pilot system configuration was based on a cordon of 18 entry points corresponding to 39 separate charge points. The figure shows the largest site, covering nine travel lanes. The site configuration includes lane-centric, laser-based vehicle detectors (center gantry) that trigger a corresponding ANPR camera (nearest gantry) as the vehicle approaches. This enables the camera to capture an image of the front license plate, while accurately truncating the image to remove information on the driver. A rear ANPR camera captures the rear license plate when the rear of the vehicle is detected by the same vehicle detector. This configuration of gantries enables highly accurate vehicle detection and high availability ANPR, and is a result of the policy requirements for the tax (not a charge) collection scheme.

The London charge point is located on a single pole/outrigger for aesthetic purposes, since many of the charge points are located in or close to residential and commercial sites. The geometric configuration of the charge point shown permits spatial matching of vehicles with their corresponding OBUs. Note that Figure 3.8 is part of a DSRC technology trial in London, not part of the operational London Congestion Charging scheme described in Chapter 8.

3.5.3 Cellular Networks/Global Navigation Satellite System

3.5.3.1 Background

The generic term for the satellite systems used for positioning or navigation is GNSS. GNSS technology within an OBU estimates position by combining measurements of

Figure 3.8 Communication and enforcement (Trial Urban Charge Point, London).

signals from a constellation of orbiting satellites, typically GPS or the Global Orbiting Navigation Satellite System (GLONASS).[1] CN refers to the bidirectional communication between an OBU and a fixed network of terrestrial transmitters, usually commercial cellular services, such as CDMA, GSM, or Universal Mobile Telephone Standard (UMTS) [third generation (3G) in Europe] mobile telephone networks [21]. GNSS-based charging also requires the creation and maintenance of a digital map of the chargeable road segments, since the position of a vehicle for charging purposes needs to be related to these segments.

In theory, positioning and communications can be continuously provided services, although in practice both are subject to the uncertainty of radio coverage (i.e., a sufficient number of satellites are not always visible, and cellular networks do not have 100% coverage). The positioning function needs to be specified (possibly with assistance and augmentation), such that it is able to accurately identify the road segment on which the vehicle is traveling, or at least flag when an accurate position cannot be determined. The reporting strategy needs to indicate that cellular network coverage is not always available (e.g., lack of coverage, loss during cell handover, or lack of available capacity). Alternative methods of reporting may need to be considered, such as batching data to be subsequently exchanged with the OBU, or requiring the user to transfer the data by memory card.

1. GLONASS is operated by the Coordination Scientific Information Center (KNITs) of the Ministry of Defense of the Russian Federation.

CN/GNSS reflects a combination of technologies, in which OBU position estimation is reported to a central collection hub site, otherwise known as a technical back office. A DSRC transceiver is usually also integrated, allowing the OBU to communicate with fixed and mobile enforcement points.

A generic positioning system uses radio transmissions to estimate position. The first areawide navigation systems used ground-based transmitters to provide reference signals for measurement. Although terrestrial positioning systems are still widely used, satellite-based transmitters are used to cover the majority of the Earth's surface, and provide positioning information with higher accuracy than from terrestrial systems. The satellites transmit timing information, satellite location information, and information that describes the health of individual satellites. The *Space Segment* is the technical term for this constellation of satellites. The most widely used satellite constellations are GPS and GLONASS, sponsored by U.S. and Russian government agencies, respectively. A third constellation, known as Galileo, funded by a consortium of member states of the European Union and others, will commence service in 2008, and will interoperate with GPS. Figure 3.9 shows the main elements of a scheme that uses CN/GNSS as the primary means of charging.

Every GNSS system employs a constellation of orbiting satellites working in conjunction with a network of ground stations. Every OBU requires a special radio receiver that is able to receive and decode the transmissions from visible satellites. This receiver uses triangulation to locate the OBU by combining information from a number of satellites, each of which transmits specially coded signals at precise intervals. The difference in time for signals to be received from the visible satellites is used to calculate the relative distance that the receiver is from each satellite. Using this information, and the fact that the receiver accurately knows the location

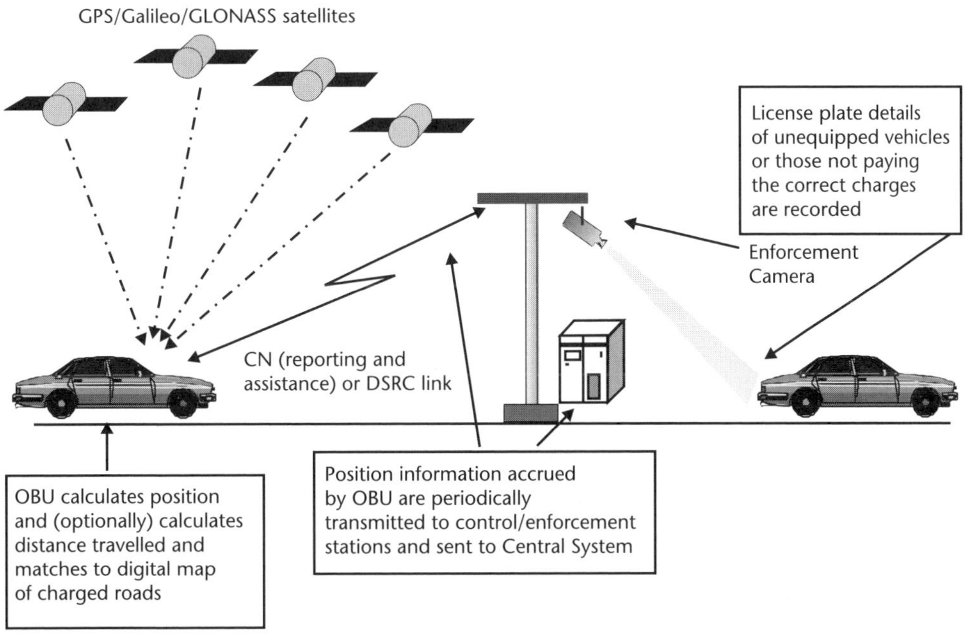

Figure 3.9 Schematic of a CN/GNSS scheme.

of each satellite and any time and point on its visible orbit, the location of the receiver in the vehicle can be calculated. The receiver converts this signal information into the position and velocity of the receiving OBU and provides an estimate of time. The OBU calculates its own position by coordinating the signal data from four or more satellites captured at about the same time. A minimum of three satellites is required to calculate location on the Earth's surface, while a fourth satellite signal enables the height above the Earth's surface also to be calculated. In practice, the receiver utilizes the signals from as many satellites as are in view practically a maximum of 12 to 13, to help overcome errors and ensure accuracy. The users (i.e., the OBUs) and their receivers are collectively known as the *User Segment*. The satellites are controlled and monitored from several ground stations, which are collectively known as the *Control Segment*. These stations monitor the satellites for health and timing accuracy, and are able to upload maintenance commands, orbital parameters, and timing corrections as needed.

It is important to note that the user does not have to transmit anything to any satellite, and that the satellites do not have the capability to track OBUs. The space segment does not need to know of the existence of the OBU, since the OBU is merely a receiver of a broadcast signal. Thus, there is no limit to the number of receivers, including OBUs, that can use the system at any one time. A typical GNSS/CN OBU for windshield mounting is shown in Figure 3.10.

The GPS and GLONASS systems each provide two sets of positioning signals with different degrees of accuracy. The higher accuracy signal was originally reserved for each country's military use, and the lower accuracy signal was available to civilian users without charge. On May 1, 2000, this restriction was removed from GPS. By comparison, the business model for the future Galileo operation is likely to be based on different service levels linked to escalating fees. The services offering the highest accuracy and availability will be charged, although general positioning capability will be offered without charge at the point of reception. Galileo will also provide an integrity indicator, so that the OBU will know whether

Figure 3.10 GNSS/CN OBU for windshield mounting. (*Source:* Siemens.)

the received signals can be trusted. This ensures that position estimates and the charges related to the vehicle's position will be credible. GPS integrity monitors are already available, although most have limited benefit.

3.5.3.2 Performance

The quality of the positioning information from a satellite radio receiver inevitably varies over time and by position of the measuring device, so we should assume that the location could only ever be regarded as an estimate. The quality of the output from GNSS depends on accuracy, yield, and latency.

- *Accuracy* is the linear offset between the actual position and position estimate, when available.
- *Yield* (0% to 100%) is the probability of providing a location estimate within a defined time period.
- *Latency* is the time from a position request to the availability of a location estimate.

Accuracy, yield, and latency are interdependent and depend on several factors:

- Time of day, since the space segment constellation geometry varies throughout the day;
- Atmospheric disturbance;
- Impact of local environment (e.g., multipath or occlusion within tunnels or urban canyons);
- Nonoptimal orientation of GPS antenna and attenuation by vehicle;
- Local multipath interference;
- Integration time of receiver;
- Instability and offset of receiver clock.

Many reports [22] into the performance of autonomous GPS in widely varying environments are based on receivers that track satellite integration times in excess of 20 minutes. However, time-critical applications, such as accurately detecting when the vehicle has crossed a tariff boundary, may require the maximum latency to be no greater than 10 seconds, and the position of the OBU relative to a charged area to be known to within 99% certainty (or better). The implementation of a charging policy may sometimes require a road segment to be identified, possibly based on several independent measurements by the same OBU over a short period, and then matched by position and direction of travel to the location and orientation of a road link that is recorded on the on-board or off-board map database. The corresponding charge can be calculated from the identity of the road segment, length, and the tariff at the time of travel.

The ERTICO-led road charging interoperability (RCI) initiative places requirements on positioning accuracy of GNSS subsystems: 95% of location estimates shall lie within 20m of the true position [23]. This technical accuracy underpins the charging accuracy based on road segment identification. Although the technical accuracy is important to ensure operational integrity, the scheme operator and road user are more interested in the billing accuracy, which depends on all road segments being correctly reported. The accuracy requirement for missed or incorrectly reported road segments creates a requirement on two parts of the operation:

- The positioning accuracy (relative to the correct chargeable road segment);
- The accuracy of the charges actually levied on the road user by the central system (see Section 6.2.3) as shown in Figure 3.11.

If the positioning accuracy is not sufficient to correctly identify the road segment (e.g., two parallel roads having different tariffs), then the final bill will be wrong. This may be mitigated by several methods, such as providing additional local augmentation at difficult locations on the road network (e.g., the German truck tolling scheme uses IR beacons to broadcast the identity of some road segments); auditing a vehicle journey to identify apparently missed or inconsistent road segments; or using the integrity information to decide whether or not to use a position estimate.

Both the Swiss LSVA and German truck tolling schemes employ GPS to provide continuous vehicle position information. As described above, the Swiss system uses the vehicle's odometer as the primary means of determining road usage. DSRC is used to enable and disable distance measurement and for enforcement. Figure 4.7 shows an example of a Swiss enforcement point. GPS provides a redundant backup to the odometer and DSRC functions, and confirms that the odometer is switched on and recording. The German scheme uses a mix of GPS to identify the road segment on which the vehicle is driving based on an on-board map database, and, where GPS is not available or where chargeable and nonchargeable roads are in close proximity, roadside infrared DSRC beacons provide localized fill-in information.

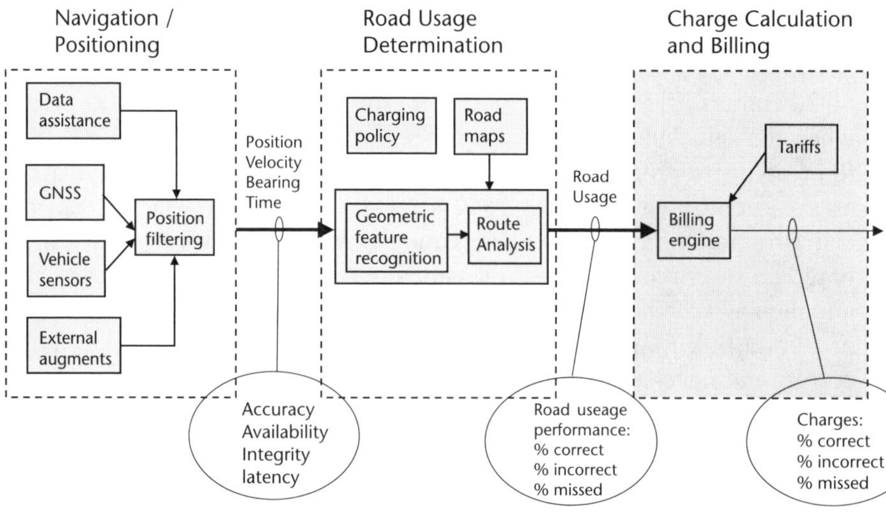

Figure 3.11 Positioning, usage determination, and billing. (*Source:* Mapflow, 2006.)

3.5.3.3 An Intelligent Client or a Thin Client?

There are primarily two types of GNSS OBUs, which differ in the division of tasks between the in-vehicle equipment and the roadside systems. The minimum requirement on the OBU is to capture satellite data and estimate a position. The minimum requirement on the central system to which the OBU reports is to allocate the total aggregate charge to the appropriate account. The ERTICO-led RCI group allocates the following tasks to either the OBU or the central system:

- Getting processed sensor data;
- Comparing data to determine location;
- Calculating charging data;
- Aggregating charge data up to thresholds.

Although the definitions of thin and intelligent have not been standardized, it is generally accepted that an OBU that estimates position and matches this to the terrestrial data of road segments is known as an intelligent client. The OBU is required to maintain a database of the road network on which the vehicle is likely to travel. The alternative approach limits the OBU to estimating its position, temporarily storing this information on-board, and subsequently reporting this data with corresponding time stamps to the central system to be matched with a map database. This is known as a thin client.

Table 3.3 compares intelligent and thin clients.

Technology vendors each make competing claims on the benefits of each system. Thin clients delegate much of their responsibility to an intelligent central system, and are the current direction of development for the delivery of location-based services for mobile phone users [24]. Thin client OBUs do not require a locally maintained map database, but still communications traffic from OBUs to the central system. 2.5G and 3G cellular networks are able to support this capacity. The same evolution in communication services benefits intelligent clients, which are chosen for both the Swiss and German truck tolling schemes.

Table 3.3 GNSS: A Comparison of Intelligent Versus Thin Clients

*Intelligent Client**	*Thin Client*
Position estimation, map matching, and reporting	Position estimation and reporting
On-board map database and tariff table, with possibility of outdated versions	No on-board map database or tariff table
Summary reports only (road segments)	Detailed reports (time-sequenced position estimates)
Potentially lower communications for reporting, offset by increased updates of map database and tariff tables	Potentially greater communications overhead and related costs, offset by no need to retain map database and tariff table updates
Near-real time display of accumulated charges	Charge only determined when report has been transmitted to central system

*Known also as a thick client to reflect its complexity compared with a thin client.

3.5.3.4 Improvement Through Augmentation

Additional factors improve the accuracy of the location estimate: data assistance from overlay services and cellular network, application augments, and complementary technologies. Each is discussed next.

Data Augmentation

Additional overlay satellite services are available to correct GPS signal errors caused by ionospheric disturbances, timing errors, and satellite orbit errors. The confidence that an OBU can attach to position estimates depends in part on the health of each satellite. Overlay services can also provide integrity information regarding this health. North American users have access to the Wide Area Augmentation System (WAAS) [25], European users have the European Geostationary Navigation Overlay Service (EGNOS) [26], and GPS receivers in East Asia have the Japanese Multifunctional Satellite Augmentation System (MSAS). Other comparable overlay services are available in India and China.

Terrestrial radio networks (i.e., a commercial GSM or CDMA network) can provide assistance to the terminal, either on-demand or broadcast periodically. This is referred to as Assisted GPS (A-GPS). The assistance data provides the OBU with knowledge of available satellites, along with corrections for time and atmospheric conditions. Assistance can therefore reduce the receiver search time, increase the number of valid observations (to increase the probability that a location can be computed with better geometry), and increase the accuracy of the observations available within the GNSS OBU.

A-GPS is a new technology that capitalizes on extensive development into the GPS network, and has driven the growth in expertise serving the emerging consumer and commercial markets for autonomous GPS terminals. These historically stable markets are vertically oriented among a very limited number of fabless[2] licensors of chipset designs/Intellectual Property, chipset vendors, and system integrators. This leads to a concentrated supplier base for GPS-based products.

A-GPS is a variant of autonomous GPS, which aims to compensate for measurement offsets, reduces the TTFF waiting time for a location estimate, and will provide a small improvement in received sensitivity to increase the number of visible satellites. Increasing the quantity of satellites that are visible to the in-vehicle receiver will improve the location geometry and reduce the error in locating a terminal that is partially or fully obscured from the sky (e.g., inside a tunnel or covered parking garage). A moving vehicle may travel down an urban canyon, where the view of the sky is restricted by tall buildings or nearby high vehicles. Poor location geometry increases the receiver's horizontal dilution of precision (HDOP). This means a higher uncertainty for each position estimate.

Visibility of more spatially distributed satellites will improve the geometry of the positioning calculation, particularly in urban canyons. The addition of more satellites (e.g., commencement of Galileo services) will have the same effect, and is expected to increase the time for which dual mode receivers are able to see more satellites (of either type).

2. Fabless literally means without fabrication, generally applied to a chipset designer that licenses designs to manufacturers.

Application Augmentation (Map-Matching and Interpolation)
The following application-level enhancements are available:

- Spatial analysis and map-matching, to snap the position or trajectory to the nearest viable road or route, often used for navigation applications (Figure 3.11);
- Knowledge of direction of travel (bearing) and logical connection between routes;
- Prediction (estimate based on fragmented data) and interpolation (estimation between data points) during temporary reporting.

The importance of estimate of position to RUC depends on the functional requirements of the system. If the charging policy is based on distance traveled according to the total length of road segments, then the accuracy of identifying the correct road segment is critical. The length, duration of time on the segment, and its directional uniqueness may be sufficient to enable the OBU to identify the road segment, even in areas of high uncertainty. Detecting the position of a tariff boundary (e.g., entering a charged area) would require higher accuracy, since the receiver is attempting to identify a transitional event at a precise location (i.e., a point), rather than attempting to identify a road segment (i.e., a line). The receiver may also be required to identify the zone (i.e., an area) in which the vehicle is located, rather than the point of transition. A scheme operator may require 99.99% confidence that a vehicle/OBU is within a chargeable area. To achieve a more relaxed confidence level of 99.9% would require an error margin of at least a 60m buffer zone in one case [27]. A thinner buffer zone would increase the probability that the OBU is not within the zone, but may be at either side of the buffer zone. To be confident that the OBU is within the prescribed area, the OBU must be positioned at least within the thickness of the buffer zone within the chargeable area—hence the term *buffer zone*.

The use of GPS in the urban environment for tariff boundary detection has currently focused on autonomous GPS [28, 29], although data and application augmentation (including with map-matching) is likely to improve performance where satellite visibility is limited.

The probability that the location estimate is close to the true position is shown in Figure 3.12, which also shows that the position error could be large for a small proportion of estimates. Figure 3.12 shows length (abscissa) as a proportion of the RMS error to illustrate the general distribution independently of the distance error. Improving the accuracy of a location estimate is not simply aimed at reducing the average or 1-sigma [63% circular error probable (CEP)] uncertainty; rather, it is aimed at reducing the area under the "long tail" in Figure 3.12, maximizing the geographic area over which the improved accuracy is available, and reducing the time taken to deliver an estimate with improved accuracy.

Accurate location estimates result in improved charge calculation accuracy and an enhancement to scheme credibility, reinforcing the need for augmentation.

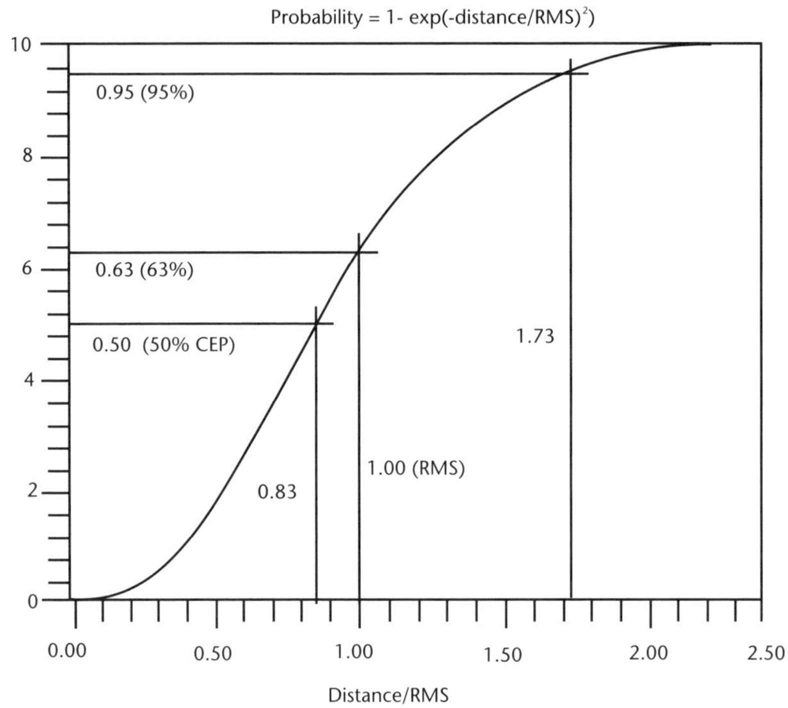

Figure 3.12 Distribution around true position.

Complementary Technology Augmentation

Map-matching and long-dwell integration are two of the methods that are known to reduce the position error from GPS. Other methods that will also improve accuracy of GPS and Galileo include:

- Dead reckoning, a proven method appropriate for vehicles traveling on a fixed route, which allows linear measurement to accurately restrain position measurements along the route;
- Direction sensors [9];
- Inertial aided technology (IAT), which allow continuous positioning despite variable satellite visibility in dense urban environments [30] (e.g., solid state angular rate sensors, and force-feedback accelerometers to provide additional information including velocity and acceleration);
- Hybridization with other terrestrial location methods, such as ground-truthed (i.e., calibrated performance), enhanced (or advanced) forward link trilateration (E/AFLT), CDMA, E-OTD, and Cell ID.

Wireless LAN receivers, such as 802.11g, can provide microcell location capability when cell ID is not available, but its usefulness is limited by the hotspot coverage in any area (currently limited mainly to areas of high population density, rather than road network density).

3.5.3.5 Long-Term Enhancements

The following improvements to the U.S.-operated GPS infrastructure are planned over the next few years, each of which will increase accuracy and geographic availability, and reduce latency:

- Signal improvements;
- New civilian frequency bands;
- Improved network stability;
- Improved network redundancy;
- Signal transmission efficiency;
- Antijamming and antispoofing (expected to be for military use only);
- Interoperability with Galileo.

Natural design improvements in GPS chipsets; increased bearer capacity to reduce opportunity cost of delivering assistance; massively parallel receiver arrays to increase the spectral window of receivers; potential deployment of "pseudolites" (i.e., fixed transmitters that provide ranging information to mobile devices, such as OBUs); and use of cellular picocells in the urban environment are all expected. If GPS receivers are built into vehicles as original equipment, then the optimum positioning of the antenna will most likely improve performance.

Galileo is expected to deliver higher accuracy and quality of service than the current version of GPS, although this may not be achievable by the free Galileo services. EGNOS commenced operation in 2006 to supplement GPS by reporting on the reliability and accuracy of GPS signals. This offers the potential that position measurements within OBUs or data correction processes within the central billing systems, will have a sufficient integrity to be usable for billing purposes. Whatever the intrinsic accuracy from a particular GNSS might be, increasing the number of satellites will be better. The situation will improve considerably with Galileo if the position measurement equipment can receive both GPS and Galileo (and GLONASS) signals. More satellites mean a high probability that enough will be visible and geographically spread in orbit to derive a location estimate with a lower error, rather than if fewer, poorly spread satellites were visible for only part of the time.

Terrestrial positioning based on cellular networks can reduce the ambiguity, or augment other methods of positioning to the resolution of a cell or cell sector, but cannot be used by itself to accurately measure distance. Terrestrial positioning methods are discussed next.

3.5.3.6 Support from Terrestrial Positioning Systems

The main methods of positioning based on 2G and 3G terrestrial cellular networks are:

- Cell ID and Timing Advance (Cell ID + TA);
- Enhanced Cell Global Identity (E-CGI);
- Enhanced Observed Time Difference (E-OTD).

The Cell ID is the identification of a cell, as designated by the network operator. This information normally defines the serving cell (connection point) of a cellular transceiver within a network. The network operator knows the coordinates of each cell site, or base transmitter station (BTS), that is used as a proxy for the estimated position of the cellular transceiver. However, cell sizes vary considerably across networks and between cellular technologies. Larger cells, referred to as macrocells, are typically tens of kilometers in radius in rural areas, while only a few kilometers in radius in suburban areas. Densely populated urban areas often deploy microcells that range from 100m to 500m to increase local capacity. Picocells can be deployed in buildings, offering a cell radius of tens of meters.

Some cells are split into three sectors, with each sector antenna pointing in a different direction, enabling a transceiver location to be estimated more accurately than from an omnidirectional cell. A parameter known as timing advance (TA) is used in normal GSM operation, and is a crude measure of the relative range of the connected mobile from the cell site to the cell boundary. This is accurate to a resolution of approximately 550m. The overall accuracy of Cell ID depends primarily on the accuracy of the BTS coordinate database, and can be improved by sectoring, use of TA, and signal strength information from more than one BTS. As a minimum, Cell ID and TA are parameters that are available for all mobiles in all networks.

The accuracy of a terrestrial positioning system depends upon:

- Density of BTSs;
- Size of cells;
- Layout of a network;
- Multipath of signals from BTSs;
- Shadowing and blocking;
- Geometry of BTSs.

An indication of the level of accuracy of a location estimate of the OBU can provide an indication of the estimated quality of the position estimate. For GSM-based positioning, [31] defines several shapes that can define the uncertainty region centered on the location estimate (see Figure 3.13).

The boundary of the shape for GSM represents the degree of uncertainty (i.e., the likelihood of the GSM receiver being within this area), at 67% or 95% confidence levels. GSM 03.32 [31] describes several shapes, including circles, sectors of a circle, segments of an arc, and ellipses. The location estimate could be weighted according to the degree of uncertainty to be used to determine the trajectory or position of a vehicle/OBU.

The accuracy of a GSM E-OTD system is between 75m and 100m 67% of the time. TOA and AOA hybrids are similar on 2G networks. Proposals were made for tolling systems based on charging for entering a radio cell, with the first trials being held on the A555 Köln–Bonn autobahn in 1996. Until recently, this option could be discounted, since this method could not offer sufficient accuracy in locating its position at any given time. This may change with the potential locating function that will be inherent in the 3G licenses for mobile terminals.

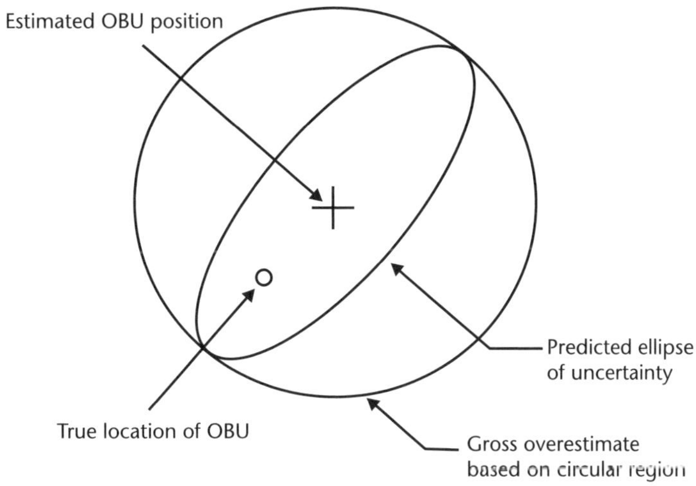

Figure 3.13 Area of uncertainty, centered on the location estimate.

3G mobile network operators claim a location service for business phone users with a 10-m accuracy, which is ample for road use charging purposes, although evidence for enforcement and prosecution may require a greater accuracy. Nevertheless, current versions of 3G phones in tests in Newcastle [32], and the extensive trials undertaken in London in 2004 to evaluate potential future technologies for an extension to the London Congestion Charging scheme, suggest that location accuracy is approximately several hundred meters [33], which is not nearly enough to operate a credible scheme and deliver credible evidence for the prosecution of nonpayers. Nevertheless, since mobile phones already have secure access and a central payment facility (as well as established interoperability), the technology needs only to provide more accurate location, and a robust and validated security and enforcement scheme, to be considered as a future contender [34].

Simple terrestrial positioning, such as Cell ID, can be used by a GNSS/CN–based OBU to request assistance data from an Assistance Server or Serving Mobile Location Center (SMLC) within the central system. The value of assistance data to an A-GPS–capable receiver in the OBU also depends on the location of the OBU, and the availability of visible satellites depends on the position of the GPS receiver. If an A-GPS receiver is capable of reporting (or allowing the cellular network to report) a coarse position based on the serving Cell ID, then the assistance data can be made more relevant, resulting in an improved TTFF and improved HDOP. This means a more rapid calculation of location from switch-on, and marginally improved accuracy.

3.5.3.7 Integration with Enforcement

The integration of a GNSS/CN–based charging solution with enforcement is similar to that for a DSRC-based solution, described in Section 3.5.2.3. The primary difference is that the calculation and reporting of road usage is physically separate from the enforcement solution; any fixed and mobile enforcement points can be independent of charging.

Regardless of technology differences, the objectives of enforcement remain the same: detecting noncompliance, providing a deterrent to nonpayment, and revenue recovery. Detecting noncompliance and capturing evidence of a vehicle's position at a specific location and time requires the vehicle to be identified, and (if fitted) the OBU to be interrogated locally to check correct functioning of the OBU, that road usage is being recorded, and that a valid means of payment is available.

3.5.4 Automatic Number Plate Recognition

ANPR systems process the video images taken by a camera in a lane, at the roadside, or on a gantry, to locate the license plate in the image and convert this into the appropriate alphanumeric characters, without any human intervention (see Figure 3.14). The significant advantage of such an approach is that it removes the need for any in-vehicle equipment to be installed, although the business case for this or any other solution needs to be justified (see Section 3.5.1.1). It also provides a solution for the occasional users (i.e., those who do not have the necessary in-vehicle equipment to automatically pay the charges), as described in Section 3.5.5. ANPR is a variation on the automatic account identification system, which relies on the vehicle's license plate as its unique identifier.

The increasing use of video cameras for road traffic monitoring has been an incentive to improve camera technology and optical processing, which is necessary to provide better contrast clearer images, even when the license plate is in a dark shadow, in the glare of low angles of sunlight, or surrounded by bright headlights

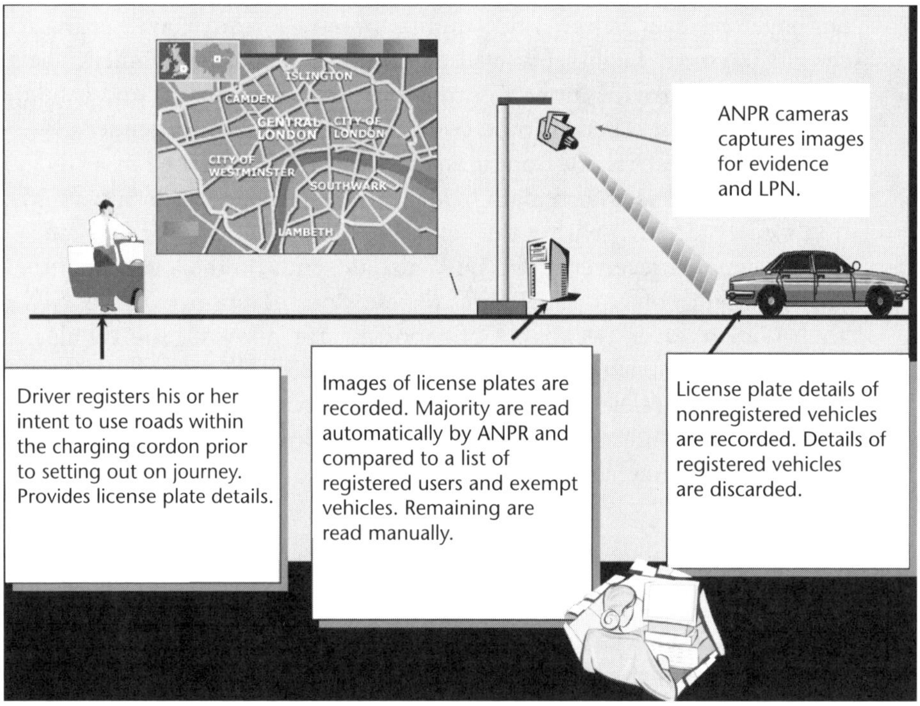

Figure 3.14 Schematic of an ANPR system.

in direct alignment with the camera. To improve accuracy and performance, the technical challenges facing ANPR technology vendors also include:

- License plates of many and different shapes and sizes due to lack of regional standardization;
- Nonreflective license plates;
- Dirt and poor weather, including rain and snow;
- Nonstandardized fonts;
- Similarities between some letters and numbers (e.g., O and D, B and 8);
- Insufficient control of ambient light at camera positions.

Some vendors capture multiple images to improve overall accuracy. If ANPR determines the same plate information for all images, then the confidence level of the data is improved and the need for manual interpretation reduced. Any discrepancies are either placed in a queue for visual inspection or treated as a "lost revenue" transaction. A Government Office for London Report [35] reviewed the road use charging options for London [the Road Charging Options for London (ROCOL) report] in 1998 and 1999. It studied the feasibility of road use pricing and workplace parking charges, as well as the likely impacts on business, traffic levels, and users' reactions. The report recommended that London should in the first instance implement a video-based road use charging system, until the results were available from the Demonstration of Interoperable Road User End-to-End Charging and Telematics Systems (DIRECTS) project [36], which would set standards for U.K. DSRC-based charging (see Section 8.7.5). In August 2002, Mayor Ken Livingstone gave the final approval to proceed toward a full-scale implementation of congestion charging in central London, using ANPR for enforcement.

If ANPR is used for enforcement, then there may be an opportunity to employ ANPR for video tolling, as described in Section 3.5.1.1. However, this apparently simple extension would still need to satisfy the benefit-cost arguments, may require additional roadside cameras at each charging point, would require new business processes and business rules, and would only be available for intermediate-use road users due to the need for manual checking before charges can be correctly allocated. Video tolling as a complement to DSRC OBUs and ANPR is used by the Melbourne City Link (Australia), the Cross-Israel Highway (Israel), and 407 ETR (Canada), and has been used on the Dulles Greenway, Virginia (United States).

There are currently no examples of video tolling in Europe for charging (with the exception of Bergen), although distance-based speed enforcement (known as section control) in the Netherlands relies on matching images captured at two separate locations to identify the same vehicle. Manual checking is still used to confirm speed offenses before enforcement action is taken.

3.5.5 Occasional Users

The vehicle rather than the user usually defines what is meant by an "occasional user." Access to the road network requires an alternate means of being charged, other than an OBU, for occasional users. In the future, it is likely that national road pricing schemes would be based on mandatory installation of OBUs regardless of the usage of the road network, in which case the definition of an occasional user becomes academic.

Section 3.5.1 outlined the economic case for developing different accounts when OBUs are not mandatory, some of which required OBUs to increase detection accuracy and to capitalize on the lower transaction costs that an automated charging process offers. It showed that the business case for OBUs (considering the operator and user costs) may not warrant that all users have an OBU-based account.

Section 3.2 identified the minimum requirements to enable a scheme to operate effectively, yet none of them specifically stated the need for an OBU; rather, it was stated that it should be possible to uniquely identify a vehicle and the road user's means of payment. ANPR can be used to read the vehicle's license plate number. However, the scalability of ANPR as an occasional user product is limited. Occasional users would need to preregister separately for multiple schemes, or the scheme operators would need to share preregistration details while meeting local data protection requirements. The handling of occasional users was regarded as technically and operationally complex in the 1990s, and, until the specific business process requirements were understood, presented a significant challenge.

The following sections outline the options available to operators of plaza-based schemes and open road schemes.

3.5.5.1 Plaza-Based Schemes

The main means of payment for occasional users for plaza-based schemes is cash, either paid to a toll officer or an ACM.

The greater the quantity of ETC-based vehicle passages, the fewer cash transactions are required, thus providing the opportunity to increasingly automate the toll collection process. As the quantity of ETC-based transactions increases, even if it varies by time of day, then the greater the opportunity to dedicate parts of the capacity of the toll plaza to ETC-only passages. There are three general approaches to the use of toll plazas, using approximate percentages of OBU usage:

- Less than 10% OBU penetration in local user population: Dedicated cash payment lanes, and mixed ETC/manual/ACM lanes for OBU-based account holders;
- From 10% to 20% OBU penetration in local user population: Cash payment in manual or ACM lanes, with ETC services in all lanes for OBU-account holders, including dedicated ETC lanes;
- From 20% to 60% OBU penetration in local user population: Cash payment in manual or ACM lanes adjacent to physically segregated express lanes or ORT lanes for OBU-account customers only.

3.5.5.2 Open Road Schemes

Examples of occasional user arrangements for nonplaza schemes are listed here.

- Melbourne City Link (Australia): *CityLink Pass* users register online or via a call center/IVR with vehicle license plate details and pay with a credit card or bank card. Each charge point is able to use ANPR to discard images from preregistered vehicles.

- LastKraft Wagen (LKW) truck tolling scheme (Germany): "Alternative user" terminals are located in truck stops and in other rest stops located at either side of the country's border. Transiting truck drivers or dispatchers are required to manually preregister a route at the roadside terminals, by contacting a call center or through the scheme operator's Internet site. Changes to the route can only be accommodated by reregistering.
- London Congestion Charging (United Kingdom): More than 5,000 retail outlets in the London area are supplemented by cash payment terminals in car parks.
- 407 ETR (Canada): No registration or prepayment is required. Vehicle is identified using ANPR, and the registered owner is identified and billed.
- Trondheim (Norway): ACMs were located in lay-bys at the toll ring, and not all entry points are manned. Over 90% penetration of OBU-based transactions occurs at peak hours. Toll collection services were completely removed on December 30, 2005, since the original purpose of the scheme, to fund road infrastructure development, had been satisfied. Ongoing road operational costs are now funded from the general taxation (see Section 8.4.1).
- Singapore ERP scheme: Installation of an OBU is mandatory for most Singapore-registered vehicles. Foreign road users planning to travel on ERP-priced roads can either get an OBU, also known as an in-vehicle unit (IVU), installed, temporarily rent a unit, or pay S$10 (approximately $6 or €5) for a daily license, regardless of the number of trips on an ERP-priced road.

The Austrian LKW truck tolling scheme offers no occasional user product. Road users of vehicles above 3.5 tons must acquire and install an OBU before using the national road network. This simplifies the business rules for enforcement, but places a greater burden on users. This also requires potential road users to be aware of the payment options, and how and where to acquire an OBU.

Other options may also be feasible where the primary means of charging is based on installation of an OBU by an accredited workshop. For example, a vehicle that does not meet the business rules based on total annual distance threshold could be regarded as occasional and therefore eligible for a simple user-installed OBU with limited automatic data collection capability. Although the installation cost would be significantly lower, the low-usage OBU would require greater effort from the road user to report usage, such as manually entering the start and end odometer readings. The data collection costs from the operating authority could also be greater in proportion.

3.6 Standards and Interoperability

3.6.1 Introduction

There are many examples where standardization has helped the competitive potential of an industry. A car tire can be bought with limited information, knowing that it will fit the wheels of a car. A GSM phone purchased in Hong Kong will function in Norway and the United States, and in any of the 860 networks and 220 countries worldwide [27]. A webcam acquired in Japan will work on a computer in Europe. The Internet Protocol (IP) can connect an FTP server in Indonesia to a client in Hungary. All this has been made possible through early cooperation between industry suppliers, leading to widespread distribution of highly differentiated, yet competitively priced products. From a user's perspective, not having to think about interoperability is a measure of the success of industry cooperation, regulatory guidance (where needed), and informed customers. However, there are many examples in which the same recipe has not led to globally interoperable products, yet consumer choice has not suffered (e.g., memory sticks for digital cameras, car entertainment systems, and electrical appliances).

There are two rules that have emerged for the selection and use of interoperable charging technologies:

1. Standards are necessary but not sufficient [37]. DSRC suppliers and road operators have shown that the variety of options defined by standards could mean that one DSRC technology uses a subset that is not compatible with another. The collective development of communication profiles, specifications, and test methods enables interoperability. This profiling is a necessary step beyond standards to enable ETC and road user charging in concentrated multiauthority road networks.
2. Multivendor interoperability may be desirable to lower the risk of technology supply and maintain ongoing competition, but the success of the scheme does not depend on it. One of the world's largest ETC schemes (measured by revenue collected) is EZ-Pass, offered by operators in the Northeast United States; it uses a single charging technology vendor. Back-office interoperability was enabled through standardizing the transaction records exchanged between operators.

Regions that aim to attract private finance to upgrade highways and infrastructure have more confidence in implementing charging if they know that specifying standards-compliant products simplifies the initial procurement, while multivendor interoperability reduces long-term procurement and operating risks. The benefits of standards and interoperability are applicable to all charging technologies, as discussed in Section 3.6.2 and 3.6.3. There may also be disadvantages if the development of standards adds delays and introduces technology development risk. This often means that debugged standards are coopted from one country to another country, since the development of a new standard for local use may make local projects less attractive to potential bidders. The alternative, with the caveats stated above, is to procure a proprietary solution, although with the significant efforts invested in standards development, this need not always be an option.

3.6.2 The Benefits of Standards

Standards designed specifically for ETC and road user charging have generally focused on the connection between in-vehicle equipment and the roadside. There is little evidence, to date, of application-specific standards being applied to enforcement, other than generally accepted methods for image format, encryption, and compression methods to maintain the integrity of evidential records.

The European Committee for Standardisation (Comité Européen de Normalisation, CEN) and its Technical Committee on Road Transport and Traffic Telematics (TC278) initiated one of the earliest standardization activities in 1991. In Spring 2004 (almost 13 years later), the completed standards defined the operation of the DSRC interface between an OBU and a roadside system. The standard is applicable to all members of CEN, including the national standards bodies of all EU member states and the European Free Trade Area (EFTA), leaving institutional barriers as the final hurdle to enable multinational interoperability.

U.S.-developed standards include Caltrans' Title 21 [38] and ASTM E2158-01 [39] for DSRC technologies in the 902-to-928-MHz band. Since the Federal Communications Commission (FCC) announced the availability of the 5.9-GHz band in October 1999, ASTM and IEEE have been developing complementary standards for vehicle-roadside communication, beginning with ASTM E2213-02 [40] in 2002 for layers 1 and 2 of the OSI model of network architecture. ASTM and IEEE are currently working on the upper OSI layers, as described in Section 3.5.2.1.

Competition for ETC projects has introduced CEN DSRC–compliant solutions in Southeast Asia, South America, and South Africa. However, CEN-compliant products do not have a market monopoly. Proprietary solutions and systems that comply with standards created in the United States and, to a lesser extent, Japan, are also being used outside Europe and the United States.

CN/GNSS generally relies on standard-bearers such as GPRS for communication of road usage information, map database updates, and tariff tables, depending on whether a thin or intelligent client is employed. Locally applicable DSRC standards and specifications apply where a CN/GNSS OBU relies on DSRC for localized communications for enforcement. Consequently, current activities are focused on the application level to ensure interoperability, as discussed in Section 3.6.3.

3.6.3 The Benefits of Interoperability

The benefits of interoperability are often treated as purely technical. The commercial benefits are far more important and include:

- Creation of multiple supply chains from multiple vendors, potentially reducing procurement risk and threat of monopoly pricing;
- Ease of technology comparison by highway operators, reducing the need to focus on technical elements, and simplifying procurement;
- Separation of infrastructure procurement (i.e., high-cost, low-volume lane equipment) from OBU procurement (i.e., low-cost, high-volume OBUs), simplifying procurement;
- Continuous competition for infrastructure expansion and new OBU business, delivering lowest cost and greatest benefits to the highway operator;

- Geographic expansion from multiple road operators without the need for coordination in technology selection, reducing procurement complexity, and simplifying expansion;
- Increased user choices among OBU supply chains, with potential for direct sales to highway users by third party outlets.

Ensuring interoperability across state or national borders with an OBU that meets minimum interoperability requirements means that road usage records (GNSS) or transaction records (DSRC) will be in a form that permits charge reconciliation between operators (or payment service providers). This ensures that road users benefit from OBU roaming, trip flexibility, continuous service provision, and a single bill, just as cellular mobile service providers routinely deliver to their customers.

Enabling cross-border usage of an OBU that complies with technical interoperability requirements depends simply on the principles of contractual interoperability, as is evident from bilateral agreements between Austria and Switzerland (currently only one-way), Denmark and Sweden, Spain and Portugal, and between other pairs of EU and European Economic Area (EEA) member states.

Increased cooperation between highway operators supported by existing standards (initially DSRC-related) has meant a power shift from suppliers to highway operators. In Europe, operator involvement in CEN TC278 was virtually nonexistent before the prEN (draft) stage of European standards. During this period, the GSS [41], A1 [42], and A1+ [43] (on board charging extension to A1) interoperability specifications were created to provide a simplified approach to specifying a useable subset of transactions, which ensured a minimum service level interoperability between different vendors' products.

However, the most prominent European interoperability programs, such as TIS (France) and the Common EFC System for Road Tolling European System [44], have been entirely driven, since 1999, by highway operators that invited DSRC vendors to participate. In addition, the Concerted Action for Research on Demand Management in Europe (CARDME) [45], DIRECTS (United Kingdom), PISTA, and the development of the WAVE Platform within the U.S. DOT–led vehicle infrastructure integration program, are all examples of interoperability initiatives also driven by highway owners and national administrations.

Nevertheless, we can already see the benefits. The first pioneering applications of ETC were initially driven by highly localized needs, and it took almost 10 years from the first use of ETC until cross-border interoperability found its way onto the agenda. In Europe, the directives that enable lorry road user charging (LRUC), and the modified directive relating to interoperability, have increased industry debate, helped form national technology preferences, and established positive support for cross-border interoperability. This process took only 5 years. By comparison, this was also the time required for Switzerland and Austria to plan, deploy, and launch national schemes.

An operator is now able to routinely procure DSRC roadside systems, OBUs, and turnkey systems from several competitive vendors. Multivendor sourcing requires standards-compliance, supported by a debugged interoperability specification. The benefit of interoperability for small-scale isolated schemes may be less important, so standards-compliance is less critical. As earlier described, the U.S. EZ-Pass, ETR 407 (Canada), and the Singapore ERP scheme are based entirely on proprietary charging technologies, although all were procured when standards-compliant products were not generally available.

Once a scheme is operational with charging technology that complies with standards and an interoperability specification, then future OBU procurements can be routinely separated from main system purchase, although many buyers have continued to depend on significant technical knowledge to ensure that vendor products comply with the local requirements for interoperability. Notably, the Chilean Ministry of Public Works (MOP) appointed the Germany-headquartered TÜV to verify OBU compliance to a local interoperability specification underpinned by CEN DSRC standards [46–49]. Similar interoperability specifications based on the same set of standards have been produced in Australia [50], Brazil [51], Chile [52, 53], Norway [54], and Sweden [55]. The French *Liber-t* project requires that all OBU and roadside systems pass a formal site acceptance test, managed by the L'Association des Sociétés Françaises d'Autoroutes et d'Ouvrages à Péage (ASFA). Standards backed by interoperability specifications, published test methods, operator-specific tests, and a willingness for scheme operators to enter into contractual arrangements are critical to ensuring a seamless user experience when roaming. The ultimate goal in Europe is the enabling of a road user to use a single OBU to travel on all charged road networks within the European Union, with few exceptions. The road user would only have to register with one organization (a payment service provider), and receive only one bill [56, 57].

The U.K. Department for Transport embarked on a program to develop a national specification for interoperable payment of road use charges, consistent with European standards and potentially enabling compliance with the European Interoperability Directive. The U.K. DIRECTS project [36], using 500 or so volunteer drivers with vehicles equipped for a trial in Leeds in the North of England, demonstrated an end-to-end solution for DSRC-based charging. The DIRECTS project is presented in Chapter 8, on international case studies.

Looking globally, ISO 17575 "provide[s] a framework for achieving interoperability between different EFC systems using satellite positioning and cellular networks and define[s] in particular a framework for on-board equipment to roam between different EFC services, even where the EFC services have different policies and charge structures" [58] applicable globally. In Europe, the Minimum Interoperability Specification for Tolling on European Roads (MISTER) initiative builds on this to guarantee technical and procedural interoperability, consistent with the aims of the European Electronic Toll Service (EETS), discussed further in Section 8.5.1.

One of the most prominent projects that aims to develop a media-independent vehicle-roadside communications approach is CALM, led by ISO/TC204 Working Group 16. It is expected that the interfaces will include DSRC (IR and microwave), millimeter wave at 63 GHz, mobile wireless broadband, GSM, and UMTS services, as a minimum. CALM will define handover mechanisms between multiple media providers to ensure service continuity that is completely transparent to the user. The multimedia expectation requires coordination with other standards bodies, including the European Telecommunications Standards Institute (ETSI) (TG37 car-to-car communications), and the Wi-Max Forum. A common global allocation of bandwidth will also need the cooperation of the International Telecommunication Union (ITU) and the Conférence Européene des administrations des Postes et des Telecommunications (CEPT), plus local spectrum regulatory bodies, such as the FCC. Further information on CALM is given in Section 9.2.5.

3.7 The Future

3.7.1 Introduction

The dominant charging policy for road use was toll collection up until the mid-1990s. This led to the emergence of products aimed primarily at ETC. Since then, new policies have evolved, and technology vendors have developed adaptations of well-understood technologies (e.g., IR, ANPR, and CN/GNSS) to meet these new policy requirements.

The future evolution of the RUC market as a whole is addressed below, based on observations of relevant global trends, market forces, and a statement of possible future scenarios. Regulatory influences are treated separately.

The most important influence on the use of charging technology and the network of technology suppliers and supported integrators continues to be infrastructure expansion driven by economic growth. "National and local Government initiatives, as well as an increasing user requirement for more convenient tolling, are the key factors driving demand for ETC systems" [59]. A shortfall in public funds and investment in highway infrastructure upgrading is also leading to growth in build, operate, and transfer (BOT) projects and commercialization of existing highways. Increased awareness of the adverse impact of economic activity on the environment, particularly among OECD nations, has led to increased political and institutional support for pay-as-you-go principles. Finally, contributors to congestion, such as population growth, increased vehicle ownership, and increased vehicle miles traveled (VMT), highlight the need for balance between capacity expansion and efficient use of existing capacity. There are highway instrumentation, telematics, and RUC solutions for either approach.

The global trends and regulatory influences described above were used to assess possible market evolution. Reports published in Brazil, Japan, and the United States describe the rapid expected growth in ETC usage:

- The private investments in concessions "have been the main factor behind the adoption of ITS in the Mercado Común del Sur (MERCOSUR) region (South American trading bloc). Because of this, the most common ITS application in the region is for highways, as in ETC and highway communication systems" [60].
- The U.S. national intelligent transportation systems program sets out a plan that ". . . advances the safety, efficiency and security of the surface transportation system, provide increased access to transportation services and reduce fuel consumption and environmental impact and [the introduction of] a single payment medium for regional and national travel" [61].
- ASECAP states that ". . . the axes that will define the future road policies that will impact its members (highway operators) include a new Infrastructure

financing framework, a common methodology for the infrastructure charging, RUC interoperability (DSRC—GPS/GSM-GALILEO), policies that differentiate between private cars and heavy lorries and between urban areas and motorways" [62].

- ITS Japan claimed recently that "It has been calculated that [the planned investment] will allow about 80 percent of total traffic on toll roads to move without stopping. In future, all toll gate booths will be fitted with a card reader capable of reading the electrically transmitted information of the IC card inserted in the on-board equipment, enabling every vehicle fitted with ETC on-board equipment to use all toll gates in the country" [63].

Other external forces that impact RUC technology developments include global decisions on radio spectrum allocation, the prominence given to large-scale projects such as Galileo, and regulatory forces at the regional and national level. The evolution of the RUC industry is also guided by forces from several directions, including: continued investment in applications trials with community funds (e.g., the Fifth Framework Programme in EU member states); technology transfer initiatives (i.e., long-term net shift of defense to civilian expenditures); infant industry protection measures through the imposition of import tariffs; and local technology transfer provisions (e.g., China and Brazil).

3.7.2 Future Scenarios

Table 3.4 describes a policy-led future scenario.

If we adopt the perspective that charging for road use is simply an application, then we have the scenario in which road user charging and tolling would reside alongside other in-vehicle applications, such as navigation, safety enhancement, and information systems. These applications are fed by sensor inputs, providing vehicle position, speed, vehicle-to-roadside communications, object detection, and other active and passive detection and measurement systems. Sensor inputs may feed one or more applications, so information sharing may drive applications to coexist on the same vehicle platform. For example, if the vehicle is equipped with more advanced methods of determining road user charges based on distance traveled, then the OBU that was adequate for interoperable charging now needs to have greater connectivity with the vehicle to be able to securely access distance traveled information. Economies of scale, security, and common information needs

Table 3.4 A Policy-Led Future Scenario

Technologies will continue to evolve as the acceptability of tolling and road user charging increases, as the complexity of charging policies increase, and as road users have increased contact with different charging policies. Many users will initially come into contact with the technology by paying a charge electronically. These users will experience technology at its most focused level: usually no more than a user-installed OBU that beeps to indicate that a transaction was performed successfully. In the long term, vehicle manufacturers will provide interfaces to retrofit devices before offering an integrated solution. Users will interact with the scheme through an intermediate service provider with whom the user has an account. The user will be able to prepay or postpay, depending on status, through a variety of channels targeted at specific user groups.

further suggest that road user charging and tolling could assume the status of an embedded application within the vehicle. This is discussed further in Section 9.3.2.

It is easy to predict complex technology scenarios where the technologies for road user charging need to encompass all possible sensor inputs to serve all possible charging policies that a user may experience in a typical journey. However, ensuring interoperability between geographic areas or road segments that have different charging policies (e.g., tolling, area pricing, cordon pricing) can be seen from three different perspectives, discussed in the following.

3.7.2.1 Home Policy Compliant

Home policy compliant relates to isolated, region-specific procurements. A user registered with one scheme would need to act as an occasional user with the other scheme. A heavy goods vehicle with a CN/GNSS/DSRC OBU would need to pay cash or register as an occasional user elsewhere. Extrapolating this scenario to the future, a gradual increase in the number of bilateral interoperator agreements would result in vehicles meeting minimum technical interoperability requirements for the bilateral agreement operators but not all regional road charging operators. The burden rests with the road user to ensure that the payment means is acceptable outside the home area.

3.7.2.2 Minimum Policy Interoperability

A more desirable outcome of the focused, home policy compliant would be where all road operators support a minimum common charging policy. For example, an OBU issued by one operator would be accepted as a valid means of recording and reporting road usage to all operators on whose infrastructure the user travels. A user registered for scheme A can participate as an occasional user in scheme B using scheme A technology. In other words, the technology issued by operator A is accepted as valid technology for occasional users on operator B's infrastructure. If the reverse also applied, then true bidirectional interoperability would be achieved, and users having either technology would be able to use either infrastructure without additional registration.

This policy is analogous to a cellular phone subscriber having a broad choice of handsets, each with different capabilities, some of which may or may not be supported when roaming (e.g., instant messaging, streaming video). However, every operator's network supports the minimum capability (e.g., voice and data).

3.7.2.3 Full Policy Roaming

This scenario states that meeting the requirements for minimum interoperability for all road operators would require a maximum capability OBU. This has no analogy in mass-market cellular communications.

This scenario would only apply if an OBU needs to meet the charging policy requirements of the operator with the most complex charging policy within the area in which the user could reasonably be expected to travel. For example, operator A would need to accept vehicles equipped with charging technology from operator B that have the capability of measuring road usage based on a multilevel, time-of-day, distance-based charging mechanism as employed by operator B. In this case, minimum interoperability means that a more capable OBU would be required, incorporating many charging technologies.

The most likely technology future will be dictated by legislative requirements, the propensity of road operators to agree on occasional user schemes, technology costs that can be borne by the road user, and the relative penetration rate of each technology choice in a retrofit and new vehicle market.

One possible impact on the course of technology development to 2010 of full policy roaming is described in Table 3.5. As earlier, this is not a forecast, but merely one of many possible future outcomes of regulatory, institutional, and technology development activities.

The integrated scenario is only applicable where the convergence of procurements, cross-border interoperability, and economies of scale drive cooperation and the emergence of new organizations dedicated to increasingly specialized parts of the road user charging and tolling value chain. Many vertically integrated scheme operators may focus on core operations, while road users benefit from a choice of payment service providers and mass customized options for payment of road user charges. Integration with other ITS services may also be possible (e.g., Japan and VII case studies in Chapter 8), including traffic information services, safety-related devices, and automatic payment for fuel and parking.

Table 3.5 Integrated Scenario

Development of hybrid OBUs supporting GNSS/CN and DSRC, where DSRC is the lowest common denominator for complex and monolithic OBUs to ensure interoperability in EU/EEA, including newly joined EU member states;
Continued routine use of DSRC technologies for highly focused, mass market applications, such as ETC;
Continued development of contractual interoperability to ensure coexistence with other forms of EFC, such as CN/GNSS and ANPR (already introduced as nationally or locally);
Evolution of charging policies from only highways towards all roads, with local differentiation based on emissions class, classification, axle weight, time-of-day, and measured congestion;
Emergence of cross-border charge clearing services, and service providers driven by economies of scale;
Further development of regional [EU, EEA and North American Free Trade Agreement (NAFTA)] contractual roaming agreements;
Broad acceptance of road user charging policies within vehicle and transport services supply chains (e.g., retrofit outlets, vehicle manufacturer options);
The development of multimode, flexible OBUs, adaptable to local RUC service requirements;
Development of pan-EU cross-border enforcement processes [e.g., based on Video Enforcement for Road Authorities (VERA)-type tools and equipment approvals], initially on a bilateral basis;
Cooperative operator-driven procurements for RUC systems;
Continued emergence of OBU-only vendors;
Scheme overlap, separating the roles of OBU issuing, account management, and RUC service provision.

Source: [64].

3.8 Summary and Conclusions

A technology perspective on tolling and road user charging reveals a long list of technology building blocks that can be combined to meet functional requirements defined by charging policies. The optimal mix of GPS and DSRC systems will be determined by national charging policies, and minimum interoperability requirements for travel on networks of regional roads that have different policies for charging and enforcement.

The choice between having or not having an OBU will depend on regulation (i.e., mandatory or voluntary installation), and the business case for scheme operators to encourage the use of OBU-based accounts by different usage categories of road users. Regulation and interoperability will blur the choice between DSRC and GNSS/CN toward OBUs that embody all technologies. We have seen that DSRC, as a technology building block, has been widely adopted for ETC. However, the introduction of distance-based charging schemes, initially for heavy goods vehicles, has already challenged the business case for discrete detection methods offered by DSRC, to also include methods that are applicable to all roads with multiple tariff boundaries. The development of increasingly accurate and reliable satellite positioning methods that depend on different forms of augmentation will increase the global applicability of CN/GNSS schemes. Regulatory pressure for distance-based charging is essential for the availability of positioning information to an OBU, whether delivered by DSRC or satellite positioning. The drive toward interoperability, underpinned by standards, will enable OBUs to roam between areas that differ in charging policy, which requires the OBUs to be capable of providing road usage information to satisfy local scheme rules. The pressure on OBUs to evolve to more sophisticated forms could be mitigated by the evolution of central systems. Chapter 6 shows that interoperability does not always require the charging technologies to meet the requirements of all schemes. Unless all schemes have the same approach and have coordinated their procurements, it is likely that the central systems should also be regarded as a critical enabler of interoperability, rather than an exclusive focus on charging technologies.

Regional solutions (defined by an economic area, such as the EU or NAFTA) will remain feasible in the future. Wide area augmentation methods and regional standards for wireless communications suggest that road user charging technologies will need to be bundled to meet regional requirements. Similarly, DSRC and ANPR provide baseline capabilities for enforcement; DSRC can interrogate OBUs to check account validity and other declarations; and ANPR allows the handling of evidential images to be highly automated. Within the confines of each scheme, ANPR also allows occasional users to be registered. For higher frequency road usage that does not warrant an OBU, the use of video tolling can reduce transaction costs for pay-per-use operations.

The common threads of RUC technology development are the continued drive towards interoperability at all levels, from technical to contractual; the trend to road use charging and tolls; and the need to find new sources of investment for infrastructure upgrade and expansion, mitigated by the institutional and organizational hurdles that need to be overcome.

Regulation is also expected to continue its impact on the development of RUC technologies. The greatest influence on the technology choice for a vehicle owner, driver, local authority, and highway operator will depend on the regulatory environment and the local or national charging policies. Distance-based charging will require discrete or continuous vehicle positioning or distance measurement capability. Toll roads will continue to maintain highly localized collection and enforcement schemes to meet long-term concession targets, but will also be under pressure to cooperate with other distance-based policies.

References

[1] Ministry of Transport, *Road Pricing, The Economic and Technical Possibilities (Smeed Report)*, London, U.K.: Her Majesty's Stationery Office, 1964.

[2] Highway Technologies Ltd., *Intelligent Vehicle Trial, Report to the Department of Infrastructure, Energy and Resources*, Australia, 1999.

[3] Underhill, P., "Road User Charging and Trucks—An Update on U.S. Programs," *Proc. IBTTA Spring Technology Workshop*, Edinburgh, U.K., June 14, 2005.

[4] Richeson, K. E., *Introductory Guide to CVISN*, U.S. Department of Transportation, POR-99-7186, vP.2, 2000.

[5] Land Transport New Zealand, *Road User Charges and Light Diesel Vehicles—Factsheet 38*, 2005.

[6] Government of New Zealand, *Road User Charges Act 1977 and Its Amendments*, 1977.

[7] UNINFO, *Italian Standard UNI 10607, Road Traffic and Transport Telematics, Automatic Dynamic Debiting Systems and Automatic Access Control Systems Using Dedicated Short-Range Communication at 5.8 GHz*, January 1998.

[8] Armstrong, L., *Dedicated Short Range Communications (DSRC) at 5.9 GHz*, Lee Armstrong Consulting, February 2003.

[9] Geovector, *Application Demonstrates the Power of the .NET Compact Framework When Combined with GPS and Heading Sensors*, 2002, http://www.geovector.com/press/downloads/mscasestudy.doc on April 16, 2006.

[10] IEEE, *IEEE 1609—Family of Standards for Wireless Access in Vehicular Environments (WAVE), Fact Sheet*, 2006, http://www.standards.its.dot.gov/fact_sheet.asp?f=80.

[11] IEEE, *IEEE P1609.1—Standard for Wireless Access in Vehicular Environments (WAVE)—Resource Manager*, 2006.

[12] IEEE, *IEEE P1609.2—Standard for Wireless Access in Vehicular Environments (WAVE)—Security Services for Applications and Management Messages*, 2006.

[13] IEEE, *IEEE P1609.3—Standard for Wireless Access in Vehicular Environments (WAVE)—Networking Services*, 2006.

[14] IEEE, *IEEE P1609.4—Standard for Wireless Access in Vehicular Environments (WAVE)—Multi-Channel Operations*, 2006.

[15] ASTM International, *E2213-03 Standard Specification for Telecommunications*, 2006.

[16] IEEE 802.11, *Official IEEE 802.11 Working Group Project Timelines, 10 April and Information Exchange Between Roadside and Vehicle Systems—5 GHz Band Dedicated Short Range Communications (DSRC) Medium Access Control (MAC) and Physical Layer (PHY) Specifications*, 2006.

[17] Jones, B., and R. Resendes, *Vehicle Infrastructure Integration Initiative*, ITS Joint Program Office, U.S. Department of Transport, 2006.

[18] Blythe, P., and P. Hills, "Automatic Toll Collection and the Pricing of Road-Space," Ch. 7, *Advanced Technology for Road Transport*, Ian Catling, (ed.), Norwood, MA: Artech House, 1994.

[19] Blythe, P., *Electronic Tolling in Europe: State of the Art and Future Trends, Operation and Maintenance of Large Infrastructure Projects*, Balkema, 1998, pp. 85–102.

[20] Neilsen, M., *DELTA Final Report, Deliverable D1.2*, copy available from http://www.ertico.com/en/activities/projects_and_fora/delta_website.htm on April 23, 2006.

[21] Elliott, K. and C. Hegarty, *Understanding GPS: Principles and Applications*, 2nd ed., Norwood, MA: Artech House, 2006.

[22] MacGougan, G., et al., "Degraded Signal Measurements with a Standalone High Sensitivity Receiver," *Proc. ION National Technical Meeting*, San Diego, CA, January 2002.

[23] ERTICO, *Road Charging Interoperability (RCI) Final Supplier Workshop*, February 2006.

[24] Mjølsnes, S.-F., and B. Forssell, *Galileo and Location-Based Services*, Norway: Norges Teknisk Naturvitenskapelige Universitet, 2001.

[25] Federal Aviation Administration, *Fact Sheet: Wide Area Augmentation System (WAAS)*, 2006, http://gps.faa.gov/Library/waas-f-text.htm.

[26] European Space Agency, *What Is EGNOS?* 2006, http://www.esa.int/esaNA/egnos.html.

[27] GSM Association, *Coverage Maps and Roaming Information*, 2006, http://www.gsmworld.com/roaming/gsminfo/index.shtml.

[28] Patchett, N., et al.. "Assessing the Use of GPS for Congestion Charging in London," *Traffic Engineering & Control*, Vol. 46, No. 3, March 2005.

[29] Kristensen, J. P., and M.-B. Herslund, "Copenhagen Trials Results," *Proc. the Progress Final Conference: Road Pricing, The Way Forward*, London, February 2004, http://www.transport-pricing.net/confppts/2B_COPEN.PPT.

[30] Applanix, *Multisensor GIS and Asset Management Applications*, 2005, http://www.applanix.com/media/downloads/case_studies/POSLV_GIS_Pasco.pdf.

[31] ETSI, *GSM 03.32 Version 7.1.0 Release 1998 Universal Geographical Area Description (GAD)*, 1998.

[32] Evans, J., "Update on the London Congestion Charging Scheme," *IEE Seminar on Road User Charging*, London, U.K., March 2003.

[33] Sharif, B., "An Investigation of Localisation Techniques Using Mobile Phones for Road Transport Applications," School of Electrical, Electronic and Computer Engineering, Internal Report, Newcastle University, U.K., 2004.

[34] Birle, C., "Use of GSM and 3G Cellular Radio for Electronic Fee Collection," *IEE Seminar on Road User Charging*, London, U.K., June 9, 2004, http://www.iee.org/oncomms/pn/auto.

[35] Government Office for London, *Road Charging Options for London*, London, U.K., Stationery Office, 2000, http://www.gos.gov.uk/gol/transport/161558/228862/228869/.

[36] Tindall, D. W., "Road User Charging Demonstration Project," *IEE Seminar on Road User Charging*, London, U.K., June 9, 2004, http://www.iee.org/oncomms/pn/auto.

[37] Pickford, A., "Chile Sauce (DSRC Grows Up)," *Traffic & Technology International*, UK & International Press, November 2002.

[38] Barclays California Code of Regulations, Title 21, Chapter 16, *Compatibility Specifications for Automatic Vehicle Identification Equipment*, §1700 to §1705.8, http://www.dot.ca.gov/hq/traffops/itsproj/Title_21/title21_index.htm.

[39] ASTM, *E2158-01 Standard Specification for Dedicated Short Range Communication (DSRC) Physical Layer Using Microwave in the 902 to 928 MHz Band*, 2001.

[40] ASTM, *ASTM E2213-02e1 Standard Specification for Telecommunications and Information Exchange Between Roadside and Vehicle Systems—5 GHz Band Dedicated Short Range Communications (DSRC) Medium Access Control (MAC) and Physical Layer (PHY) Specifications*, 2002.

[41] GSS Group, *Global Specification for Short Range Communication*, Version 3.0, December 2001.

[42] Alcatel, Kapsch, Combitech, CSSI, A1: TR4001 A1, "Interoperable EFC Transaction Using Central Account Based on DSRC," Version ER9_1.3, June 1999.

[43] Telematics Application Programme, TR 4001 A1 project, "Interoperable EFC Transaction Using On-Board Account Based on DSRC," Version IR9_1.3, June 23, 1999.

[44] PISTA, *Transaction Model, Project IST-2000-28597, Deliverable 3.4*, Version 6, November 2002.

[45] CARDME, *The CARDME Concept, Project IST-1999-29053, Deliverable 4.1*, June 1, 2002, http://www.cardme.org.

[46] CEN, *prEN 12253: 2002 Dedicated Short Range Communication, Physical Layer Using Microwave Medium at 5.8 GHz*, Brussels, Belgium: CEN Central Secretariat, 2002.

[47] CEN, *prEN 12795: 2002 Dedicated Short Range Communication, DSRC Data Link Layer: Medium Access and Logical Link Control*, Brussels, Belgium: CEN Central Secretariat, 2002.

[48] CEN, *prEN 12834: 2002 Dedicated Short Range Communication, Application Layer*, Brussels, Belgium: CEN Central Secretariat, 2002.

[49] CEN, *prEN 13372: 2002 DSRC Profiles for RTTT Applications*, Brussels, Belgium: CEN Central Secretariat, 2002.

[50] Standards Australia, AS 4962, *2001 Interim Australian Standard—Electronic Fee Collection—Transaction Specification for Australian Interoperability on the DSRC Link*, December 11, 2001.

[51] Minstério dos Transportes, *Brazilian DSRC-EFC—Specification for Interoperability*, Version 01, November 19, 2000.

[52] Ministry of Public Works, Transport and Telecommunications Ministry (MOPTT), *Specification for Interoperability in the Beacon—Transponder Transaction*, Version 1.25, July 15, 2002.

[53] Ministry of Public Works, Transport and Telecommunications Ministry (MOPTT), *Conformance Tests to the Specification for Interoperability in the Beacon—Transponder Transaction*, Version 1.05, July 15, 2002.

[54] Norwegian Public Roads Administration, *Autopass Suite—Specification for Norwegian Electronic Fee Collection Systems*, February 4, 1999.

[55] Swedish National Road Administration, *Basic Requirements Specification for Interoperable EFC-DSRC Systems in Sweden*, Version 0.5, http://www.viv.se/pga/betalsystem/efc-dsrc/.

[56] European Commission, *Directive 1999/62/EC of the European Parliament and of the Council of 17 June 1999 On the Charging of Heavy Goods Vehicles for the Use of Certain Infrastructures*, 1999, http://europa.eu.int/comm/transport/infr-charging/library/directive1999-62.pdf.

[57] Catling, I., "Minimum Interoperability for Tolling on European Roads (MISTER)," *Proc. European Standards and EFC*, June 28, 2005.

[58] European Commission, *Directive of the European Parliament and of the Council on the Widespread Introduction and Interoperability of Electronic Road Toll Systems in the Community*, 2005.

[59] Hughes, C., *ETC Market Status, Strategy Analytics*, 2003.

[60] ITS America News, *Intelligent Transportation Systems in Brazil and First ITS Brasil Congress*, 2002.

[61] ITS America, *Delivering the Future of Transportation—The National Intelligent Transportation Systems Program Plan: A Ten-Year Vision*, January 2002.

[62] ASECAP, "Round Table Discussion with Kallistratos Dionelis (Secretary General, European Association of Tolled Motorway Companies ASECAP)," *7th IRU East-West Road Transport Conference*, Budapest, May 15–16, 2003.

[63] ITS Japan, *Guide: ITS Deployment Progress in Japan*, Section 4, 2003.

[64] Pickford, A., *Road User Charging: Market Development Scenarios*, Transport Technology Consultants, 2005 (unpublished).

CHAPTER 4

Technology Options for Enforcement

4.1 Background

Enforcement is the term used for systems and procedures to ensure that road users follow scheme rules. Vehicles not equipped with an appropriate charging device (e.g., an OBU) and/or that have not paid the charge are detected and fined (or penalized). The real meaning of enforcement will depend on the perspective adopted by a highway operator or any other stakeholder in the process, but as a minimum, an enforcement strategy needs to be based on three fundamental objectives:

- *Compliance*, to ensure the charging policies and payment rules are followed by all road users;
- *Deterrent to nonpayment*, to inform and raise awareness of scheme requirements to reduce the temptation to evade payment;
- *Revenue recovery*, to ensure that the fees that are due are paid by road users and protect the revenue stream.

Enforcement is not an event but a process—a mix of technologies, well-defined procedures, human resources, and enabling laws—to deter any attempts at noncompliance with the charging policy, and when noncompliance is detected, to take steps to secure revenues as a priority.

A review of the technologies for enforcement must consider the underlying reasons, objectives, and strategies of the charging regime. It is clear that without an enforcement strategy (in its broadest sense), a charging regime cannot be operated on a sustainable basis.

We saw in Chapter 3 that there are different types of road user charging systems. A road user may choose not to pay charges directly for road usage by selecting public transport, for example. A road user may choose to pay a lower direct fee by selecting a vehicle that is subject to a reduced tariff, such as a low emissions vehicle, or by using a vehicle with a lower maximum gross weight, such as a two-axle rather than a three-axle vehicle. Traveling at certain times of the day may also lower the rate. Some road users may be eligible for a discount or exemption because of their status, including disabled drivers who may depend upon specialized modes of transport.

The overall charges for road usage may depend upon the classification of the vehicle, the purpose of the trip, the status of the road user, or a mix of these. Ensuring compliance with the rules that govern the charging scheme is necessary to ensure that the correct fee, discount, or exemption is paid, whatever the vehicle type, purpose of trip, or status of road user.

An enforceable event (often known as a violation or exception passage) may need to be pursued as a criminal or a civil offense, depending on the purpose and legal basis of the charging scheme. If sufficient evidence cannot be captured to meet the evidential test imposed under civil or criminal law (whichever is locally applicable), then the charge will be unenforceable.

Section 4.2 considers the importance of relying on data presented by an OBU (if used) to help assess the level of charge to be applied. These statements made by an OBU at an enforcement point are known as declarations. Section 4.3 outlines the basis of enforceability and the importance of preserving the integrity of the charging scheme by confirming the validity of the declarations.

Section 4.4 defines the enforcement strategy options. If the strategy is simply to deny access to, or exit from, a charged road network, then the enforcement process can rely on physical constraints, such as barriers. However, the introduction of charging on roads without a toll plaza means that enforcement needs to rely upon capturing evidence of a vehicle's presence at a specific location and at a specific time. Installing physical constraints is not an option on most existing roads.

Section 4.5 describes the end-to-end enforcement process, with an emphasis on front-end technologies within the vehicle and on the roadside. Section 4.6 focuses on several short examples to highlight the different issues faced when enforcement schemes are designed. Section 4.7 considered cross-border enforcement, and Section 4.8 considers emerging trends and innovations in enforcement techniques aimed at improving the effectiveness of the enforcement process. Chapter 7 extends the approach to enforcement by considering a hypothetical charging regime. In this case, enforcement is put into a policy context that inevitably faces all road user charging systems to some degree.

4.2 Declarations

Many parts of the transport network rely on a person making a declaration to gain permission to use a facility or to assert a status that affords some other benefit. This benefit could include a discount or exemption from charges. This is analogous to using a paper or magnetic card permit to gain entry to a secure building.

The importance of declarations on which to base enforcement decisions is best illustrated with an example. In-building security systems are often configured for closed user groups, where the access permit is already known to the entry system, and it is normally assumed that the access permit is being used by the assigned cardholder. Photo ID helps manual checking, and in the future, biometric methods that are based on prerecorded characteristics of the individual will help further automate access control. Small, closed user group schemes are dominated by proprietary techniques, so every new user needs to be locally registered, and, if needed, issued with an access permit that is also locally accepted. The security and integrity of these schemes usually depends on the proprietary designs of access permits, the structure of data held on the access permits, the business rules used to grant or deny access, and the flow of data through the system.

Large-scale schemes, such as credit card payment networks, mobile telephone networks, and interoperable road user charging systems, are operated as open user groups, since there are many service providers and virtually an unlimited number of users associated with one or more of the service providers. Authentication of an access card, cell phone, and PINs for credit cards can ensure a tighter relationship between the electronic means of declaring a privilege and the authenticity of the device or person requesting the privilege. The same technique can be applied to vehicles equipped with the necessary electronic credentials provided by OBUs.

An open user system could not operate if every access point or payment point is required to have a priori knowledge of every account, all users, and all of their privileges. If this analogy is applied to charging and enforcement systems, then we can see that the information flows within a small, closed, and centralized scheme will be entirely different from the flows within a scheme that is part of a region with many chargeable roads, each operated by a different organization. Closed user schemes include isolated toll roads that are often managed by operators who manage all parts of the road user charging process, including charging, enforcement, and customer relations management. For this reason, these operators are known as vertically integrated. Open user schemes that are applied on dense road networks with many operators are becoming increasingly prevalent in Europe (e.g., France, Spain), the United States (e.g., New York, New Jersey, and California), and Asia (e.g., Hong Kong, China).

As local and regional economies become increasingly connected, then so do the road networks that connect them. This greater interdependency places pressure on road and transport service providers to agree on common rules and approaches to reflect the common user base. Embodying all charging and enforcement functions within each organization makes less sense, since the trend towards interoperability and realization of the benefits of economies of scale suggests that some functions should be shared between organizations. Chapter 6 further explores this subject.

Europe, the United States, and Japan have been the source of some of the most active, original standardization work for charging technologies, although attempts to harmonize approaches to enforcement in these regions are still relatively immature. There are as many different approaches to enforcement as there are legal jurisdictions, even if the principles of effective enforcement are becoming increasingly well known, as described in Section 4.4.

The following illustration highlights the differences between closed user group and open user group schemes, which both depend on OBUs. Declarations of a vehicle class made by an OBU issued by authority A at a charge point (or toll lane) managed by operator B need to be common, or at least meaningful, to operator B. If operator A's OBU declares "class 2," then this should be consistent with the same class defined by operator B. Any inconsistencies can cause unnecessary enforcement events, and the road user could be unfairly penalized due to the declaration of a class that is not acceptable to a third-party operator. A vehicle that provides an electronic declaration of class 2 (e.g., passenger car) should ideally be understood by all operators. A class 2 declaration (e.g., passenger car) should always mean the same thing to all operators. Technical misunderstandings can be eliminated through technical agreements between operators that serve the open user group. A better solution is not to depend upon multiple bilateral agreements, but instead, aim for regional agreements, which standardize the format and content of declarations within an economic region, such as the European Union or the United States. Chapter 5 explores this issue further.

Small-scale charging schemes (e.g., up to several hundreds of thousands of vehicles) can depend entirely on reading the vehicle's number plate or a code read from a simple tag fixed to the vehicle's windshield. The number plate can be used as a pointer into a database containing information on the account to which the vehicle is associated [automatic account identification (AAI)]. A simple fixed code AVI tag could provide a similar pointer or, in the future, EVI or ERI (see Section 4.8).

Larger interoperable systems need to rely on electronic declarations provided by an OBU, such as the status of the user and vehicle characteristics. Increasing the information delivered by the OBU to the roadside system at the point of enforcement enables localized decision-making. However, it also increases the complexity of the logistics to encode, distribute, install, and maintain the validity of the data provided by the OBU and its association with roadside databases. Electronic declarations generally have a higher detection accuracy rate than do license plate–based systems, although this bears little relationship to the level of compliance that can be achieved. The choice between OBU-based and license plate methods is generally defined by the charging strategy (see Chapter 3).

Declarations can relate to the vehicle, purpose of trip, or status of user, as discussed in the following sections.

4.2.1 Vehicle Type

The OBU is encoded with the type of vehicle, such as its taxation class. A manually operated switch can also be provided on the OBU itself, so that driver can dynamically modify the class definition (e.g., to declare whether or not the vehicle is towing a trailer), as used by the Austrian LKW (heavy goods vehicle) charging scheme. A simple keypad/display on the OBU could allow the modified vehicle configuration (e.g., floating axle status, trailer configuration) to be changed by the driver.

Many schemes provide discounts for taxis and hybrid fuel vehicles, although this concession often depends on whether the charging scheme is to collect tolls or to manage demand.

4.2.2 Usage/Purpose of Trip

The OBU can be encoded to highlight the purpose of the trip, such as a military vehicle being used on military service, an emergency vehicle dealing with an incident, or a bus driving on its regular route. If the purpose changes, then the declaration needs to be updated. A simple keypad and display on the OBU would be an adequate interface for this declaration.

Usage-based discounts or exemptions can be difficult to enforce, since there may be nothing that can be measured externally to suggest the purpose of the trip. Historical and institutional discounts are often used, despite the enforcement burden. Sections 4.3 and 5.2 further explore the measurability problem.

4.2.3 Status of Road Users

If the charging policy provides discounts for selected categories of users, such as registered disabled person and local residents, then the OBU could indicate this category. User declarations could vary through a simple interface on the OBU or integrated card reader, although enforceability may be problematic if users are required to perform some action to declare their status—some will forget and others will not fully understand their obligations.

However, the flexibility of electronic charging systems to accommodate many differentiated charges can sometimes make it difficult to be accurate when making a decision on whether or not the vehicle passage should be enforced.

4.3 Measurability and Enforceability

If the charging structure is based on vehicle characteristics, then it should be possible to automatically or manually observe or measure these characteristics in order to enforce the charging scheme. This is also true if the charging structure includes differentiated charges that depend on the purpose of the trip or the status of the user. Some of these characteristics can be observed and checked by a manual toll lane operator or mobile enforcement officer, and others can be automatically measured or checked against a vehicle registration database for charging in a toll lane or on an open highway.

Some vehicle characteristics cannot be easily measured. Table 4.1 lists the three main bases of charging, and the degree to which they can be automatically measured at the roadside.

Since an effective enforcement strategy is critical to support a charging regime, then vehicle usage and user characteristics need to be measured if the charges depend upon these factors. The enforcement strategy does not need to depend on automatic methods if compliance can be assured by other means, such as by dedicated vehicle patrols or curbside checks by authorized enforcement officers (see Section 4.4). Where these characteristics cannot be automatically measured, the other options include manual enforcement, or using the vehicle's license plate number to check against the operator's local database. Interrogation of the Department of Motor Vehicles' database may be necessary if the operator's database is inadequate. Evidential records in these cases could be retained until these checks have been completed.

Innovation in on-board and roadside sensor technologies aims to improve the accuracy of automatically measured characteristics, although methods of determining the purpose of the trip and status of the road user are either nonexistent, expensive, complex, or still under development as early stage technologies (see Section 4.8).

The charging policy usually defines vehicle characteristics that cannot be directly measured, including the vehicle taxation class (e.g., maximum vehicle gross weight and public service vehicle with over 12 seats). The origins of these class definitions are as old as the principles of vehicle taxation, and have slowly evolved as new vehicle types are introduced. Consequently, few of these class structures readily conform to automated measurement techniques, but nonetheless are often adopted

Table 4.1 Automatic Measurability and the Main Bases of Charging

Basis of Charging	*Specific Differentiator*	*Automatic Measurability*
Vehicle characteristics	Taxation class	Cannot be measured, but may be inferred from axle count and vehicle profile
	Plated weight	Generally known by the driver, a static parameter, may be inferred by volumetric measurements
	Dynamic weight	Can be measured on the vehicle itself, or using in-ground weigh-in-motion sensors
	Number of axles	In-lane treadles (plaza-based schemes), or combination of in-lane loops and video analysis
	Engine size and power	Not externally measurable
	Low-emission or dual fuel vehicle	Emissions could identify fuel being used, but is not a reliable measure of the capability (on which discounts or exemptions are usually given) for low emission or alternate fuel
Usage	Military purposes, buses in service, emergency vehicles	Cannot be measured
	Occupancy	On-board sensors or face/skin/heat detectors (see Section 5.5 on measurement trends and emerging technologies); presently unknown viability
User status	Registered disabled, local resident	Cannot be measured

by road user charging operators due to their legal basis. These class structures are often defined to meet national vehicle taxation preferences, so that divergence between adjacent countries makes cross-border interoperability more complex. Harmonizing vehicle taxation policies across borders is unlikely to be an option for a single scheme operator, although cooperative efforts have made some progress (see Section 4.7).

Chapter 5 explores the capability of a variety of on-board and roadside sensors for vehicle classification, and how well they can be used to infer the vehicle taxation class.

Road users in barrier-controlled lanes at toll plazas expect that the charging transaction will be rapidly completed, without requiring the vehicle to stop. If the information provided by an OBU needs to be checked against some remote database, including the balance of funds in a prepaid account, then the vehicle could be kept waiting until the remote database confirms the vehicle status or its class. It may be satisfactory to wait 60 seconds for a credit card to be checked in a store, or for a mobile phone to register with a third-party mobile network, but it would be unacceptable to have the same delay to pay a toll electronically at a foreign toll plaza while the road user's credentials are being checked.

The decision on whether or not to enforce needs to consider the integrity and content of declarations, the measurability of the declared characteristics of the vehicle or user, and other information that may be held by the roadside system.

All networks must be aware of road users that are known to be persistent violators or are associated with other account anomalies. As stated earlier, the enforceability of any charging system is paramount. Sharing information between operators for the purposes of enforcement is an inherent requirement of road operators within all road networks, particularly on travel corridors or within the same economic area. Mobile communication network operators and credit card payment service providers already apply the same principle. The principles are therefore already proven in other mass-market applications.

The enforcement system must process information collected in a few seconds in a toll lane, or a few hundred milliseconds on an open highway. Complex charging structures based on several categories of vehicle, combined with usage and user discounts, places a burden on the charging and enforcement systems to collect and properly assess the reliability of the declarations. Where there is some doubt about the accuracy of the measurements, the enforcement procedures need to depend on manual support.

The most effective solution is to measure selected vehicle characteristics, such as length, width, height, and number of axles. Independent measurements that have sufficient resolution to differentiate between each category can check the validity of the declarations. The differences between the class definitions should ideally be matched by the ability of automatic vehicle classification systems to resolve the same differences (see Section 5.2.2). Vehicles also tend to cluster at class limits (e.g., maximum gross weight limits), so the measurement methods must accommodate this nonuniform statistical distribution to minimize classification errors. Chapter 5 further explains this. The class definitions also may not be based on physical characteristics, such as size.

The accuracy of capturing OBU data from all vehicles ranges from 99.5% in an urban environment, where OBUs may be obstructed, to 99.995% in a toll lane or on an open highway, where optimal measurement geometry is possible. However, the accuracy of vehicle classification systems typically varies from 60% to 95%, assuming class definitions are measurable (see Sections 5.2.1 and 5.2.3).

To ensure payment process efficiency, OBU declarations of vehicle type, usage, or user status should be used to trigger the charge, rather than the measured vehicle characteristics. The relatively high error rate of vehicle classification systems could cause overcharges or undercharges, depending on the tariff structure. The increased cost to resolve errors could worsen the economic feasibility of the operation, create customer dissatisfaction, and stimulate adverse media interest for highly public schemes. The overall legitimacy of the charging system could be undermined, and could cause the operator to face penalties. In the worst case, repeated errors could force the termination of the operating concessions and takeover by the public governing authority.

If the observed or measured results differ from the declared results by more than a preset margin, then the business rules used in the toll lane or MLFF charge point will trigger the enforcement process. For example, if the OBU declares the classification to be a passenger vehicle, but in-lane sensors detect three axles, then the barrier could remain closed to permit a plaza attendant to resolve the difference. Immediate resolution is not possible on the open highway, so the evidential strategy would depend upon the capture of images of the suspected vehicle, together with the measured and declared classification. The content of the images must satisfy legal requirements for evidence, comply with relevant privacy laws, and (ideally) allow an operator to manually confirm the true classification.

Table 4.1 provides some examples of what is, and is not, measurable. It would not be possible for an operator viewing images to confirm that an engine size declaration is correct. The solution is either to conduct a roadside check, or not to differentiate charges based on engine size.

The importance of ensuring that the majority of the payments due are collected cannot be emphasized enough. The reasons for this depend on the purpose of the charging systems, and may not always be obvious to the road user. For example, a private toll road operator needs to ensure the effectiveness of the fee (revenue) collection process to reassure the providers of the investment funds. Private capital providers regard the effectiveness of the revenue collection system as extremely important to meeting long-term financial targets. On the other hand, a congestion charging scheme imposes fees (charges) as a means of demand restraint, and may use the fees to fund complementary measures, such as enhanced public transport. The accuracy of the charging system need not be as high, since it is the perception by road users of the existence and integrity of the enforcement regime that induces behavior change and maximizes compliance.

4.4 Enforcement Strategy Options

4.4.1 Considerations

The main methods of enforcement are either physical or evidential. Enforcement actions can be based on either a physical denial of service (see Section 4.4.2), or the capture and processing of evidential records (see Section 4.4.3). Physical, historic, legal, data protection, and economic constraints (see Section 4.4.4) often limit the range of enforcement strategies. The tendency for road users to evade payment depends on their assessment of the likelihood of being detected, and the benefits of not been detected. Section 4.4.5 provides a brief discussion on strategies to reducing this tendency.

4.4.2 Physical Methods

There are several physical approaches to deterring evasion, including:

- In-lane vehicle height restrictors;
- Lane exit barriers at toll plazas;
- Hydraulic or pneumatic bollards/ramps, rising curbs, or similar devices;
- Manual enforcement.

4.4.2.1 In-Lane Vehicle Height Restrictors

If a vehicle taxation class is used to differentiate between vehicles, then it may be possible to link physical characteristics, such as a vehicle's height, to one or more classes. This may be crude, but height restrictors in toll lanes can effectively filter vehicles above a prescribed height into manual lanes, where the vehicle class can be more accurately determined by visual inspection.

Figure 4.1 shows a plaza on the N1, north of Pretoria, South Africa—one of the many plazas worldwide that employ height restrictors to stop high vehicles (i.e., mainly commercial vehicles) from entering unmanned ETC lanes. Many plazas in France and Spain also employ this method. Vehicles can also be filtered by clear and understandable approach signage, backed up by plaza attendants if a vehicle enters a lane intended for vehicles of a lower class.

4.4.2.2 Lane Exit Barriers and Rising Bollards

The second physical method in road user charging systems is the barrier or gate arm (see Figure 4.2), which acts as a visible, physical deterrent to evading payment. Some city center restricted area schemes can use hydraulic bollards (Figure 4.3) or rising curbs.

When the charging event has been completed based on declarations made by an OBU and confirmed by measurements, the barrier is lifted to allow the vehicle to continue. Typically, revenue loss from automatic barrier-controlled systems can be as low as 0.1%, mainly due to tailgating; that is, charging can sometimes be evaded by closely following the vehicle in front, such that any vehicle detection method is unable to distinguish the second vehicle from the first and allows the second vehicle to clear the barrier before it is lowered. Tailgating, and its complement frontgating, potentially represents a source of systematic evasion, although

Figure 4.1 Height restrictors (N1, South Africa).

Figure 4.2 Rapid action barrier. (Courtesy of Tecsidel.)

Figure 4.3 Rising bollards, being lowered for a bus (Cambridge, United Kingdom).

the unpredictable behavior of the unwilling partner vehicle means that it is often difficult to repeat the evasion action on every occasion.

The simple function, familiarity, and direct action of the barrier has led its adoption by the majority of worldwide toll collection systems, despite its obvious flaws. Barrier-controlled toll plazas are not always desirable or not an option in many cases. MLFF or hybrid plazas that combine traditional lanes with ORT lanes have thus been introduced. Since the passage of a vehicle on an open highway cannot easily be denied if there are no physical barriers, the only solution is to employ evidential methods, as described in Section 4.4.3.

4.4.2.3 Manual Enforcement

Local regulations may permit a vehicle to be restrained if there is a reasonable expectation that the vehicle has violated a local road user charging scheme. If the vehicle has an OBU, then it can be interrogated with a handheld reader (see Figure 4.4) to check consistency of any declarations for discounts or exemptions.

Manual enforcement may mean on-foot enforcement officers checking OBUs and vehicle license plates in random areas. "Recent camera system developments, requirements for enforcement and pressure on the budgets to pay for enforcement have all combined to provide incentives for innovators . . . to create new tools and processes for enforcement" such as vehicle-mounted ANPR cameras that can rapidly identify violators from the license plates of vehicles parked at the curb or in parking lots [1].

The cost of manual enforcement for a single offense would probably warrant apprehension of a vehicle only if it were associated with a persistent violator, defined as a person who has evaded payment on several occasions at one or more road user charging schemes. The charging regulations will also state whether the driver of the vehicle at the time of the offense or the owner of the vehicle is liable for payment of the original charge and any additional payments that have accrued.

Figure 4.4 Handheld DSRC reader. (Courtesy of Q-Free.)

For example, the Stockholm congestion charging trial made the registered owner liable for payment and subject to criminal proceedings if payment was not made. Depending on the regulations, it may be possible to restrain or remove the vehicle to force compliance, and, if payment is not made, to dispose of the vehicle to recover any lost revenue.

Finally, enforcement may permit mutual recognition of offenses between regions, that is, an offense committed in one jurisdiction may be enforced in another jurisdiction (see Section 4.7).

4.4.3 Evidential Methods

4.4.3.1 Principles

If physical methods cannot be used, then evidence of the suspected violation is needed. The evidence needs to meet the criteria defined by local guidelines that unambiguously show the following three characteristics:

- The vehicle's identification (usually its license plate);
- The location of the vehicle relative to its surroundings;
- The time and date of the vehicle's presence at the point of detection.

Some jurisdictions may require the capture of the color of the vehicle, front and/or rear images, side images showing the number of axles, and an overview image showing the vehicle in context. The image needs to be secured by using mechanisms to detect/resist tampering, prevent interception, and avoid unintentional deletion. The Stockholm congestion charging pilot uses evidential enforcement, but does not allow the image to contain the driver or passenger.

There are a few jurisdictions that permit unattended automatic capture of evidence. The standards that apply to the equipment and the capture and management of evidence acceptable for enforcement need to be well defined [2]:[1]

> It is of paramount importance that this evidence is of such unquestionable accuracy and quality that it is readily accepted by the courts and public . . . While any medium that can record the evidence with sufficient quality to meet the above requirement may be used, to meet the aims set for this system, the data needs to be recorded and stored in electronic form . . . [and, if images are employed] . . . in general terms, acceptable image quality is that which most people would feel clearly portrayed all the information relevant to proving the offence.

In general, jurisdictions that employ safety (speed detection) and red light cameras may already have suitable legislation (under civil or criminal law) to allow evidential enforcement for road user charging. Evidence usually means an image of the vehicle meeting the quality criteria listed earlier. Additional information, such as data read from an OBU, an axle count (measured by an in-ground detector), and license plate number (from an ANPR system) would not be regarded as evi-

1. In the United Kingdom the London Congestion Charging Scheme relies on an external adjudicator rather than the courts to assess claims.

dence, but instead as supporting information or metadata. Figure 4.5 shows an example of evidence captured from one camera, showing the vehicle in context (on an anonymous section of road) and an enlarged view of the license plate. Figure 4.6 provides the same evidence, but shows a vehicle in clear context with the specific road segment and all relevant metadata collected at the time of the suspected violation.

The automated capture of the evidence needs to rely on business rules that can be effectively and reliably applied, based on other information known at the time. For example, the charging policy may require the vehicle's passage to be linked to a payment or a means of payment. If this association cannot be established for a specific vehicle (either by reading its OBU or the license plate), then the enforcement process decision logic needs to capture, secure, and retain evidence for later processing (see Section 4.5.2). If postpayment were permitted, then the evidence would need to be retained until the payment period time limit has expired.

4.4.3.2 Permanent Enforcement Sites

Image-based evidence is generally accepted worldwide, since it allows an independent observer to see the offense as if it were witnessed at the roadside. This natural acceptance of image-based evidence, the availability of low-cost cameras, and new video recording technology stimulated the development of video enforcement

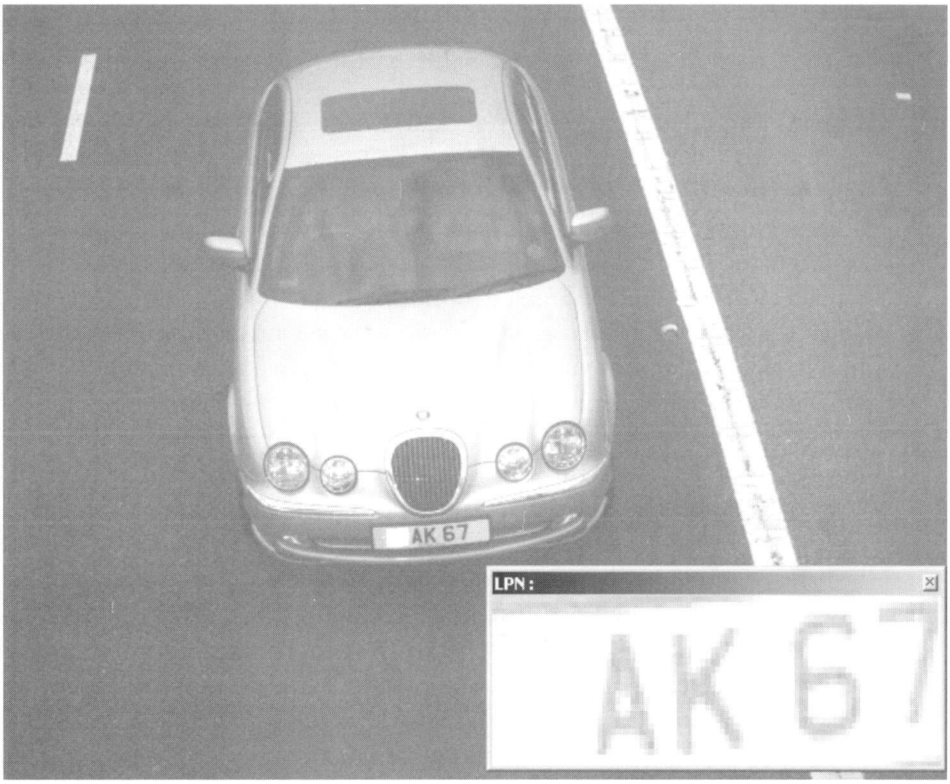

Figure 4.5 Evidential images (number plate and overview image).

Figure 4.6 Evidential record and metadata. (Courtesy of Kapsch TrafficCom AG.)

systems (VES) by many toll system integrators. Early systems were based on time lapse video recorders triggered by in-lane logic, although current MLFF solutions rely on asynchronously triggered high resolution digital imaging cameras located above the road (see Figures 4.7 and 4.8). Cameras may be mounted on the side of a lane or in the canopy on toll plazas. Secure digital records are increasingly replacing analogue recording, as confidence in digital records increases and as local laws permit. This gradual conversion to digital imaging process offers both the opportunity to implement low-cost existing solutions for small toll roads, and the efficient mass management of images in large-scale systems.

A typical small-scale VES includes functions that capture, store, and allow the images to be reviewed off-line by a trained operator. VES can also supplement barrier-based tolling schemes to capture evidence of tailgating. More sophisticated solutions extend the definition of VES to include image compression, encryption, transfer, filtering, automatic extraction of a vehicle's license plate number, temporary bulk storage, preview, and mass mailing of infringement notices. Encryption mechanisms, such as DES and AES-128, can be used to protect images from being interpreted if intercepted unlawfully while being transferred over public telecommunication networks. Watermark methods can also help authenticate an image, for example. The physical, operational, and legal framework together help define the evidential strategy.

Figure 4.7 MLFF enforcement (Switzerland). (Courtesy of Kapsch TrafficCom AG.)

Figure 4.8 Enforcement point (German truck tolling scheme). (Courtesy of Vitronic Dr.-Ing. Stein Bildverabeitungssysteme GmbH.)

4.4.3.3 Mobile Enforcement

A fleet of mobile enforcement vehicles (see Figure 4.9) can be deployed at any random strategic location to make compliance checks in moving traffic or as stationary checkpoints [3]. In moving traffic, the enforcement staff can remotely check the functionality and status of a suspected OBU, and can access application data (i.e., declared vehicle parameters) from the OBU that are relevant for compliance checks. If an OBU is not installed, then the enforcement staff will capture images of the vehicle in context and its license plate.

Automatic compliance checks can be performed at appropriate locations on the road network where the mobile enforcement vehicle can be safely parked, such as a service lane or a rest area. The additional equipment needed for fixed automatic checks could also include a vehicle detector, vehicle classifier, and an ANPR camera to capture images.

To deliver mobile and fixed enforcement functions, the equipment installed within the mobile enforcement vehicle (see Figure 4.10) could include:

- A fixed-mounted DSRC transceiver;
- Internal DSRC controller;
- Handheld DSRC OBU reader, for portable use;
- Portable classifier for classification/axle-counting, for stationary use;
- Notebook PC with the mobile enforcement application and local database of vehicles of special interest;
- ANPR camera(s) to capture context and vehicle license plate images;

Figure 4.9 Mobile enforcement vehicle (ASFINAG). (Courtesy of ASFINAG.)

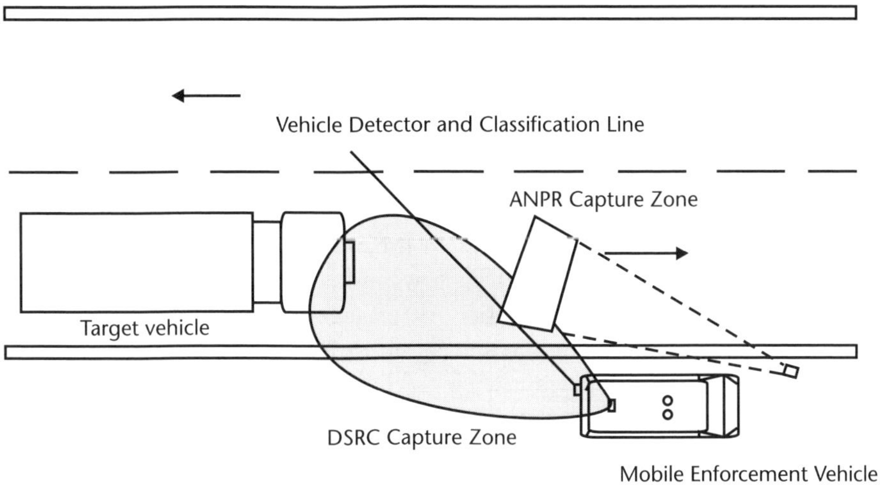

Figure 4.10 Mobile enforcement equipment (stationary operation).

- GSM/GPRS modem to interrogate a remote vehicle database, and to transmit evidential information to an enforcement site;
- Printer (to issue receipts).

A GPS receiver and externally synchronized clock can be installed within the mobile enforcement vehicle to provide geographical coordinates for automatic logging and incorporation within all evidential records generated, ensuring that the vehicle's position is accurately recorded.

If the charging scheme applies only to specific categories of vehicles, then a preclassification system (roadside scanning laser detector) can be used to detect vehicles that meet an equivalent profile description (e.g., number of axles). An image of the front of the suspected noncompliant vehicle is captured, showing its color, position, and license plate. The images are typically encrypted and incorporated into an evidential record with other metadata, such as the license plate number, measured vehicle dimensions, detected presence of a trailer, and measured number of axles. The images are cryptographically secured to prevent tampering. The OBU (if installed) also provides data, including its ID, mileage declaration, operating status, and tamper status. If noncompliance is determined (using the same business rules as a fixed enforcement point), then the OBU data can also included in the metadata with the evidential record.

The legal basis of witnessed capture of evidence may require the evidential quality to be relaxed to a less onerous burden of proof (e.g., use of a civil rather than criminal evidential test), thus allowing lower capital cost equipment to be used. Mobile enforcement can also compress many stages of the enforcement process into a few process steps that can be completed by the mobile operators.

4.4.4 Constraints

There are many constraints on the development and operation of an enforcement strategy. The five most common constraints are physical, historic, legal, privacy-related, and economic.

4.4.4.1 Physical

Direct manual collection of fees may be the most economically advantageous or practical method of charging. Enforcement is immediate and localized, although plaza-based tolling occupies land space and requires vehicles to stop or slow down. Channeling vehicles into separate traffic lanes also requires traffic management, road user familiarity, good signing on the approach to the toll plaza, and physical infrastructure specifically designed for manual and automatic payment.

Barriers slow down traffic flow and can increase emissions as compared to emissions from moving traffic. An efficient rapid-action barrier lane can allow vehicles to travel through at speeds up to 20 km/hr, although good signage on the approach to the toll plaza is critical. Higher throughput can be achieved by holding barriers in the open position during the morning and evening peaks when the proportion of ETC users is greatest, although this may increase the incentive to evade being charged.

4.4.4.2 Historic

The first and still dominant application of ETC, as measured by the number of operational sites, is on toll plazas. The ETC system need not make any complex decisions on the position of the vehicle or its OBU. It simply needs to detect the presence of an OBU, and, by ensuring that the DSRC footprint is highly localized and stable, the system is able to associate the tag with a vehicle detected by an in-ground loop or optical light curtain (see Chapter 5). If the ETC communication zone is not localized, then cross-lane reads may occur, in which tags are read in adjacent lanes and may cause an OBU in one lane to be associated with a vehicle in an adjacent lane. ETC communication transceivers are increasingly able to electronically limit detection into a narrow, well-defined zone, or to actively localize the OBUs to within a few decimeters.

The use of ETC as an alternative to payment by cash, tokens, or cards means that plaza-based ETC often uses the same enforcement system. ETC enables charging at higher vehicle speeds, so some tolling operators (e.g., in France and Spain) have attempted to deliver this benefit in barrier-controlled lanes by placing barriers at a further distance from the OBU detection point, and, in some cases, providing road users with an opportunity to change to a non-ETC lane in case of detection of a violation or some other anomaly (e.g., the user forgets to install the OBU). A high penetration of tags in the traffic at the plaza may mean that a lane can be partially or fully reserved for ETC accountholders, hence the term dedicated lane.

The use of toll plazas has traditionally linked charging with payment: the charge is settled immediately with cash or tokens. The introduction of ETC breaks this historical dependency. The passage of a vehicle through a toll lane or at a charge point generates a charging event that needs to be matched off-line with the appropriate payment. Depending on the account, the charging event can be used to trigger a payment from a prepaid account, or a liability to be settled later (i.e., postpayment) by cash, credit card, or direct bank transfer.

Familiarity and institutional inertia, coupled with the need for low cost, generally points toward traditional toll plazas enforced by manually or automatically operated barriers. Incentives to reduce the use of land for traditional plazas, development of charging and enforcement technologies, central systems, and enabling laws all supported the emergence of MLFF/ORT. Truck tolling schemes in the United States, Germany, Austria, Switzerland, and Australia depend on permanent and mobile enforcement systems approaches. Finally, the introduction of high-occupancy vehicle (HOV) and high-occupancy and toll (HOT) lanes currently depends on manual enforcement, although the there is the possibility of automatic counting of vehicle occupancy as part of evidence gathering in the future (see Section 5.5.2).

4.4.4.3 Legal Basis for Enforcement

The legal basis for enforcement is typically either civil or criminal law. The use of a barrier for enforcement generally ignores this distinction, but the capture of evidential information for an MLFF system cannot. The evidential strategy needs to provide enough information to unambiguously identify the vehicle and its license plate to meet the civil or criminal law test for evidential quality. Any supplementary information, such as the location, color, make, and model of the vehicle, and the time and date of the violation, would strengthen the evidence.

If there is no reliable database of vehicle registrations, then it is unlikely that a specific vehicle could be enforced solely on the basis of an image of its license plate. In this case, the enforcement strategy would need to depend upon immediate denial of service methods, such as barriers. This means that plaza-based fee collection is the only feasible method that could be enforced until a reliable, reasonably complete database of registered vehicles is created. Lack of such a database may not only prevent a charging system from being employed, but may prevent the construction of a road if the fund providers perceive the enforcement risk to be too high. For example, the need to use MLFF for ETC on urban highways in Santiago de Chile stimulated the need to centralize the various vehicle registration databases in Chile.

The legal requirements for toll collection in some countries may require the vehicle to stop, which is a legacy of legislation written specifically for plaza-based toll collection. The same law would actually prevent enforcement on the open highway at MLFF charge points. Since charging cannot be applied without effective enforcement, the lack of a legal basis for enforcing moving vehicles would prohibit the use of MLFF for road user charging. In France an offense is committed by violating a traffic signal in a toll lane rather than paying a toll. This currently precludes MLFF in France.

It is generally assumed that evidence needs to rely on one or more images, but the development of electronic number plates and other forms of identification that are part of the vehicle may offer new forms of evidence that could be acceptable in the future (see Section 4.8). Of course, the OBU used for charging can also provide vehicle, usage, and user-specific information to the enforcement system. As long as the OBU is not an integral part of the vehicle, it could never be proven that the data originated from a specific vehicle, in the absence of an image. It could only be proven that the data originated from a specific OBU, but business rules may make this good enough to pursue payment.

The integrity of the OBU installation process may provide this reassurance. For example, the Singapore ERP scheme, managed by the LTA, required most of the city-state's 600,000 vehicles to be equipped with OBUs (see Section 8.2.1). An OBU is physically bonded to the inside of each vehicle's windshield with UV-curable adhesive. The OBU is installed on behalf of the LTA, which means that the identification of an OBU at the charge point is directly equivalent to the detection of the vehicle. In this example, the physical, relatively secure link reduces the tendency for OBUs to be switched between vehicles of different classes. The LTA requires that OBUs be installed or removed only at authorized inspection or other appointed centers.

The Stockholm pilot congestion charging scheme that commenced on January 1, 2006, imposes fees as taxes (see Section 8.2.4). As mentioned earlier, nonpayment of taxes is regarded as a criminal offense, as reflected in the Swedish enforcement process.

Local regulations may prevent a vehicle from being relicensed if any payments are outstanding. This is known as plate denial, and used in the United States and Canada to ensure high levels of compliance (see Section 7.3.6).

4.4.4.4 Data Protection and Privacy

Local data protection and privacy laws may affect the collection and use of customer data, and any other information that may relate to an individual, such as images captured for enforcement purposes.

The Stockholm Congestion Charging system evidential strategy required images to be captured that, legally, could not include the driver. This required the capture of images to be triggered precisely by vehicle position. As the front of the vehicle crossed a defined photo line, an image is captured of the front of the vehicle containing the number plate. The camera truncates this image by simply removing all information above a predefined position.

Local laws may also limit the maximum period for which images can be retained. The London Congestion Charging scheme operator retains images of a vehicle's presence at a defined location and time until the vehicle's number plate (as recorded) can be associated with a payment for the specific charging period. The maximum time that images can be held is generally until the purpose for which they are held has been satisfied. The enforcement process starts by notifying the registered owner by issuing a penalty charge notice (PCN). The ensuring process provides sufficient time for the recipient to respond, and, if needed, an appeals process to be concluded.

4.4.4.5 Financial Necessity and Revenue Assurance

The funding for road construction (e.g., government finance, bonds, a syndicate of investors, one or more banks, and so forth), and the provision of these funds, will be linked to legal rights and obligations, which define the purpose of the charging scheme, enable charges to be levied, and enable enforcement actions to be taken in the event of nonpayment. The level of detail and the need for specific laws will depend upon the host country.

The providers of funds for road infrastructure may also impose an enforcement strategy that minimizes revenue risk. The charging and enforcement technologies and processes must offer an acceptable level of financial assurance to the providers, and therefore, a minimum acceptable technical performance of the end-to-end enforcement process, including:

- The vehicle detection rate: the percentage of vehicle passages that are detected;
- The capture accuracy: the percentage of detected vehicle passages that result in the capture of one or more images;
- The read rate: the percentage of images that contain the vehicle's number plate that is automatically readable;
- The false positive rate: the percentage of automatically generated results that have a high confidence, but which are nevertheless incorrect.

The accuracy of any automatic process to extract the characters from a license plate relates more to process efficiency rather than to revenue assurance. An efficient, highly automated enforcement process has lower marginal costs and therefore lower costs of operation (see Section 4.5). Typical single point detection rates measured over a 24-hour period should exceed 85%. The enforceability of a charging policy also needs to be economically viable. The enforcement process should not penalize the wrong person, which requires a low false positive rate. Safeguards (e.g., manual checking) need to be used before issuing an infringement or penalty charge notice.

The performance capture and read rates should be applicable for all charging schemes. Some manufacturers depend on a "lane-centric" approach, which assumes that vehicles will primarily travel in the center of each lane. This may apply in free-flowing interurban highways where driver discipline is high, but will be less applicable at low vehicle speeds, such as in the city environment. If traffic flow is undisciplined and chaotic, then the image capture and ANPR read rates should still apply, regardless of the vehicle's position on the road. Since charging and enforcement are interlinked, the enforcement cameras ideally should provide spatial information on the vehicle and its license plate, to distinguish between vehicles with a specific OBU and vehicles without any OBU. Section 4.5 discusses further the minimum dynamic requirements for MLFF/ORT systems.

Finally, the integrity of any charging scheme needs a credible enforcement process to recover funds and (ideally) pays for the operating cost of the recovery process. The objectives of enforcement may also include providing a viable, secondary source of revenue, rather than merely covering its costs. The enforcement revenue in some schemes can be many times the amount of the revenue lost, although this depends on the number of detected violations, the proportion of the value of the fine/penalty recovered, and the existence of regulations that allow penalties to be levied.

4.4.5 Tendency to Evade Payment

The principles of road user charging depend on maximizing compliance, deterring nonpayment, and enabling the necessary action to recover payments and additional fees. The fee structure needs to be understandable to road users, and it must be accessible and require low effort to pay any charges. The scheme operator also must signal the existence of enforcement, and have the authority to take direct actions that may be escalated over time, to ensure that revenue and additional fees are collected.

The tendency for road users to evade payment depends on the probability of being detected, the scale of the charges being avoided, and the scale of the penalty. However, this tendency cannot be measured in absolute terms.

- An information campaign can help inform road users to consider the probability of detection of evading payment as a deterrent, especially if periodic reminder campaigns are used rather than continuous broadcasts. See Figure 4.11.
- Schemes that are based on high charge rates, such as distance-based truck tolling, will represent a greater benefit to the road user if payment could be avoided. The incentive can be reduced through deterrents such as mobile enforcement, which can be used as an effective visible deterrent by increasing the perceived probability of being detected. See the mobile enforcement vehicle in Figure 4.8.
- The scale of the penalty may be set by the scheme operator to act as a deterrent. It may not be legally possible for the operator to impose punitive rates to deter enforcement. Instead, it may only be possible to charge the toll rate plus a reasonable administration fee.

If the scale of the penalty is greater than the charge avoided multiplied by the perception of probability of detection, then evasion is suppressed. If the relationship

Figure 4.11 London congestion charging information campaign poster.

is reversed, then the tendency to evade payment is increased. Increasing the perception of the probability generally offers a better deterrent to evasion than simply increasing the scale of the penalty charge.

4.5 The Enforcement Process

4.5.1 General Outline

Section 4.1 emphasized that enforcement is not an event but a process. Effective enforcement depends on business rules that process data from multiple sources, including electronic declarations from an OBU (if equipped), measurements of vehicle characteristics (if possible), and use of the vehicle registration mark (VRM) as a pointer into a local or remote vehicle database.

The process is described in two stages; at point of detection, and in the back office (addressed in more detail in Chapter 6).

4.5.2 Image Capture and Interpretation

4.5.2.1 Principles

The relationship between the vehicle, its OBU, and its license plate are shown in Figure 4.12. This relationship exists, regardless of whether the violation detection is automatic or manual.

The vehicle is physically linked to its license plate. The logical relationship is retained by the local department of motor vehicle registrations, and probably by the road operator with whom the road user has an account. A third-party operator could learn more about the vehicle by contacting the department of motor vehicle registrations, if the license plate provides sufficient information for the correct jurisdiction to be identified. In Europe, every vehicle is required to display the country of registration, with a decal either on the vehicle or on the license plate, although in practice this is not widely accepted. If the vehicle is equipped with an OBU that is readable at the point of enforcement, then it should be possible to identify the issuer of the OBU (interoperability rules will define these data requirements), and, if issued locally, the identification of the road user's account. The business rules at the enforcement point would include interrogation of immediately accessible databases and consistency checking with any measured attributes. These checks will decide whether or not the vehicle is a potential violator, regardless of the success of the charging transaction. It may be impossible to immediately verify this, so an evidential record may be created in some schemes, and later deleted when all business rules have been applied to the satisfaction of the scheme operator. If the vehicle is in a toll lane, the business rules may require the barrier to remain lowered to enable a toll lane attendant to resolve the apparent violation.

4.5.2.2 The Technology

The performance of optical character recognition (OCR) systems that can automatically read the alphanumeric characters from a vehicle's license plate should not be

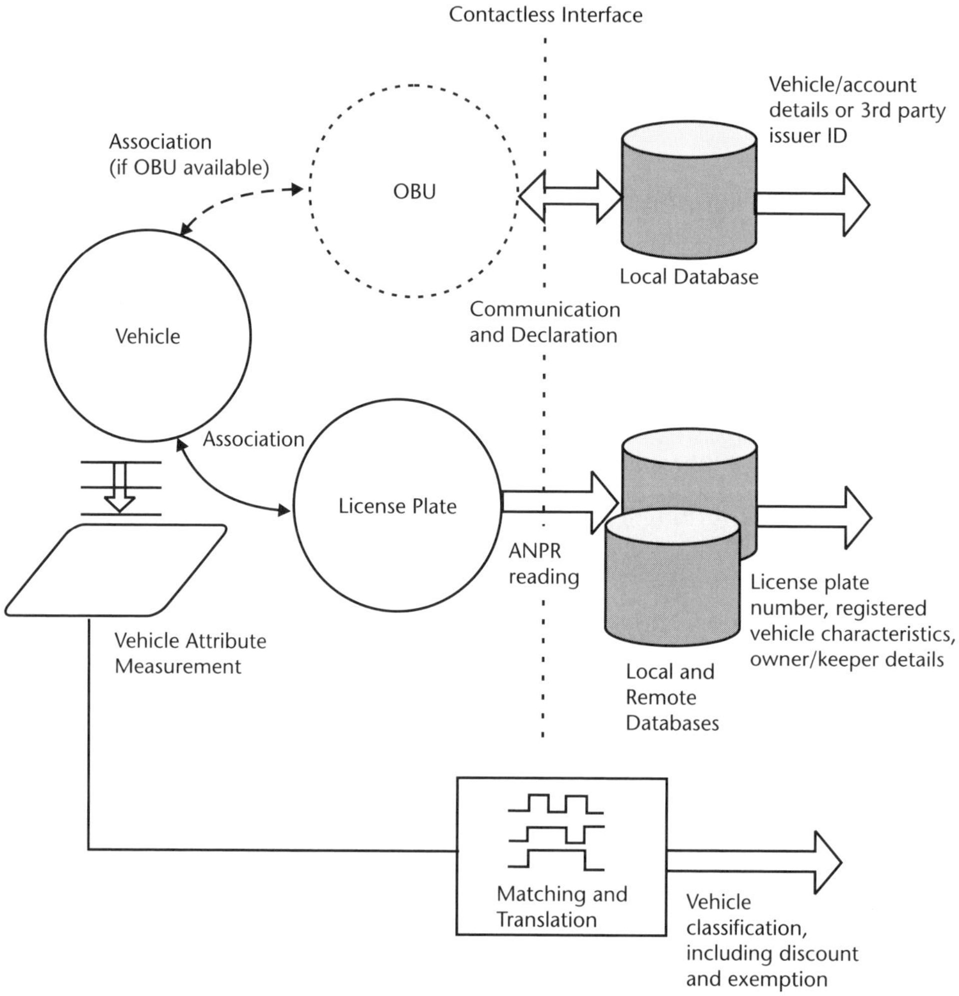

Figure 4.12 Relationships at the point of enforcement.

confused with the minimum technical performance requirements for an end-to-end evidential strategy. The simplest image-based enforcement schemes do not need OCR. A low volume of daily enforcement events can be interpreted by an operator who will apply business rules and exercise discretion in deciding whether or not to pursue the violation. Higher volume schemes necessarily depend on a higher level of automation, which assists manual operators in deciding whether or not an offense has been committed. Manual image checking is a source of process costs, so it should be minimized without compromising the effectiveness or error rate of the enforcement scheme.

ANPR systems that employ OCR techniques are able to process image-based records with almost no human intervention, and are therefore essential to high process efficiency. However, a computer record of the vehicle's registration number currently cannot act as a legal substitute for the images captured for enforcement. The vehicle's registration mark extracted from the image therefore needs to be seen only as an interpretation of an image (metadata). The OCR reading is used in the enforcement process and can be used to trigger automated enquiries to remote databases (e.g., the department of vehicle registrations). An image can be used by an operator to resist a challenge made by a road user that a charge was erroneous. Evidence of a vehicle's presence can show that a charge triggered by an OBU was correct. Thus, an image can be used to strengthen a charging event and support an enforcement action.

ANPR systems usually also indicate the confidence of the character sequence, and, for some camera systems, a confidence value for each individual character. If the confidence of the whole character string is high enough, then it can be used to compare with a preregistered list of vehicles. If the confidence is too low, then the image can be discarded or passed to manual checking (a simple example of a "filter"). The level of confidence of the interpreted character sequence can also drive the business rules for the enforcement process, although in practice there are other, more sophisticated methods that are used.

The London Congestion Charging system uses ANPR cameras for enforcement of its area charging scheme. A typical configuration in an urban context, including context cameras and dedicated ANPR cameras, is shown in Figures 4.13 and 4.14.

Figure 4.13 London Congestion Charging camera site (configuration).

Figure 4.14 London Congestion Charging camera site (urban context).

A typical range of specifications offered by ANPR cameras, comprised of a camera and an integral OCR engine, are given in Table 4.2. The OCR engine may also be located in an external controller fed by two or more cameras, depending on the approach to OCR. The optimal configuration depends on the preferences of a system integrator, although the lanes in toll plaza systems usually operate independently of each other. An infrared illuminator may also be needed, and can either be integral to the camera (for reflective plates) or separate (for nonreflective plates). Context images can either be monochrome or color (depending on local regulations), so additional white light illumination may be required after dark, if accurate color information is required. The London charging scheme determined that there would be sufficient illumination, and that additional lighting would not be necessary.

MLFF/ORT ANPR systems usually have a single camera mounted so that its field of view (FOV) lies centrally on the lane. The camera is mounted at a vertical angle that is steep enough to read plates on vehicles separated by a few meters,

Table 4.2 Typical ANPR Camera Specifications

Parameter	*Value*
Sensor	IR sensitive at 850 or 950 nm
Field of view	Vendor-specific, generally from 2.0m to 3.5m
Image format	8-bit JPEG (or 12-bit JPEG for a higher dynamic range)
Accuracy	Capture and correct read accuracy from 85% to 95% of readable plates, depending on maximum font size and variation of the legally accepted order and content (syntax) of the alphanumeric sequence
Font support	Vendor-specific: Latin, Korean, and possibly Arabic can be supported
Power	Typically 12V to 18V dc, although project-specific
External interfaces	TCP/IP (IP addressable), GPRS
Local service access	RS232 or LAN with password-controlled Web client access
Packaging	Rugged and waterproof to IP67, in some cases nitrogen filled
Maintenance	Remote diagnostics, periodic lens cleaning
Illuminator	Integral (range up to 30m), synchronized with frame rate of camera or trigger to external, separately powered IR illumination, lifetime up to 10 years; alternatively, external illuminator
Other	Integral clock; integral image capture/encryption/compression; alternatively, split camera/image capture and OCR (dynamic control of aperture and gain often not offered)

but shallow enough to ensure that the height of the characters (measured in pixels) is sufficient for accurate OCR.

Two cameras may be used to increase the probability that an OCR reading that has a high confidence and therefore believed to be correct is actually correct. The Melbourne City Link applies this principle as part of its video tolling product. Assuming that vehicles are required to have license plates on the back and front, a double camera enforcement point can reduce the probability of false positives, since it would be expected that the front and rear license plates have the same number. Two ANPR cameras with different geometries record the passage of each vehicle from the front and rear.

The cost of the site increases by the additional integration and installation cost of a single camera, additional maintenance, and a marginal cost of developing and proving new business rules. The false positive rate also declines, meaning that a greater reliance can be placed on the combined output, although a manual check will be required to ensure that the license plate has been correctly decoded before targeting the evader. ANPR is also a technology option for occasional users as described in Sections 3.5.1 and 3.5.4.

4.5.2.3 Fixed Enforcement Site Deployment

Fixed enforcement site planning and selection of optimum locations must consider many factors, including:

- Land ownership and rights of way;
- Proximity to utilities (power and communications) and cable routes;

- Crash protection structure requirements;
- Physical/geotechnical feasibility;
- Accessibility for maintenance personnel;
- Road user safety assessment;
- Aesthetics and environmental impact;
- Road closure and traffic management constraints;
- Maximizing detection of violators, based on preferred local and regional travel routes.

The selected site is then subject to a topographical survey and ideally an assessment of the environmental impact of the proposed infrastructure.

A large-scale project would require a specific organization to plan and construct enforcement points. The project leadership would include a project director, a program manager, an engineering manager (systems integration), an operations manager (implementation), and a risk manager. The tactical staff would include operations and technical/engineering, comprised of regional installation managers, each with their own installation and commissioning team. Following site acceptance and any silent running period, enforcement scheme operations would include performance monitoring, reporting (summaries, service level agreements, issues, and assumptions), and general systems monitoring.

A small-scale deployment of barriers and/or enforcement systems at a toll plaza would not specifically depend on an enforcement deployment team; instead, enforcement systems would form part of the general systems integration, installation, testing, and maintenance.

4.5.3 "The Funnel" and Back-Office Procedures

The enforcement strategy and the necessary laws and regulations to recover lost revenue, or to otherwise ensure compliance, underpin an enforcement scheme. The complexity of the central system (described in Chapter 6) generally depends on the geographic scale of the enforcement regime and the forecast number of violators (i.e., the level of compliance). The part of the central system that manages the flow of evidential records would include several validation checks to guarantee the consistency and integrity of the information, and to prioritize the evidential records before manual review. The records and related metadata are presented to the enforcement center operator for consistency and accuracy checking. These checks are critical to enforcement service delivery, and maintain the evidential quality for the entire end-to-end process.

Some vehicle images may have poor quality, making them unusable. For example, the vehicle's number plate may be dirty, obscured by a trailer hitch, outside of the FOV, or hidden by another vehicle. These images may be rejected, or, if enough of the vehicle is visible, it may be possible to match an image with another image of the same vehicle captured elsewhere [4]. Many images may be rejected if they do not meet evidential quality requirements, such as the availability of a context image, and a license plate image that is clearly and unambiguously readable by manual operators. This successive reduction of the list of possible violations is known as a "funnel."

The enforcement system back office performs the following functions:

- Workflow management: automatic analysis of evidential records, and prioritization for manual inspection;
- Information warehouse: access to customer registration data and history;
- Reporting: validation and compilation of ad hoc and periodic reports;
- Image preview, interpretation tools, and enhancement of evidence images;
- Monitoring of process performance and generation of summary statistics;
- System configuration;
- General systems monitoring and health checks;
- Enforcement points: security key distribution, software configuration, and updates;
- Data security: management of user privileges and rights;
- User management: prevention of unauthorized system access, and logging of access to meet audit trail requirements;
- Communication with external service providers: use of digital signatures to ensure transaction integrity and nonrepudiation, transaction and digital signature logging.

Depending on the scale of the operation, these tasks may be handled by a single part-time enforcement officer, or a full-time team of operators. Several operational assumptions determine the size of the system, including a noncompliance rate in steady state, and expected peak loading at the start of scheme operations. The expected image size and quantity of images generated each day (before the application of any business rules) will define the image storage capacity. The data storage strategy may require images to be retained for longer than 12 months, in which case the image retrieval time is important if a case is being built against persistent offenders. Typical enquiry functions should provide a response within 2 minutes for evidential records less than 12 months old, and 24 hours for evidential records older than 12 months.

The enforcement solution will also include disaster recovery provisions (e.g., second site infrastructure, handover procedures, and related management structure). The repetitive nature of manual checking warrants attention to occupational health and safety, workstation ergonomics, and appropriate shift patterns that include adequate operator breaks to avoid fatigue and to ensure good performance.

4.6 Examples

The following examples highlight a small selection of different issues that face enforcement scheme designs for toll plazas and MLFF/ORT-based schemes.

4.6.1 Example 1—OBU Association with Vehicle

The German heavy goods vehicle charging scheme [5] employs accredited workshops to install OBUs that are programmed with vehicle characteristics, and that can be interrogated by fixed or mobile enforcement equipment. Truck operators may also decide to manually register for a specific route without needing any in-vehicle equipment. In this case, the vehicle's number plate is read as part of the process to confirm that the initial declarations are correct.

The opportunity for tag swapping can be minimized by employing locally accredited tag installation workshops to securely attach the OBUs to each vehicle [6]. This installation provides some confidence that the electronic declarations made by an OBU accurately reflect the characteristics of the vehicle to which it is fitted. In the future it may be acceptable that OBU data would be regarded as proof of vehicle presence, if it could be assured that the OBU could not have been in another vehicle at the time of the violation. Section 4.8 describes a complementary technology known as EVI, which potentially provides a method to electronically identify a vehicle.

4.6.2 Example 2—Discount for Residents

A charging scheme may require local residents applying for a discount to select a specific vehicle that is registered to their address. This enables the discount to be associated with the vehicle rather than its driver or passengers. It could be assumed that a resident would only own or have access to one or two vehicles. This makes it easier to ensure compliance with the privilege, since the identification of the vehicle is easier to confirm than the identification of the driver or passengers while the vehicle is moving through an MLFF or ORT enforcement point. The association between the vehicle and user (in this case) improves the measurability and therefore the enforceability of the discount for residents.

4.6.3 Example 3—Poor Measurability

A charging scheme for truck tolling employs a charge that is based on distance traveled, the quantity of vehicle axles, and emissions class. The number of axles and emissions class can be electronically declared, but only the number of axles can be measured with any certainty. The scheme operator has the following two options:

1. Random, manual off-line checks of images may be captured at each enforcement point, to determine whether the electronic declaration of the number of axles matches the physical vehicle characteristics. Further enforcement action would only be taken if a mismatch could be clearly verified.
2. Each vehicle's number plate may be captured and the automatic ANPR record may be used to request a copy of the vehicle's registration record. This record would be compared with the electronic declaration provided to the enforcement system by the OBU. If there is any mismatch, a further manual check of the image is conducted, to confirm that the ANPR system read the number plate correctly and that the context image unambiguously showed the correct vehicle and number of axles (if used as a differentiator). Images of all vehicle passages would need to be retained to permit off-line checks. Unused images are deleted.

4.6.4 Example 4—Vehicle Segregation at Toll Plazas

An electronic toll collection operator differentiates charges based on taxation class of the vehicle, which is in turn based on a combination of maximum gross weight and design purpose (e.g., carrying goods on a commercial basis). Each of the toll plazas employs physical height restrictors to limit HGVs to lanes that combine automatic ETC with manual lanes. As the vehicle enters the manual lane, the toll lane attendant manually classifies the vehicle through visual inspection. If the vehicle has the means to electronically declare its classification (e.g., an ETC tag), and if this matches the classification determined by the toll lane attendant, then the ETC transaction proceeds normally. If there is a mismatch, then the attendant is able to resolve the difference with the driver in the toll lane.

4.6.5 Example 5—Manual Enforcement

An urban charging system based on an area charging policy employs in-zone camera-based enforcement points to check compliance, based on measurability and declarations. All parameters cannot be measured, so manual inspectors and clearly marked enforcement vehicles that are equipped with the means to electronically access OBUs (e.g., using a handheld DSRC reader) increase the deterrent to road users. Each OBU is securely labeled with a subset of the declared parameters (excluding any personally identifiable information), to enable manual inspection without any equipment. The OBUs are also color-coded to reflect the intended vehicle class and application of any exemptions or discounts. The Dartford Thurrock Crossing in the United Kingdom, among other operators, employs color-coded OBUs that can be manually inspected in the toll lane if needed.

4.6.6 Example 6—National Vehicle Database

The national department of transport may decide to update its legal definition of classification, based on parameters that can be measured in lanes at toll plazas (e.g., number of axles and vehicle profile). Enforceability of charging schemes would also benefit from updating the national vehicle registration database, but inaccuracies will continue to arise if owners do not notify the transport departments of the sale or disposal of vehicles, change of address, or modification to any other measurable vehicle attributes. The enforcement policy would dictate whether the national database was used for enforcement, or, if the vehicle is already known to road operator, then the local database may be used in the first instance.

4.6.7 Example 7—Nonregistered Vehicles

A charging regime may allow road users to use the road without registering for an OBU. Each vehicle detected without an OBU would not be assumed to be a violator, but may be a registered OBU customer that is not displaying an OBU, or a customer that does not have an OBU (e.g., 407 ETR, Canada). The ANPR system reads the license plate, and (to an accuracy of about 90%) generates an evidential record with metadata based on any data held by the road operator. If the vehicle is associated with an OBU customer, then an administration charge is applied to cover the additional cost of enforcement. If the vehicle is not known, then the license plate number is submitted to the department of motor vehicle registrations. A manual check is made to confirm that the plate number and other vehicle details (e.g., vehicle type, color) match before issuing a bill to the registered owner. The amount of the bill covers all of the additional costs incurred by the operator, plus a premium to encourage preregistration of the vehicle on its next journey.

4.7 Cross-Border Enforcement

Differences in vehicle registration policies, privacy and data protection laws, the definition and prosecution of traffic offenses, and evidential quality requirements all mean that cross-border enforcement of road user charges is potentially complex and expensive to operate (see Section 7.3.8). The procedures required to identify the party responsible for a vehicle and to serve a notice requesting payment differ widely between EU member states.

The interest in cross-border enforcement is a key issue in an open market like the European Union and led to the creation of the VERA initiative in 1998, the creation of which was supported by representatives from Europe, the United States, and Asia. The VERA program aimed to "develop the technical tools necessary to support cross-border enforcement and define relationships between enforcement agencies to govern the use of these tools" [7]. The program included an assessment of legal, operational, and organizational issues in enforcement [8]; best practice in enforcement, using digital imaging enforcement systems [9]; and a common functional specification for enforcement systems, based on digital imaging techniques [10]. VERA2 extended this to develop and test a common format for the exchange of data, an operational framework within which cross-border enforcement could be managed, approval of enforcement equipment, and a memorandum of understanding (MoU) on cross-border enforcement. The VERA3 program aims to validate the technical approach to data sharing, and to identify minimum legislative requirements. This legislation to enable enforcement on nonresident violators of traffic offenses does not yet exist. The related CAPTIVE [11] initiative examines the current state of legislation and suggests solutions, including mutual recognition of traffic offenses, including nonpayment of charges.

Section 5.2.4 details current attempts at regional harmonization of class structures to assist with the enforceability of vehicles across borders, and Section 7.3.8 outlines the challenges faced with charging and enforcing road users registered in another jurisdiction.

4.8 Innovation and Trends

We have seen that the complexity of the enforcement process may range from an in-lane barrier to a comprehensive digital imaging system supported by an extensive administrative process to recover revenues.

Innovation is apparent at all stages in the enforcement chain, including the recent development of combined imaging cameras that include pulsed IR illuminators and improved image authentication techniques to highlight image tampering. Other ongoing developments are aimed at improving OCR engines to deliver higher accuracy, with a reduction in false positives, using higher resolution cameras that can operate with a high dynamic range in difficult lighting conditions, and cameras that can resolve characters on license plates at difficult angles (using CMOS technology).

Other recent developments, some of which are now commercially available, include:

- Cameras that include IR illumination (optimal for reflective plates);
- ANPR systems that are capable of supporting effective matching processes required by ORT/MLFF enforcement systems;
- Improvements in automatic triggering by ANPR cameras, to reduce the variability of the captured vehicle position on the road;
- Cameras that provide additional information on the trajectory of a vehicle (or its plate), to assist in matching vehicles with their OBUs in ORT/MLFF charging systems;
- Improved OCR engines that can interpret scripts on foreign number plates;
- Faster, more secure encryption mechanisms, such as AES 128 or AES 256, that are less processor intensive [12];
- Improved lossless compression methods, to enable more efficient transfer of images without compromising evidential quality;
- Microfluidic lenses that can change the viewing direction of fixed-mount cameras through remote control, potentially reducing on-site maintenance time to readjust a camera's geometry [13];
- ANPR combined with vehicle fingerprinting, to improve matching of subsequent images [14];
- New methods of image-based vehicle identification, based on extraction of distinct characteristics and matching with subsequent vehicle detection [4];
- Handheld imaging cameras integrated with OBU readers and localization, to permit aiming with the same accuracy as fixed imaging systems;
- New methods to accurately measure vehicle characteristics and occupancy;
- Convergence of legal descriptions of vehicles with automated measurement methods;
- Low-power, rapid action barriers coupled with vehicle separators, to reduce tailgating;
- Mass-market OBUs with integral smart card readers [15], to allow user declarations at the point of charging made by smart card.

In addition, ERI [16] and EVI [17] both offer the possibility of unambiguously and securely identifying a vehicle without depending on optical methods. These technologies include radio frequency tags that are securely fixed to vehicles, which allows the remote reading of license plate or vehicle identification numbers (VINs). The contents of the tag will depend where in the manufacturing and supply chain it is installed. A vehicle manufacturer–led scheme could provide tags that provide the VIN. In addition, the vehicle's registration process could require tags to be encoded with other static information, including emissions characteristics, engine size, and quantity of axles. At the time of this writing, there is no firm acceptance of the installation and tag-encoding process. Installing the tag rigidly to the vehicle's structure means that the ERI tag identification acts as a proxy for the vehicle identification. The installers of the tag may include the vehicle manufacturer, distributor, authorized workshop, or any entity that can be regarded as trusted and confirmed by an accreditation process.

The choice of the encoding method for future ERI and EVI technologies requires institutional and legal support, and offers the opportunity to enhance automatic enforcement systems for road user charging on the open highway along with or as an eventual replacement for depending on automatic reading of license plates.

In the meantime, the use of images and supplementary information will still dominate evidential strategies on worldwide MLFF and ORT schemes. We may progressively depend less on images as primary evidence as the evidential strength of the electronically captured information from ERI/EVI devices increases, as trusted entities are defined, as privacy concerns are met, and as nonimage data becomes more acceptable. Acceptance will require images to be regarded merely as one class of electronic fingerprint that also includes nonimage-based sources, such as secure data packets provided over an RF interface from an ERI/EVI device located in the vehicle.

Enforcement technology is worthless without the appropriate legal context. Developments in this area include regional and national legislation that is recognized across jurisdictional borders, which specifies equipment approval and evidential quality requirements.

4.9 Summary

Every charging scheme needs an effective enforcement strategy, which ensures compliance with payment rules, deters nonpayment, and provides the means to recover revenues and fees.

Enforcement can generally be physical or evidential. Physical methods use a barrier to restrain every vehicle suspected of violating the charging policy. Toll collection attendants then resolve the violation or inconsistency with the driver of the vehicle. Physical restraint is not possible on the open road, so evidential methods, usually based on the capture of images or witnessed by a traffic officer, are used. Road users will tend to evade payment if the scheme rules are not understandable, if the perceived risk of being detected is low, and if the penalty is less than the road user charges.

Enforcement is not an event, but a well-defined process (based on business rules) that comprise automatic and manual process steps to efficiently reduce possible violations to a subset that meets quality thresholds sufficient to identify evaders using the vehicle's license plate. Evidential methods depend on ANPR cameras located at strategic locations on the road network to capture high-quality images and decode the vehicle's license plate. Fixed enforcement sites can be supplemented by mobile enforcement, and, if needed, manual enforcement. Making road users aware of the enforcement scheme through the use of highly visible enforcement vehicles, targeted advertising, and other means can provide a sufficient deterrent to nonpayment and can raise the credibility of the enforcement scheme.

Enforcement within a jurisdiction is less complex than cross-borders enforcement. The increasing interdependency between regions means that enforcement increasingly needs to extend across jurisdictional borders. This requires that violations be mutually respected in other areas. Legislation will support this interdependency, and will help increase the use of images for enforcement and MLFF/ORT charging. Other innovations relate to the process and the detection technologies themselves.

References

[1] Percival, M. E., A. Sedgwick, and T. Ellis, "Street Enforcement Applications for Mobile ANPR Systems," *Proc. 12th IEE Intl. Conference on Road Transport Information & Control—RTIC 2004*, London, U.K., April 20–24, 2004, pp. 211–213.

[2] Lewis, S. R., *Home Office and ACPO Traffic Outline Requirements and Specification for Automated Traffic Enforcement Systems*, March 1996.

[3] Jung, S., "The German Heavy Vehicle Tolling System," *Proc. IEE Road Transport Symposium*, December 5–6, 2005.

[4] Sines, T., and J. E. Hedley, "Video Enforcement—The Road to Electronic Tolling," *Proc. IBTTA Technical Workshop*, Edinburgh, U.K., June 11–14, 2005, http://www.ibtta.org/files/PDFs/HedleySines.pdf.

[5] Egeler, C., and M. Bibaritsch, "Enforcement of the Austrian Heavy Goods Vehicle Toll," *Proc. 10th World Congress on Intelligent Transport Systems*, Madrid, Spain, November 2003.

[6] Singapore Land Transport Authority, *Electronic Road Pricing Authorised Inspection Centres*, 2006, http://www.lta.gov.sg/motoring_matters/motoring_guide_centres.htm.

[7] VERA Project Team, *Video Enforcement for Road Authorities, VERA TR4027, 4th Framework Research Project*, Cordis, http://www.cordis.lu/telematics/tap_transport/research/projects/vera.html.

[8] VERA Project Team, "Legal, Operational and Organisational Issues in Enforcement," *Video Enforcement for Road Authorities (VERA), Deliverable D3.2*, 1999, http://www.cordis.lu/telematics/tap_transport/research/projects/vera.html.

[9] VERA Project Team, "Report on Best Practice in Enforcement Using Digital Imaging Enforcement Systems," *Video Enforcement for Road Authorities (VERA), Deliverable D3.3*, 1999, http://www.cordis.lu/telematics/tap_transport/research/projects/vera.html.

[10] VERA Project Team, "Common Functional Specification for Enforcement Systems Based on Digital Imaging Techniques," *Video Enforcement for Road Authorities (VERA), Deliverable D5.1*, 1999, http://www.cordis.lu/telematics/tap_transport/research/projects/vera.html.

[11] CAPTIVE Project Team, *CAPTIVE Overview*, 2004, http://www.veraprojects.org/CAPTIVE/CAPTIVE_overview.html.

[12] Lewis, S. R., *Home Office Requirements for the Protection of Digital Evidence from Type Approved Automatic Unattended Traffic Enforcement Devices*, Home Office Scientific Development Branch, October 12, 2005.

[13] Bains, S., "Going with the Flow," *The IEE Review*, U.K., March 2006.

[14] Bureau Verkeershandhaving Openbaar Ministerie (the Netherlands), Section Control, Verkeershandhaving Dossiers, 2006.

[15] Q-Free, *Breakthrough for Q-Free in Turkey*, Media Release, February 22, 2006.

[16] International Standards Organization, CEN ISO/TS 24534-1 to 4—Road Transport and Traffic Telematics—Automatic Vehicle and Equipment Identification—Electronic Registration Identification (ERI) for Vehicles—Parts 1 to 4, 2005.

[17] ERTICO, *ERTICO Deliverable D2 (Final Requirements), D3 (High-Level Architectures), D4 (Final Assessment), D5 (Conclusions and Recommendations)*, EVI Project Consortium, 2004, http://www.ertico.com/en/activities/projects_and_fora/evi_website.htm.

CHAPTER 5
Vehicle Detection and Classification

5.1 Background

Vehicle detection and classification are functions used to determine the relevant attributes of a vehicle for charging and enforcement. The traditional view of these functions has been based entirely on measurement of external characteristics of vehicles for traffic monitoring, intersection control, and traffic surveys.

Accurate vehicle detection and classification for road user charging and the management of tolled highways is needed for statistical reporting, supporting manual toll lane operation, and as an integral part of open road charging and enforcement schemes. Shadow tolling is a mechanism by which a road operator is paid by the highway owner according to the number of vehicles using the road, and the quality of the road in terms of speed flow and lane availability. Measurements are taken of each of these parameters, and payment is made off-line without any participation of road users. No toll is actually levied from the driver, since shadow tolling is a business relationship between the road operator and owning authority. Accurate vehicle counting and classification is critical to the payment mechanism.

Classification methods aim to measure vehicle parameters in order to infer the classification of the vehicle, usually for the purposes of enforcement.

Road user charging is developing as an integral part of worldwide pay-per-use transport policies, so the charging policies and the differentiation of charges relating to vehicle class become increasingly important. Each road operator must ensure compliance with the rules and laws that govern the road user charging scheme. If the charging policy defines a tariff table that is based on vehicle class, then correct execution of the policy depends upon accurate resolution of the vehicle's classification to ensure that the correct charges are applied, and that the appropriate enforcement actions are taken against road users that violate the regulations.

The scope of the vehicle detection and classification process is expanded here to reflect advances in external measurement methods, developments of in-vehicle technologies, and the use of vehicle-to-roadside communication. Section 4.3 further extends this discussion in the context of enforcement.

Capturing the vehicle attributes at the point of charging, enforcement, or monitoring could include any of the following four approaches:

- Use of one or more sensors for the measurement of attributes that directly relate to the defined vehicle classification (e.g., number of axles and height above first axle);

- Measurement of attributes, such as a vehicle's height, width, length, and cross-sectional profile, which permits the vehicle classification to be inferred if it cannot be directly measured;
- Reading of declarations of vehicle classification from a security label or other equipment fitted to a vehicle, such as a sports utility vehicle (SUV), 12-ton medium goods vehicle (MGV), or motorcycle;
- Unique identification of a vehicle to enable vehicle attributes to be read from a local database (e.g., account information read from an OBU or a vehicle's license plate).

These approaches show that the classification process (see Figure 5.1) is not only the capture or measurement of information from a vehicle, but also includes two additional phases: matching the reported attributes to the vehicle (for enforcement purposes), and translating them into the locally applicable classification definitions.

The end-to-end process needs to ensure that the classification objectives are met. The classification definition is often based on vehicle taxation classes that were derived many years ago as part of the vehicle licensing policies of a country or state, which were then used to determine the appropriate annual fee or import duty based on the engine's maximum power output, seating capacity, and maximum gross weight. The legal, financial, and operational pressures on the automotive industry competing in a global market have since increased the complexity of the vehicle design process and means that vehicle manufacturers now must consider emissions categories, safety requirements, target market segment preferences, product differentiation, and the need for local variants of the same vehicle platform.

As regulatory authorities within economic regions harmonize class definition for heavy goods vehicles, and as vehicle manufacturers aim to capture the economies of scale from platform sharing and common body styles, the specifications of vehicles are increasingly clustered into engine size, maximum gross weight (MGW), and dimensional categories. At the other end of the scale, designs for passenger vehicles increasingly result in hybrids and crossover formats that make it difficult to accurately classify a vehicle, even when inspected by an expert. This also makes automatic classification more difficult. The regulatory definitions often develop more slowly than the product lifecycle of new vehicles.

There are increasing pressures to establish charging regimes based on vehicle characteristics that are not easily measurable (e.g., emissions class or type of engine configuration). Classification systems that intend to confirm the correct fee has been paid may also need to determine the number of occupants, as discussed in Section 7.5.7. For example, some HOT lanes in the United States can be used by single-occupant vehicles if they pay an additional charge, or if the vehicle has a hybrid (electric/gasoline) engine. Vehicle taxation in New Zealand is primarily

Figure 5.1 The classification process.

based on a vehicle registration tax plus a fuel tax. However, diesel engine vehicles must pay a fee based on distance traveled [1, 2]. Some countries, including the United States, have defined a list of possible vehicle classifications, from which each regional transport authority (e.g., the state DOTs in the United States) can make a selection. These classifications provide some guidance when a classification strategy is being developed, although in many countries, the vehicle classifications primarily exist for vehicle taxation rather than road user charging.

All road user charging schemes depend on effective charging and enforcement measures, underpinned by regulations. A vehicle classification strategy cannot be developed for road user charging independent of these regulations. The development of regulations also must recognize the capabilities of technologies for vehicle classification, charging, and enforcement. There is no point in having an elaborate, fine-grained classification regime if the classes cannot be readily measured.

There are many examples worldwide where regulatory policies have been developed to reflect the capability of the measurement methods used to determine a vehicle's classification [3–5]. Other schemes are based on MGW [6] or vehicle type [7]. Highway operators therefore face the challenge of developing classification, enforcement, and charging schemes consistent with classification tables that were developed for another purpose. Any inconsistencies between the regulatory definitions and actual measurements may result in unfairly targeting road users for declaring a class that varies from the measured class. A partial remedy would be to manually check an image of the vehicle at the enforcement point and compare it with the vehicle registration records.

The number of available measurement methods has significantly expanded since defense manufacturers turned to the civilian market as a potential outlet for sophisticated target detection and tracking and vehicle measurement technologies. Operators of toll plazas and open highway schemes are faced with a broad array of classification techniques, often delivered as part of fully integrated tolling systems by system integrators serving regional or global markets. The challenge is to ensure that the appropriate choice of attribute capture, matching, and translation techniques are used, and that the road operator is fully familiar with the available choice. Modifying the underlying vehicle taxation scheme to align with available classification techniques is often not an option (or at least could take many years). Any known shortcomings of the classification process need to be identified early in the design stage, and accommodated in the workflow of the charging and enforcement processes. This ensures that the correct charges are levied, road users are not wrongly charged, and the operator is able to employ efficient business processes that are not entirely focused on compensating for operational shortcomings in the classification method. Figure 5.2 shows the relationship between charging and enforcement processes.

Operators of new roads may be able to specify an approach to classification that reflects the enabling charging regulations. For example, Midland Expressway Limited (MEL), operator of the M6Toll (the United Kingdom's first private toll motorway), specified a classification scheme that is not only simple for road users to understand, but that can be automatically measured in each of the ETC lanes at all of MEL's six plazas (see Figure 5.3).

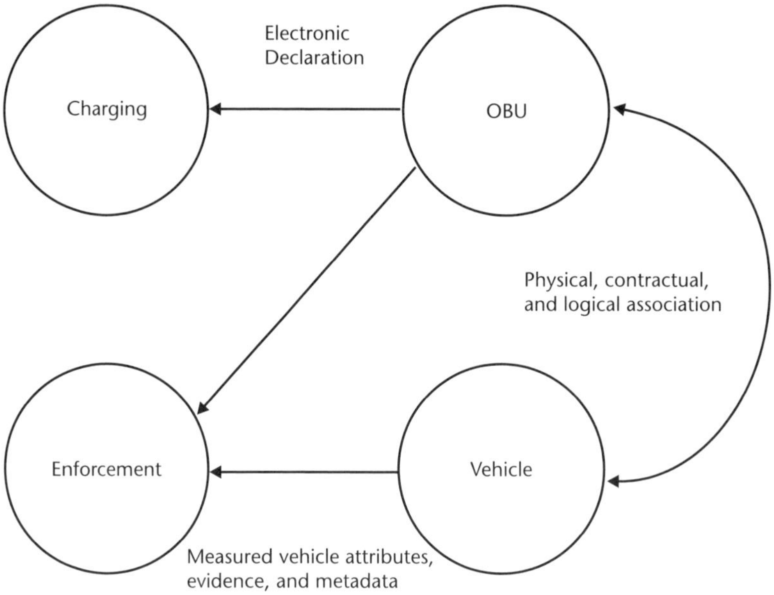

Figure 5.2 Classification for enforcement.

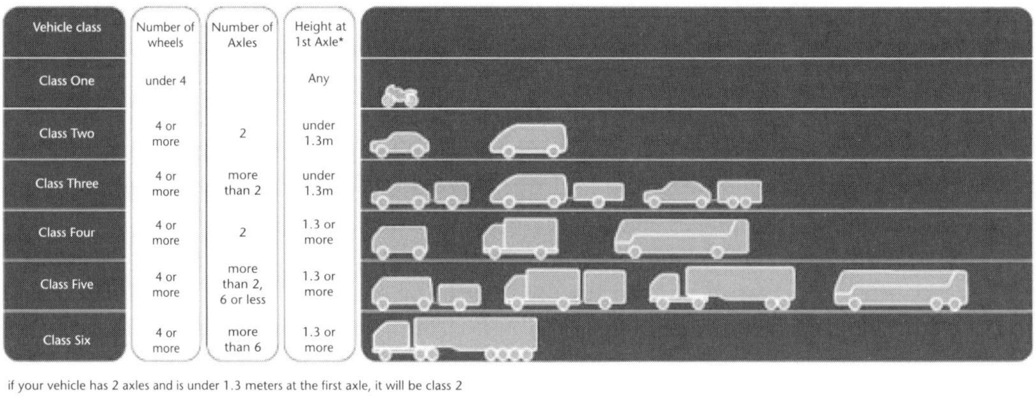

Figure 5.3 M6Toll classification scheme. (*From:* [3]. © 2006 Midland Expressway Limited. Reprinted with permission.)

Section 5.2 elaborates on the four approaches described above. Road user charging systems use either plaza-based toll collection systems or open highway, multilane free-flow systems. Collecting charges at a toll plaza or enforcing a road user charging scheme on the open road requires specific vehicle classification techniques, which are described in Section 5.3. Section 5.4 provides examples that reflect the variety of measurement and classification challenges facing operators of toll plazas, urban charging schemes, or interurban fee collection systems. Section 5.5 considers the future evolution of sensor development to improve vehicle detection, separation, classification, and translation.

5.2 Approaches to Detection and Classification

5.2.1 Context

The technological leap from vehicle detection and classification systems that create aggregated statistical reports, to becoming an integral part of an enforceable road user charging scheme, highlights the range of requirements facing sensor developers, system integrators, and road operators. There is a range of different environments, from manual toll lanes at toll plazas to unconstrained interurban highways, and from slow moving (or stationary) undisciplined traffic in an urban congestion charging scheme to high-speed traffic flow.

As well as the four approaches outlined above, the requirements on the detection and classification system also depend on its intended purpose:

- Auditing and performance monitoring (e.g., counting and classifying for shadow tolling, or audits of private operators);
- Enforcement (e.g., deterring nonpayment or wrongly equipped vehicles);
- Charging (e.g., calculating the charges due).

Tolling and road user charging policies have gained widespread use in developed and developing nations, and the enforcement regimes have also developed to the point where the detection and classification processes are integral to image-based enforcement, as elaborated in Chapter 4. The vehicle detection function may be independently provided, or as part of another function, such as vehicle separation or classification.

The accuracy of a classification system depends on several factors, including:

- Underlying technologies and their combinations;
- Quantity of vehicle classifications (fewer categories often means easier resolutions);
- Mix of vehicles (error rates often depend on the type of vehicle and their distinctiveness);
- Type of road (the classification mix varies between road types);
- Flow rates (low-speed, stationary, or irregular flows can be more difficult);
- Taxation regime (this impacts the classification mix of locally registered vehicles).

The achievable accuracy also depends on the requirements. It may be sufficient for auditing purposes to only count vehicles and allocate them to general or tariff categories, each of which could include several classifications. Enforcement requires capturing evidence of a vehicle's presence at a specific location and time (see Section 4.4.3), and classification confirms that the correct fee has been assigned to the vehicle passage. This invariably means that measurement-based methods are typically not used to calculate the charge. Instead charges can be based on the electronic declaration made by an OBU, since this can be accurately captured (approaching 100% accuracy), or by using the vehicle class declared at the time of registering the charging account. The matching process (see Section 3.5) will determine whether or not the vehicle is likely to have an OBU. If not, the matching process will have gathered enough information from available sensors and classifiers to determine the vehicle's position and trajectory, and to determine the best time to capture images for enforcement purposes.

Classification is therefore part of the charging and enforcement processes that rely on vehicle positioning, image capture, vehicle attribute measurement, processing/inference, and vehicle-roadside communications. For plaza-based tolling schemes, it is necessary to determine the position of the vehicle to operate the lane exit barrier (and to prevent it from being lowered onto the vehicle), and to trigger video enforcement cameras. The classification process must resolve every vehicle regardless of its speed, to allocate the measurements to the correct vehicle, and, for in-lane systems, to trigger the barrier for the correct vehicle. The sensors and other systems increasingly serve multiple functions. The vehicle's classification is only one of many intermediate outputs used for charging and enforcement.

The requirements on accuracy must be put into context. The requirement to accurately measure a vehicle's width on open highways may be relaxed without compromising the ability to correctly classify the vehicle, especially if the width adds little or no value to the accuracy of the matching process. Measurement of the number of axles may be a critical determinant of whether or not the vehicle should be placed into the heavy goods category, but may still lead to ambiguities if several heavy goods vehicle definitions have the same number of axles. Therefore, a combination of measurement technologies and inference may be the only option, as described below. The overall accuracy of an automatic classification system is usually greater than 60%, but can be 85% if the class definition is externally measurable, and can be up to 99% if the classification regulations can be easily mapped onto the method of measurement (e.g., number of axles for in-lane measurement at a toll plaza). This also means that any claimed accuracy is meaningless unless it is qualified.

On toll plazas, a mismatch between the measured class and the class declared by an OBU caused by a measurement error would result in the barrier remaining closed. Manual intervention would then be required. On an open highway, the image evidence needs to be manually checked to if an electronic declaration did not match the measured classification of the vehicle. Incorrect classifications therefore result in a higher volume of manual handling events, so the capacity of the enforcement process workflow for charging schemes based on classification must accommodate this increased volume.

An error can be either one of underclassification or overclassification. The charging policy could define an offense as a vehicle that declares a classification that is in a lower tariff category than the true classification. Declarations in higher tariff categories would result in a higher charge than appropriate for the vehicle, and depending on the scheme, a notification to the charge payer/accountholder.

5.2.2 Direct Measurement

This section deals with direct measurement of a number of attributes that directly relate to the defined vehicle classification.

Automated classification methods were first developed to help resolve some of the constraints of manual counting methods for statistical surveys. Early methods were based on laying a hollow, airtight rubber tube laterally across the highway. Driving an axle over the tube creates a pressure shock that can be detected by a roadside pressure switch linked to a mechanical or electronic counter. A vehicle's axle spacing can be determined by combining the time separation of axles detected by a pair of tubes (separated by a known distance) with the vehicle speed measured by adding a third tube at a known distance from the tube pair. The multiple detection events help build the axle configuration of the vehicle, which can be compared with reference classification models stored as a class table within the measurement device or processed off-line. This method may be technologically simple, but nevertheless highlights the three steps in all direct measurement processes:

- Capture (e.g., air pressure pulse detection);
- Matching (association of the measurements with a single vehicle passage);
- Translation (comparison with an internally held class table).

This example is strictly suitable only for temporary monitoring applications, since the surface-mounted air pressure tubes are not robust enough for use in toll lanes. Other proven surface-mounted or in-ground detectors are applicable for measurement of vehicle presence, such as electromechanical treadles or capacitive strip sensors coupled with inductive loops.

The U.S. Federal Highway Administration (FHWA) defines 13 vehicle categories (known as the FHWA 13 Categories Classification System) for reporting purposes, including seven classes for trucks and one definition for passenger vehicles [8]. The appropriate class for a truck mainly depends on the number of axles in contact with the road at time of classification. The FHWA requires that floating axles (i.e., axles that can be lifted off the ground) need not be counted. Some vehicle configurations are classified according to the number of axles on the tractor (pulling unit), regardless of the number on the trailer unit. A direct measurement classification system compliant with the FHWA Traffic Monitoring Guide Section 4 Category Definition [8] therefore must discriminate between tractor units and trailers and count the number of axles in contact with the road. If the classification scheme is intended for trucks, then it obviously also needs to discriminate between trucks and nontrucks, although the attributes that describe a truck differ from that of a passenger vehicle.

The externally measurable attributes of a moving vehicle generally include:

- Position (lateral and longitudinal);
- Instantaneous speed and direction;
- Number of axles (including dual tires) in contact with the road;
- Dynamic weight of each axle in contact with the road surface;
- Wheelbase (distance between axles);
- Track (distance between wheels on the same axle);
- Height (maximum height, height over first axle);

- Overall length, including trailer(s);
- Overall width, including or excluding mirrors;
- Presence of one or more trailers;
- Side profile (visible);
- Inductive profile or magnetic permeability.

The matching process also requires a single vehicle to be resolved at the time of measurement, regardless of its speed or proximity to other vehicles. The external measurement system must also provide sufficient information to separate vehicles without losing the ability to detect a trailer (usually by the presence of a towing bar linking the towing vehicle to the towed unit).

As of March 2006, the New York State Thruway's vehicle classification definition [4] was based upon height (two categories: above and below 7 feet, 6 inches, or 2.3m), and number of axles (greater than two). These attributes are measurable both in toll lanes and on the open highway. This potentially enables a migration to ORT. Other road operators in the Northeast United States offer EZ-Pass ETC services based on similar (but not identical) class definitions. The vehicle types sometimes cannot be distinguished completely without knowing the intended purpose of the vehicle. This may or may not be important, and entirely depends on whether accurate classification is needed to ensure consistency of statistical reporting based on local vehicle taxation categories or to enforce the correct payment that may require additional information about the vehicle that cannot be measured.

In this example, without harmonization of class definitions, local solutions to classification will remain, and, in some cases, direct measurement will not be enough to accurately resolve the vehicle class.

Section 5.3 describes available direct measurement technologies for plaza-based and open road schemes.

5.2.3 Translation and Inference

This section deals with the measurement of attributes that permit the vehicle classification to be inferred (e.g., by measuring height, width, length, and lateral cross-sectional profile).

The direct measurement example above applies to trucks categorized according to the FHWA [8]. Passenger vehicles are defined, irrespective of the number of axles, as:

> All sedans, coupes, and station wagons manufactured primarily for the purpose of carrying passengers and including those passenger cars pulling recreational or other light trailers.

This classification system needs to confirm the described purpose for which the vehicle was designed, namely "[for] carrying passengers."

Direct measurement of the vehicle profile rather than its height or number of axles could be used to infer that the true classification is a passenger vehicle. Since there are several body styles for passenger vehicles, several combinations of length and shape are required to cover the majority of passenger vehicle types. The classification process requires direct measurement, followed by successive comparisons of business rules (e.g., template matching) to select the most likely vehicle class.

Figures 5.4 to 5.7 highlight the use of profiling of passenger cars and trailer detection. The profiling is based on an overhead IR scanner that has a fixed scanning rate of 300 Hz, which provides vehicle detection, separation, and classification functions.

The SUV with trailer shows that slow-moving vehicles are stretched. The vehicle speed can be measured to improve the accuracy and to allow restoration of the profile prior to the successive comparison process. Alternatively, the comparison process can be repeated for different assumed vehicle lengths, although this is less accurate than including a true length and speed measurement from a secondary loop, or a light curtain based on a dense matrix of IR beams through which the vehicle passes.

Vehicle manufacturers are developing vehicles that span two or more categories, such as passenger vehicles and taxis that may be physically identical but have different purposes and tariff classes. Inference and secondary sensors would be unable to assist in this case. Measurement methods cannot generally determine the purpose of a vehicle's use, so it may not be possible to accurately infer the category for all vehicles based only on direct measurements. Several methods could improve classification accuracy:

Figure 5.4 Passenger vehicle. (Courtesy of Tecsidel.)

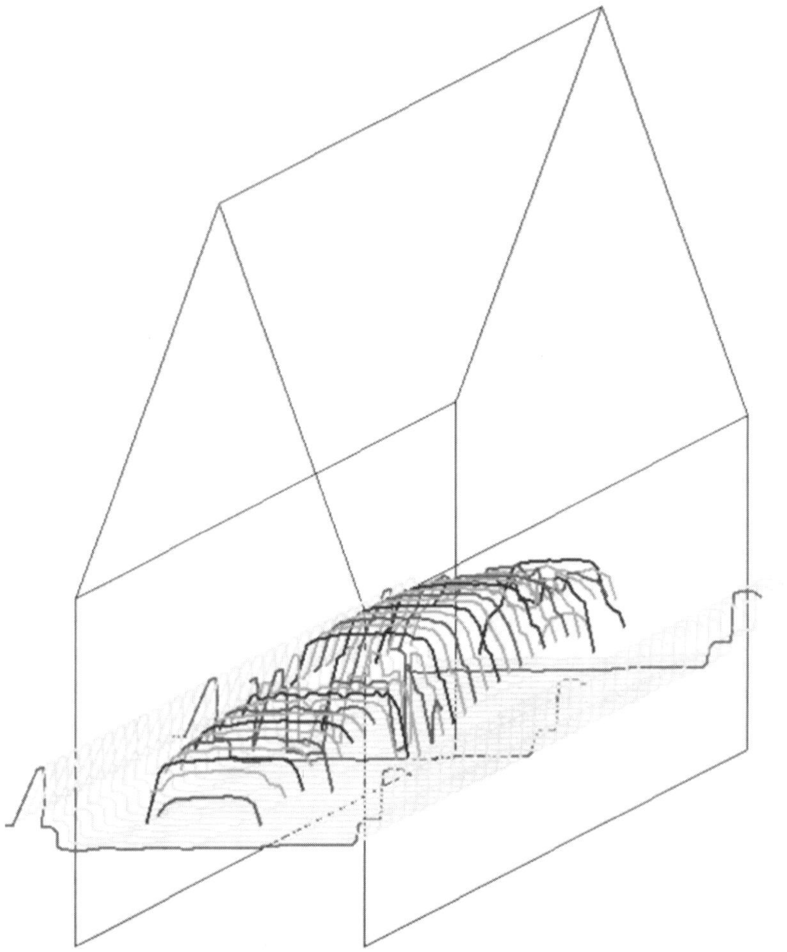

Figure 5.5 Passenger vehicle—laser profile. (Courtesy of Tecsidel.)

- Modify the classification definitions to ensure that all of the vehicle characteristics are directly measurable with limited use of inference (technically feasible, but often regulatory changes are not possible);
- Require the OBU to be programmed with the vehicle's classification before installation [e.g., attaching it securely to the vehicle by an approved authority—the Singapore electronic road pricing (ERP) and the German LKW heavy truck tolling schemes], which can then be regarded as a reliable substitute for the measured classification whenever the OBU is challenged at the enforcement point;
- Filter vehicles into specific lanes according to their class (e.g., use mechanical height limiters in lanes dedicated to vehicles with a class below a specific class-related height threshold, as described in Section 4.4.2);
- Install signage informing vehicles to travel into lanes that relate to the general class division (e.g., typical ETC schemes allow passenger vehicles to use ETC lanes without classification equipment);

Figure 5.6 SUV with trailer. (Courtesy of Tecsidel.)

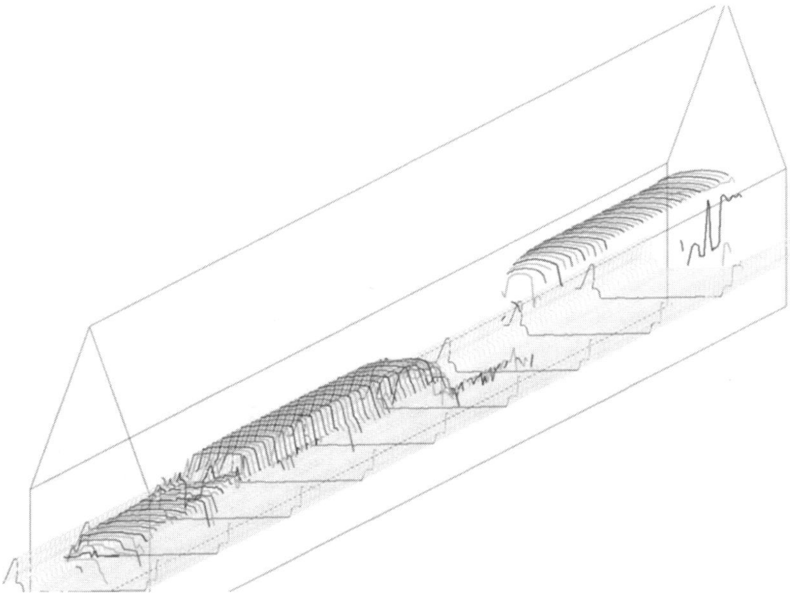

Figure 5.7 SUV with trailer—laser profile. (Courtesy of Tecsidel.)

- Define an enforcement process that accommodates measurement inaccuracies (i.e., operator support in toll lanes). For open highway charging, ensure that the workflow includes manual validation of all evidence, to confirm that the ETC account matches the true vehicle classification shown on any images collected.

A classification process that works well in one region may not work well in another. For example, local vehicle taxation and local market preferences may change the relative proportions of vehicle types, so any translation engine that uses statistical inference to resolve ambiguities could introduce errors. Vehicle designs are often motivated by the regulatory class definitions themselves, so vehicle types tend to cluster at the upper end of each vehicle category, particularly the maximum gross weight for trucks. Any classification strategy that assumes that vehicles are evenly distributed throughout the spectrum of vehicle classes is therefore likely to be inaccurate near classification boundaries, although local adjustments to measurement and translation processes can be used and will probably continue for as long as there are regional differences in vehicle purchasing and usage preferences.

Several multilane free-flow schemes, including the Cross Israel Highway, Electronic Toll Road 407 (Canada), Costanera Norte (Chile), and Melbourne City Link (Australia), use overhead-mounted cameras to measure vehicle dimensions. Stereoscopic imaging systems [9, 10] shown in Figures 5.8 and 5.9 provide a measurement of length, width, and height (i.e., volume), which is matched against criteria predefined in three dimensions to estimate the most likely class of vehicle.

There are several possible processes that translate volumetric measurements into vehicle classification. The configuration of the classification system requires each tariff class to be translated into one or more volumetric descriptions (length, width, and height). The two-dimensional nature of the imaging systems means that the output of the stereo classification system also includes the instantaneous vehicle position and velocity. The number of axles cannot be solely determined from overhead imaging methods, so supplementary buried sensors to count axles would be needed if inference is not adequate. Manual checking of an image of the side view of the vehicle (to show the number of axles), or automatic enquiry to a

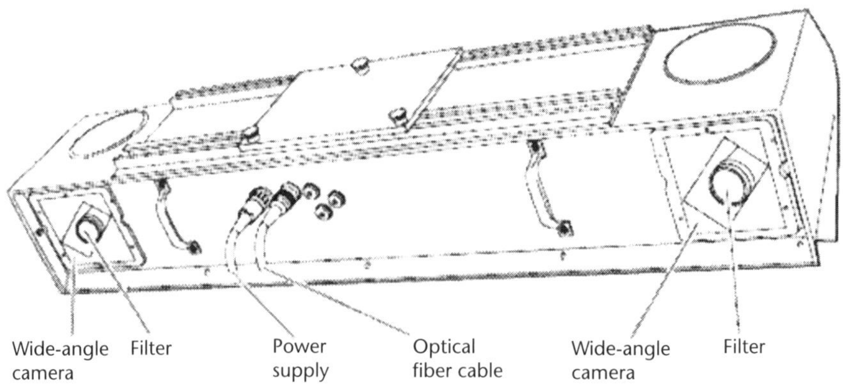

Figure 5.8 Stereoscopic vehicle profiler. (Courtesy of Kapsch TrafficCom AB.)

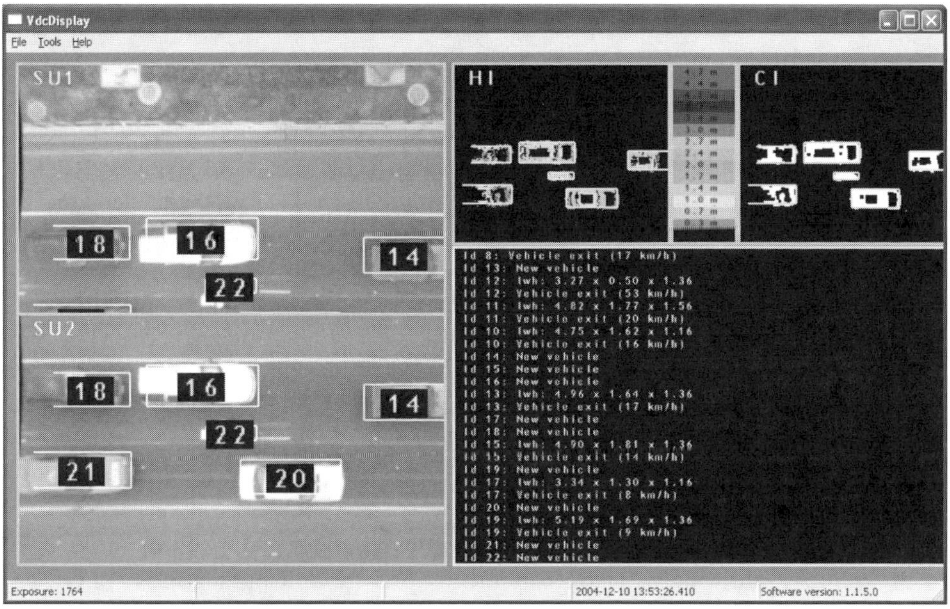

Figure 5.9 Stereoscopic vehicle profiler—captured image. (Courtesy of Kapsch TrafficCom AB.)

vehicle registration database, would be needed to confirm the measured class for enforcement purposes. The coverage of a discrete array of overhead dimensional measurement detectors in an MLFF environment also generally depends on the height of the vehicle and the spacing of the detectors, as shown in Figure 5.10.

5.2.4 Electronic Declarations

This section deals with the reading of vehicle classification declarations from a security label or in-vehicle equipment fitted to a vehicle (e.g., SUV, 12-ton medium goods vehicle, or motorcycle).

Highway operators use electronic declarations to capture information from an OBU, which enables the appropriate charge to be determined for enforcement purposes. The charge amount is not fixed, but will vary, depending on the toll

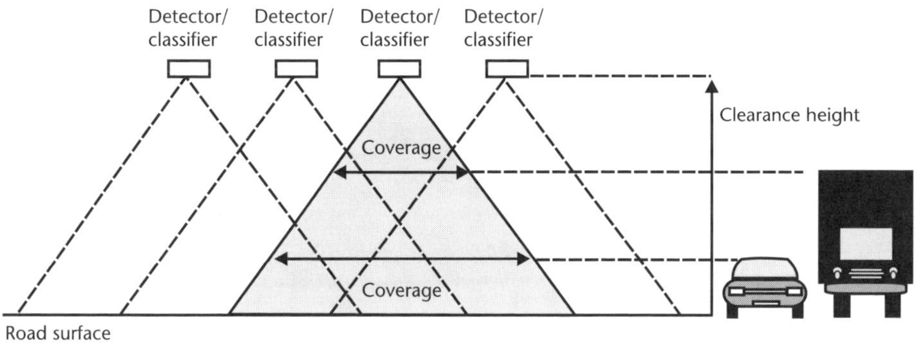

Figure 5.10 Overhead detection (MLFF).

plaza or road segment, or the tariff class of the vehicle. The tariff class usually depends on the vehicle classification, other vehicle attributes (e.g., maximum gross weight), emissions class, and any applicable exemptions or discounts.

Class tables from different operators' networks may not be aligned. The same class code could mean a totally different vehicle class in the absence of a regional agreement on class definitions. Harmonization of class codes declared by an OBU is one of the enablers of interoperability in a road network with multiple OBU issuers and multiple highway operators.

The Expert Group 2 on Vehicle Classification, established by the European Commission, created a harmonized classification code based on recommendations of vehicle parameters to be stored in OBUs. Expert Group 2 used vehicle characteristics defined in ISO 14906:2004 [11, 12] and the United Nations Economic and Social Council (UNECE) [13] to develop a common set of vehicle groups [14], which could be used as the basis of electronic declarations within the European Union, as shown in Table 5.1.

The Expert Group 2 recommendation defines groups 0, 1, and 2 as light goods vehicles, although strictly this means private and light goods (PLG), and defines groups 3 and 4 as heavy goods vehicles. With reference to UNECE classes, additional weight-based definitions can be used. For example, light goods vehicles have a MGW of 3.5 tons or less; large passenger vehicles (group 3) weigh less than 5 tons (UNECE class M2) or more than 5 tons (UNECE class M3); N2 vehicles in group 4 weigh up to 12 tons; and N3 vehicles weigh more than 12 tons.

Harmonization generally requires an agreement between highway operators, or a directive either from a national highway authority (e.g., FHWA) or a supranational regional authority (e.g., European Commission), which provides a list of vehicle descriptions applicable for road user charging applications. The UNECE or ISO descriptions can be used as the source [12, 13]. Ideally a highway authority should then select the most relevant subset of the standard classes and cluster them into enforceable categories applicable to the highway segment.

The accuracy of the measurement process is often specified when considering the direct measurement of vehicle attributes. Electronic declarations require a reliable communications path between the roadside charging or enforcement system and the OBU. Enforcement systems, including electronic tolling and truck tolling systems (e.g., Germany and Switzerland), usually rely on direct local communications over a DSRC interface to capture electronic declarations. However, the OBU may

Table 5.1 Proposed European Vehicle Groups

Group	*Description*	*Characteristics*	*UNECE Class*
0	Motorcycles	2 or 3 wheels	L
1	Small passenger vehicles	Seats ≤ 8 + driver	M1
2	Light goods vehicles	Weight ≤ 3.5 tons	N1
3	Large passenger vehicles	Seats > 8 + driver	M2, M3
4	Heavy goods vehicles	Weight > 3.5 tons	N2, N3
5	Not used	—	—
6	Not used	—	—
7	Other vehicles	—	—

be obstructed by another vehicle, in which case a declaration may not be captured or may be corrupted. The likelihood of the declaration using DSRC not being captured depends on the communication system geometry. In a toll lane (overhead or roadside DSRC transceiver) or open road (portal gantry), the capture accuracy can be as high as 99.99% for vehicles properly equipped with an OBU. It may be 99.5% for properly equipped vehicles in an urban environment. Once the declaration is captured, the likelihood of it containing an undetected error can be as low as 1 in 10 million. Finally, using the vehicle's license plate to determine its classification (with reference to a database) will be as accurate as the underlying ANPR technology, typically 85% to 90% for a single point of detection. Thus, classification based on electronic declarations is very accurate, which is why DSRC charging systems use declarations programmed into an OBU rather than physical measurements of a vehicle's attributes. However, electronic declarations from OBUs are not currently regarded as having sufficient evidential strength for enforcement purposes. Images must be used as primary evidence for automated enforcement points in express lanes or on the open highway. Direct measurement (with inference, if needed) therefore supports the enforcement process rather than the charging process, as shown in Figures 5.11 and 5.12.

Future electronic declarations of a vehicle's identity and static characteristics may be provided by an EVI device (see Sections 4.8 and 5.5.1). However, until

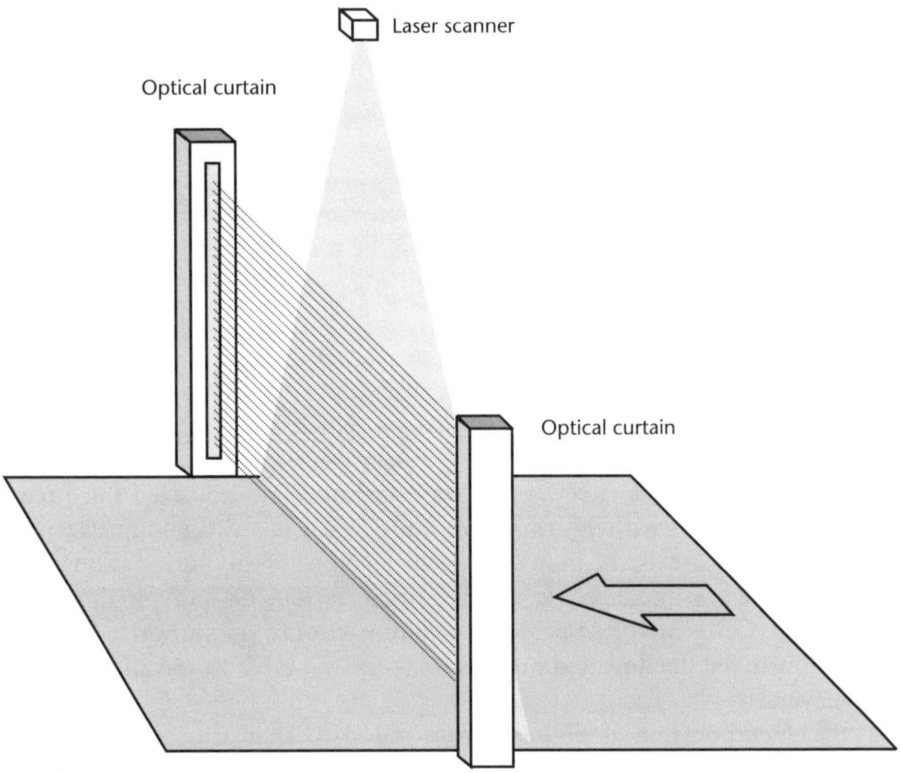

Figure 5.11 Capturing vehicle attributes for charging and enforcement: Type 1—line measurement.

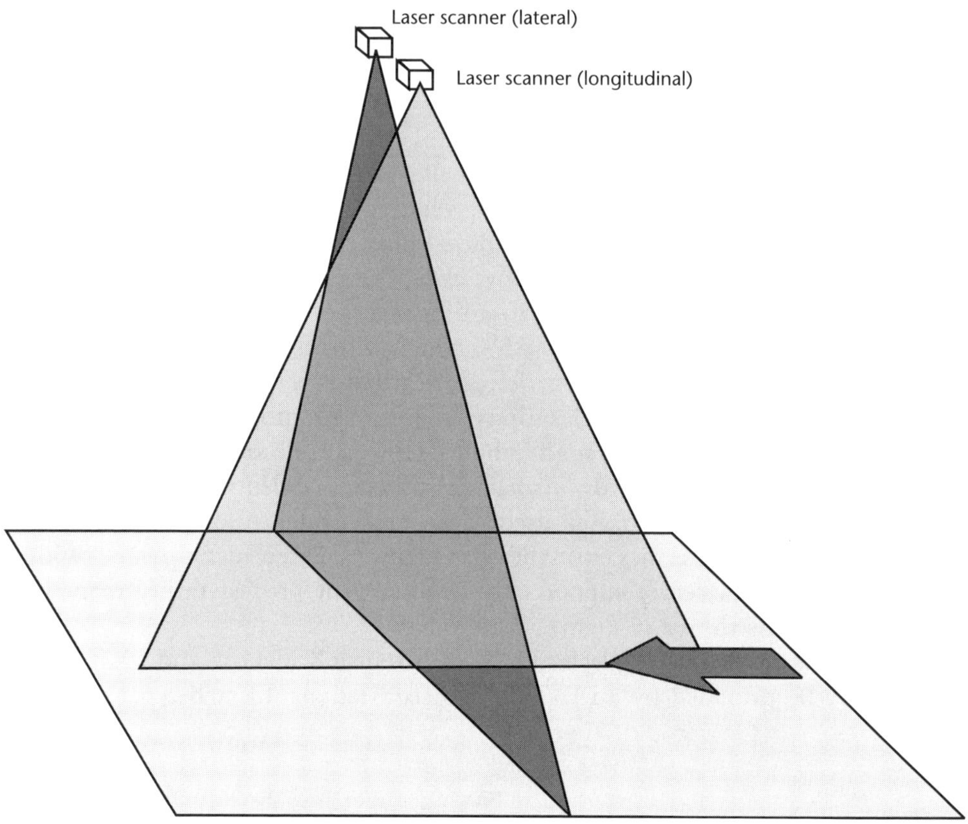

Figure 5.12 Capturing vehicle attributes for charging and enforcement: Type 2—plane measurement.

the evidential quality of the EVI data meets the requirements of the enforcement regime, image-based evidence will still be critical.

5.2.5 Indirect Capture

Indirect capture means uniquely identifying a vehicle to enable its attributes to be read from a local database (e.g., using a vehicle's license plate or account information read from an OBU).

An OBU issuer (e.g., a highway operator or authorized third party) can preprogram OBUs with information to sufficiently identify a unique account, including the issuer and classification of the vehicle. None of this information need be logically associated with a specific vehicle, organization, or individual, until the OBU is issued. OBUs are often personalized in batches by an authorized body that follows a controlled documented process to assign the OBU to specific vehicles and charge payers.

Incorporating attributes within the OBU that are specific to a vehicle (e.g., VIN or license plate number) means that the personalization process can only be completed when the information is known. Instead, vehicle-specific information can be logically associated by linking the unique OBU ID read from an OBU to the specific vehicle, the unique ID of the charge payer's account, and the organization or individual responsible for the vehicle using a central system database. Electronic declarations should be sufficient to determine the OBU issuing authority, tariff category of the vehicle, and other status indicators, such as public service vehicle, diplomatic vehicle, or security services vehicle.

5.3 Detection and Measurement Technologies

Sensor technologies can measure one or more vehicle attributes. The choice of technology depends on the vehicle descriptions used to calculate the tariff class, and can be put into two categories (in-lane and open road), as Table 5.2 shows.

If the tariff class for a vehicle relies on the number of axles and height, then the most common choice for toll plazas is the in-ground inductive loop or the IR curtain. Both technologies can detect the gap between vehicles and perform the role of matching described above. A mechanical treadle can count axles, but must be supported by a vehicle separator (e.g., in-ground loop, single-point IR detector, or light curtain). If the toll operator prefers above-ground equipment, then an IR light curtain for axle counting and separation could meet the requirements. If the number of axles does not resolve the vehicle class, then an overhead IR profiler coupled with an IR curtain provides a vehicle profile with axle position. Combining this with length measurement using a second profiler can provide a complete three-dimensional model. Inference (e.g., based on length assumptions and height above first axle) can improve the classification accuracy. Figure 5.13 shows an IR profiler located above a DSRC transceiver to help match the profile with the electronic declaration for enforcement purposes in a toll lane.

Figure 5.14 shows an array of piezoelectric strip sensors laid to count axles and measure the track (i.e., distance between tires) of the vehicle. Approximately

Table 5.2 In-Lane and Open Road Direct Measurement Technologies

In-Lane (Toll Plaza)	*Open Road (Interurban, Urban)*
In-ground inductive loop (ferrous object detection)	In-ground inductive loop (ferrous object detection)
Capacitive or piezostrip sensor (axle detection, axle counting, double wheel detection, and wheelbase measurement)	In-ground inductive loop arrays
Contact treadle (axle detection, double wheel detection, axle counting, and wheelbase measurement)	Overhead scanning lasers (lateral vehicle profile, direction, speed)
Single-point IR sensor (separation)	Offset, overhead scanning lasers (side and partial top view of vehicle profile, direction, speed)
Laser curtain (vehicle profile, height above first axle, length, separation)	Overhead stereoscopic imaging (volumetric profile)
Overhead Doppler shift sensor (profile, separation)	Offset overhead stereoscopic imaging (volumetric profile)
Overhead scanning laser curtain (profile, separation)	Roadside or surface-mounted wireless magnetic sensors
Magnetic permeability sensor	Overhead Doppler shift sensor (profile)
	Overhead acoustic/vibration sensors
	Magnetic permeability sensor

Figure 5.13 DSRC transceiver and overhead laser profiler. (Courtesy of Tecsidel.)

Figure 5.14 Piezodetectors and optical curtains. (Courtesy of Tecsidel.)

1.5m downstream of the piezosensors is an IR curtain, comprised of two 1.6-m-tall pillars located on either side of the toll lane. The IR curtain can separate vehicles, create a side profile, count axles, measure the height above each axle, and detect a trailer and its towing bar. The sensor outputs are combined and key parameters compared with a set of preprogrammed vehicle descriptions to find the closest match.

Many factors affect the accuracy of the direct measurement methods, including the vehicle speed, accuracy of measurement of distance between axles, sensor lag (e.g., sensitivity, hysteresis), false detection rate (missing or false reports), and the validity of assumptions (e.g., a single vehicle, constant velocity at measurement point) in the translation algorithm. A combination of sensors can build a more accurate picture of the vehicle, and mitigate the effects of speed, increase the detection rate, and increase the overall classification accuracy.

The vehicle's lateral position is known with great certainty in a toll lane but not on the open highway. Therefore, a combined charging and enforcement point in an urban environment cannot assume that all vehicles travel along the same trajectory. In slow-moving traffic, the driving patterns may be undisciplined and chaotic. New imaging techniques based on roadside or overhead stereoscopic imaging [9, 10] have shown that vehicles can be detected and profiled regardless of their position on the road. These techniques would be suitable for locations prone to congestion and chaotic traffic flows. An overhead laser profiler is usually adequate for more predictable traffic flows, as used at some of the Austrian heavy truck tolling scheme managed by ASFINAG, and this technology may also be applicable in the urban environment [15] although may have to face more unpredictable traffic flows.

5.4 Worked Examples

Five examples of the relationship between vehicle detection, classification, and enforcement are provided next.

5.4.1 Example 1: Sydney and Melbourne (Australia)

Some harmonization has been achieved in Australia, although vehicle classifications and charges differ from state to state. The Sydney Harbour Bridge and Harbour Tunnel charge a flat fee per trip, regardless of the vehicle's classification, so measurement or electronic declaration of classification is not used. The charges for the New South Wales M1, M2, M4, and M5 motorways, Queensland Gateway, and the Logan Motorway are all based on the OBU declaration.

The operator of the Melbourne City Link (MCLP) (Melbourne, Australia) calculates the charge based on the vehicle's OBU declaration, not the measured class. The enforcement system estimates the vehicle classification by comparing the vehicle's volumetric profile (captured by stereoscopic camera pairs located over the road) with a volumetric class definition that includes height, length, and width. The declared class is then compared with the measured class. If there is a discrepancy, then the relevant images are extracted from a short-term image cache and an evidential record is created for later processing.

Table 5.3 shows the tariff classes (based on vehicle classifications) for the Melbourne City Link, which are resolved using volumetric measurement and inference.

5.4.2 Example 2: LKW Maut (Germany)

The German LKW Maut Heavy Truck Tolling Scheme requires trucks above 12 tons to pay a fee related to the distance traveled on Germany's federal autobahns. The scheme was introduced as pay-per-use substitute for the previous paper vignette scheme. Vehicles can prepay based on a manual declaration before entering the autobahn network, or pay a charge calculated by an OBU that can identify the segment of road on which the vehicle is traveling.

GPS estimates the vehicle's position, uses map-matching to identify the road segment on which the vehicle is traveling, and then calculates the charges that are due. No roadside infrastructure is provided for charging, although IR broadcast beacons are used to provide position information where chargeable and nonchargeable roads are in close proximity, or where GPS coverage is poor. More than 300 enforcement gantries are distributed throughout the autobahn network. A pair of IR curtains on each gantry profiles vehicles from the side and partially from above to build a time-slice image of each vehicle [16]. The enforcement point includes a set of business rules that compares the volumetric time-sliced profile with a set of signatures for eligible vehicles, and determines whether or not a truck is likely to be above the 12-ton MGW limit. The ability of the classifiers to resolve tow bars and separate closely following vehicles must be matched by robust business rules that are able to infer the vehicle's classification. If the vehicle is classified as being above 12 tons, and the vehicle is not equipped with a working OBU, then an image of the front of the vehicle showing its license plate is retained for later processing. The next stage of filtering removes vehicles that had previously been manually declared for the route, and vehicles that had been incorrectly classified. German enforcement points use classification measurement to detect chargeable vehicles, although this needs to be confirmed by referring to the German vehicle registration database. Foreign-registered vehicles are processed and enforced separately.

This example highlights a MLFF classification scheme that, in its simplest form, aims to determine whether or not a vehicle is chargeable (i.e., whether or not it is in the 12-ton MGW category). A laser profiler is used to construct a volumetric signature of the vehicle that is compared with volumetric thresholds that are likely to relate to vehicles that are in the 12-ton MGW category.

Table 5.3 Tariff Table (MCLP)

Vehicle	*Description*
Passenger car	Includes cars towing a trailer or caravan
Light commercial vehicle (LCV)	Any cab chassis, from 1.5 to 4.5 tons gross vehicle weight, two axles
Heavy commercial vehicle (HCV)	Rigid trucks with three or more axles, or over 4.5 tons gross vehicle weight; buses with 13 or more seats, including driver; articulated trucks

5.4.3 Example 3: Dartford Thurrock Crossing (United Kingdom)

Dartford Thurrock Crossing is one of Europe's busiest toll facilities and has employed ETC since 1991. A twin-bore tunnel and bridge carry more than 150,000 vehicle passages every day across the Thames on the east side of London. ETC charging services, branded as DART-Tag, are available in each of the 27 barrier-controlled lanes at two plazas on the south side of the estuarial crossing.

The tariff table includes passenger cars, two-axle goods, two-axle goods and trailer, multiaxle goods, and multiaxle goods and trailers. Cars are permitted to use the dedicated ETC lanes, and are directed by signing on approach to the toll plaza. Other classes, including heavy goods vehicles, must use the manual lanes, whether or not they are equipped with a DART-Tag. The attendant manually classifies all vehicles as they enter these lanes, and displays the appropriate tariff to the driver. If a DART-Tag is detected, then the declared classification is compared with the manual class, and, if they match, the automatic lane controller raises the barrier. If there is a classification mismatch, or if a tag is not detected, then the toll lane attendant resolves the discrepancy directly with the driver. All tags are color-coded, so it is visually possible for the attendant to confirm whether the correct tag has been installed before any electronic checks on the tag are conducted. The charge payer is not required to provide the vehicle's registration details at the time the account is opened, and the tag may be exchanged with other vehicles, but only if they are of the same class [17]. Buried vehicle detection loops are used to lower the barrier as the vehicle exits the toll lane.

This example illustrates how classification is used in a combined barrier-controlled ETC/manual lane, with minimal impact on manual lane processes. Color-coded tags simplify manual checking if any discrepancy is detected.

5.4.4 Example 4: EZ-Pass (United States)

EZ-Pass is the largest ETC scheme in the United States, as measured by vehicle transactions and revenue collected. Operators in the Northeast United States, including The New York State Thruway Authority, MTA Bridges and Tunnels, New York State Bridge Authority, Port Authority of NY & NJ, Delaware DoT (DelDOT), Atlantic City Expressway, Massachusetts Turnpike Authority, New Jersey Turnpike Authority, New Jersey Highway Authority, operators in Delaware, and the Pennsylvania Turnpike Commission offer EZ-Pass services based on a common, single-sourced ETC technology.

Charge payers are contractually permitted to transfer a tag between vehicles of the same classification, although the vehicle's registration details have to be given to the road operator that manages the account to which the tag is linked. The monthly statement lists all ETC transactions, including the time and date, toll plaza, charges, and vehicle classification for all EZ Pass toll facilities that are used.

The New Hampshire DOT defines vehicle classes for EZ-Pass by the type of vehicle, number of axles, and number of dual tires [18]. Each of the EZ-Pass lanes includes a two-contact treadle to determine the total number of axles, and an optical scanner to separate and profile each vehicle passing through the lane. A class mismatch is deemed to be a potential violation. Images of the front and rear license plates of the vehicle are captured for later manual processing. According to the terms and conditions of the EZ-Pass account [19], if the tag is moved to a vehicle of another classification for which the tag is registered, an administrative fee of up to $50 per occurrence may be charged. The tag may be revoked for repeat occurrences.

This illustrates how users are encouraged through contractual terms and a "tag misuse administrative fee" to ensure that the tag remains in the same category of vehicle for which it was registered. The charge payer must notify the account holding authority (toll road operator) if the tag is moved to another vehicle, which itself is not an offense.

This example shows how ETC gives users the flexibility to move tags between vehicles, and, through charging an administration fee, discourages moving of tags to vehicles of a different classification. It could be practically difficult to force users to register the vehicle to which the tag is used, since legally (assuming the classification is correct) the correct charge would be paid.

5.4.5 Example 5: Stockholm (Sweden)

From January 3 to July 31, 2006, the City of Stockholm operated a pilot cordon charging system [20, 21]. The pilot scheme was based on a single zone, with a boundary that encircled downtown Stockholm. Each of the 18 entry points to the city was equipped with DSRC-based charging and enforcement systems that operated every weekday from 6:30 a.m. to 6:29 p.m. The toll rates were defined independently of the vehicle classification and charged for peak and "shoulder" period travel, rising from SEK10 ($1.28), to SEK15 ($1.92), to SEK20 ($2.56). The maximum charge per day per vehicle, regardless of the number of times the vehicle has crossed the cordon, was SEK60 ($7.50). The charge was treated as a tax to which the vehicle owner becomes liable when the cordon is crossed. Within 4 weeks of commencement of scheme operations, the payment compliance rate was 95%.

Exemptions are granted for emergency vehicles, vehicles with disability permits, certain vehicles whose owners are exempt from taxation in Sweden, buses on scheduled routes, environmentally friendly vehicles as defined by the City of Stockholm, taxis, transport services for the disabled, school buses, and motorcycles.

An array of downward-facing IR scanners were located at each charge point to detect the transition of the front then rear of the vehicle as it passes beneath the gantry. The vehicle detectors trigger enforcement cameras, and, when an anomaly was detected, retain the images for automated (then manual) processing. The enforcement camera cropped images of the front of the vehicle to ensure that an image of the driver was not retained. Only the number plate and the front of the vehicle were captured. The front and rear images provided sufficient evidence (in this case) to prosecute for evasion of tax (a criminal offense), which required three gantries at the largest charge points (front and rear camera, detector, and DSRC antennas).

This example shows that vehicle detection is not always linked to classification. The scheme does not depend on classification for charging or enforcement, and the accurate vehicle detection function discards the part of the image that is likely to contain the driver.

A referendum was held on September 17, 2006, to decide whether to abandon the congestion charging pilot, or whether it should be fully adopted and maintained in operation.

5.5 The Future

5.5.1 New Forms of Vehicle Identification

Electronic vehicle identification (EVI) was defined by the ERTICO-hosted EVI Project Consortium as ". . . a system that uniquely identifies a vehicle electronically [and] as an electronic device that allows the unique, remote and reliable communication of one or more identifying parameters of a vehicle" [22].

In its 2-year study concluding in 2004, the Project Consortium identified several application areas that would benefit from EVI, including crime prevention, access control, electronic road user charging, vehicle registration ownership identification, and enforcement of traffic regulations. The Consortium suggests that ". . . it should be possible to replace the existing systems for classification and identification . . . by storing a minimum set of vehicle-related data in the in-vehicle EVI components" [23].

The ISO standard for electronic registration identification (ERI) [24] defines an electronic registration tag (ERT) that optionally can be personalized with read-only vehicle data records, including the vehicle's classification.

The usefulness of EVI or ERI information, even if accurately read, would depend on how reliably it can be associated with a vehicle. An on-board data carrier can be linked physically, contractually, or logically to the vehicle, but unless it is secure and part of the fabric of the vehicle, then it could be difficult to prove the presence of the vehicle at a specific time and date, in the absence of an image of the vehicle. The acceptance of electronic identification, whether or not an EVI or ERI perspective is adopted, will depend on the intended application and its acceptance as a reliable form of evidence for enforcement purposes.

5.5.2 New Sensors

Much of the current research that is likely to provide innovations in vehicle detection and classification could emerge from the following areas:

- New short-range wireless communication media;
- High-performance/low-energy microcontrollers combining analog and digital circuitry;
- Three-dimensional visual spectrum imaging;
- Multispectral imaging;
- Fault-tolerant communications networks;
- Embedded, pervasive nanosensors;
- Emissions measurement.

Innovations in policy, particularly in HOV and HOT lanes, have triggered development in novel methods for external [25, 26] and internal [27, 28] vehicle occupancy measurement. The migration to ORT and MLFF approaches to charging and enforcement has increased the need for improving the evidential quality of images. Recognition that heavier vehicles cause greater damage to roads than do lighter vehicles has also led to research into on-vehicle measurements of a vehicle's weight [29].

Compact, internally powered devices can be equipped with a variety of sensors and rapidly deployed onto the road surface. For example, battery-powered magnetic sensors [30] offer the possibility of "low-cost, ease of deployment and maintenance . . . [and] . . . more detailed information provided by these sensor networks, suggest that they can serve as a foundation for an accurate, extensive, and dense traffic surveillance system." The sensors are able to detect the effect that a ferromagnetic material with a large permeability, such as a vehicle, has on the Earth's magnetic field. It is claimed that a triaxis magnetometer can classify five classes of trucks with an accuracy of over 80% [30].

5.5.3 Distributed Sensor Networks

Mobile ad hoc networks (MANETs) are self-configuring wireless communications networks that link multiple, discrete, autonomous processors known as nodes. Physically, a node is a small, self-contained, self-powered computer with one or more sensors and a short-range wireless interface [31]. In principle, MANETs can act as a communication subsystem for urban road user charging for two types of charging policy: cordon or passage-based charging, and area pricing. Multiple, low-cost, fixed nodes could be used to create a thick cordon to maximize the detection accuracy of vehicles equipped with a mobile node (mote). For cordon charging, clusters of fixed nodes could hypothetically be spread on either side of the boundary that separates different charging tariff zones. This would confirm the vehicle's passage across the cordon.

A node that can sense local magnetic disturbances can transfer this information via other nodes to a fixed roadside location. A dense matrix of nodes deployed on the road surface at an enforcement point can collectively measure the characteristic signature as a vehicle locally disturbs the Earth's magnetic field. Failure of one sensor would automatically cause the measured data to be transferred by a different communications path. Other nodes, known as access points (APs), could be dedicated to other tasks, such as collecting the measured data from the network and transferring it to fixed networks located at the roadside for remote processing.

MANETs technology is currently in the research stage, but offers the potential for rapid deployment, low acquisition and installation cost, network resilience, and flexibility. The flexibility of the approach to untethered communications [32] between members of a cluster of nodes offers a robust communications technique between sensors, each of which provides information that can detect a vehicle, and, if the node were suitably equipped, determine the local magnetic or inductive profile. Collating these localized measurements could determine a complete vehicle profile.

The ultimate physical form for a node would be no larger than a grain of sand; thus, the term "smart dust." A carpet of motes (nodes) could be installed by painting an emulsion containing hundreds of motes directly onto the road surface. The vibration generated by passing vehicles or the cycling between day and night temperatures could provide a source of power that could be used by each mote.

Chapter 9 provides further information on MANETs technologies.

5.6 Summary and Conclusions

The misalignment between the measurement of a vehicle's classification and its taxation class is expected to continue. However, this has not precluded the development of increasingly sophisticated road user charging systems, and the migration from a dependency on toll plazas to open road methods. Both public and private sector investments are being made in infrastructure expansion and improved use of existing infrastructure. The need for accurate, robust means of measuring a vehicle's attributes will continue as the use of road user charging grows.

The degree of alignment of vehicle classifications with measurable classifications will dictate the extent to which automatic vehicle classification can be confidently used as part of an enforcement scheme. Increasing the resolution by which a vehicle's length can be measured will have little bearing on the accuracy of the vehicle classification, if length cannot be related to vehicle class. As the charging tariffs become more complex, it is likely that a combination of measurement methods will be used along with inference to confidently accept the vehicle's declared classification, and whether further manual validation will be required. Future developments are expected to reduce the cost and increase the density of sensor deployment. Innovation in design will partly be led by innovation in charging policy, particularly related to vehicle occupancy counting, dynamic vehicle weight, and emissions.

Automatic classification technology, combined with vehicle detection and the use of electronic declarations, will continue to form a critical part of the enforcement process for worldwide tolling and road user charging schemes.

References

[1] Land Transport New Zealand, *Road User Charges and Light Diesel Vehicles—Factsheet 38*, 2005.

[2] Government of New Zealand, *Road User Charges Act 1977 and Its Amendments*, 1977.

[3] Midland Expressway Limited, *M6 Toll Pricing Chart*, 2006, http://www.m6toll.co.uk/pricing.

[4] New York State Thruway Authority, *Vehicle Classification Information*, 2006, http://www.thruway.state.ny.us/tolls/classes.html.

[5] Øresundsbron (Denmark), *Standard Prices*, 2006, http://osb.oeresundsbron.dk/documents/document.php?obj=3080.

[6] Tate's Cairn Tunnel (Hong Kong), *Toll Table*, 2006, http://www.tctc.com.hk/eng/toll.html.

[7] Kesas (Malaysia), *Toll Fare Table, Lebuh Raya Shah Alam (LSA)*, 2006, http://www.kesas.com.my/TFT.htm.

[8] Federal Highway Administration, *FHWA Vehicle Types—Traffic Monitoring Guide*, Section 4, Appendix 4-C, 2001, http://www.fhwa.dot.gov/ohim/tmguideindex.htm.

[9] Sieber, K., "Enforcement in the Austrian Truck Tolling System," *Proc. ITS World Congress*, San Francisco, CA, November 2005.

[10] Patno, B., "Toronto 407 ETC," *Proc. IBTTA Technical Workshop*, Miami, FL, June 2004.

[11] International Standards Organization ISO, ISO/EN 14906:2004 *Road Transport and Traffic Telematics—Electronic Fee Collection—Application Interface Definition for Dedicated Short-Range Communication*, 2004.

[12] ISO, ISO 3833:1977 *Road Vehicles—Types—Terms and Definitions*, 1977.

[13] United Nations Economic and Social Council (UNECE), *Consolidated Resolution of the Construction of Vehicles, Annex 7 (Classification and Definition of Power-driven Vehicles and Trailers)*, April 16, 1999.

[14] Perrett, K., "Common Approach to Vehicle Classification in Support of the European Electronic Toll Service," *Proc. ITS World Congress*, San Francisco, CA, November 2005.

[15] Neuhaus, F., "Future Proof Enforcement of Free-Flow Tolling and Congestion Charging Schemes," *Proc. ITS World Congress 2005*, San Francisco, CA, November 2005.

[16] Stein, Dr.-Ing., "Toll Checker, Enforcement of the GPS/GSM Truck Tolling in Germany," *Proc. IBTTA Technical Workshop*, Edinburgh, June 2005.

[17] Le Crossing, *Terms and Conditions of Use of DART-Tag*, 2005, http://www.dartfordrivercrossing.co.uk/dart-tag/terms.htm.

[18] New Hampshire DOT, Chapter Tra 700 *EZ-Pass Electronic Toll Payment*, Part Tra 701 *EZ-Pass Violations.*

[19] New Jersey Turnpike, *EZ-Pass Customer Agreement Terms and Conditions (Individual Terms and Conditions)*, 2003, http://www.ezpassnj.com/static/terms/index.shtml.

[20] City of Stockholm, *The Stockholm Trials Start on 22 August and 3 January*, June 2005.

[21] Trivector, *Evaluation of the Congestion Charge Trial in Stockholm (Summary)*, February 16, 2006.

[22] EVI Project Consortium (ERTICO), *EVI Requirements and User Needs*, Work Package 2, Version 3.0, EVI Project Consortium (Ertico), October 2003, at http://www.ertico.com/download/evi_documents/2_EVI_D2_V3.0.pdf on March 2, 2006.

[23] EVI Project Consortium (ERTICO), *Feasibility Assessment of EVI with Respect to Requirements, User Needs and Economic Aspects*, Work Package 4, Version 2.0, August 2005, http://www.ertico.com/download/evi_documents/2_EVI_D4_V2.0.pdf on March 2, 2006.

[24] CEN, CEN ISO/TS 24534-2 *Road Transport and Traffic Telematics—Automatic Vehicle and Equipment Identification—Electronic Registration Identification (ERI) for Vehicles*, Part 2: Operational Requirements.

[25] Pavlidis, I., et al., *Automatic Passenger Counting in the HOV Lane*, Minnesota Department of Transportation, 1999.

[26] Tyrer, J., and A. Andrew, "Automatic Occupancy-Based Tolling for the Forth Road Bridge," *Proc. IBTTA Spring Technology Workshop*, Edinburgh, June 2005.

[27] Electronic Design, *Sensors Measure Up to Emerging Automotive Safety Standards (Occupancy Seat Sensors and Angular-Rate Sensors Enhance the Effectiveness of Airbags and Vehicle Dynamic Controls)*, September 2000.

[28] The Auto Channel, *Bosch Intelligent Bolt Seat Occupancy Sensor System Helps Protect Vehicle Occupants*, September 16, 2004.

[29] Dodoo, N., and N. Thorpe, "Towards Fair and Efficient Charging for Heavy Goods Vehicles," *Proc. IEE 12th Intl. Conference on Road Transport Information & Control*, April 2004.

[30] Cheung, S. Y., S. C. Ergen, and P. Varaiya, "Traffic Surveillance with Wireless Magnetic Sensors," *Proc. ITS World Congress*, San Francisco, CA, November 2005.

[31] Blythe, P., A. Tully, and G. Martin, "Next Generation Wireless Technologies to Deliver Pervasive Road User Charging and Other ITS Services," *Proc. ITS World Congress*, San Francisco, CA, November 2005.

[32] Vollset, E., and P. D. Ezhilchelvan, *Design and Performance-Study of Crash-Tolerant Protocols for Broadcasting and Reaching Consensus in MANETs*, University of Newcastle upon Tyne, School of Computing Science, August 2005.

Selected Bibliography

Garner, J., C. Lee, and L. Huang, *Infrared Sensors for Counting, Classifying, and Weighing Vehicles*, University of Texas, Report #FHWA/TX 91+1162-1F, December 1990.

Kansas Department of Transportation, *Accuracy of Automatic Vehicle Classifiers*, July 1999.

Mohottala, S., M. Kagesawa, and K. Ikeuchi, K., "Vehicle Class Recognition Using 3D CG," *Proc. 10th World Congress on Intelligent Transport Systems*, November 2003.

Wyman, J. H., G. A. Braley, and R. I. Stevens, *Field Evaluation of FHWA Vehicle Classification Categories*, Maine Department of Transportation, Final Report for Contract #DTFH-71-80-54-ME-03 for USDOT, 1985.

Yoshida, T., et al., *An Investigation in the Use of Inductive Loop Signatures for Vehicle Classification*, California PATH Research Report UCB-ITS-PRR-2002-4, 2002.

Yoshida, T., et al., "Vehicle Classification System with Local-Feature Based Algorithm Using CG Model Images," *IEICE Trans.*, Vol. E85-D No. 11, November 2002, pp. 1745–1752.

CHAPTER 6

Central System

6.1 Context

We now live in a world where we can buy a mobile phone and use it wherever we travel. We expect our credit cards to be accepted for goods and services purchased in stores, over the Internet, and by telephone. We expect to receive a monthly bill detailing every purchase made. We routinely require the ability to pay a utility bill by sending a check in the mail, by credit card over the phone, or by Internet banking. We expect to be able to register our complaints by phone, by letter, or using a company's Web site. We expect that a wide range of payment options will back up the annual ritual of declaring earnings and income tax. Even if the principle may sometimes be painful to accept, paying tax should be easy. Paying taxes is harder to avoid, due to an effective combination of deterrents to nonpayment and effective revenue recovery methods, supported by knowledge of the account history.

Almost every contact that we have with a service provider, large or small, is with the customer interfaces of numerous systems that underpin the service delivery operation. Some work well, but many are poor and inefficient. The physical infrastructure may be distributed across several states or countries, and connected by low-cost, high-bandwidth data links serving independent operations that meet cross-border data protection requirements. Alternatively, the functions may be centralized in a small operation managed by a handful of people who are able to accept cash or credit card, can directly deal with complex inquiries and complaints, and are able to reconcile accounts on a daily basis.

A central system is therefore not a complex distributed hierarchy of computer systems. Instead, it represents a bundle of functions and administrative processes that follow prescribed business rules, to create predetermined outputs meeting quality of service expectations. For the purposes of this chapter, the term central services reflects an array of functions that may be located in one place, or distributed across many physical sites and between several service providers.

The central system is defined as the IT and core services on which charging, enforcement, and all external interfaces depend. The scale of the underlying road user charging or electronic tolling infrastructure is based on the quantity of events that need to be managed, such as the number of road users, variety of accounts, payment transactions, charging transactions, and target level of compliance. Small barrier-controlled toll plazas may need only a handful of staff, a single server, and a couple of workstations. A nationwide lorry road charging scheme or citywide pricing scheme would typically be based on distributed functionality, proven technology components, proven operational processes, adherence to internationally recognized standards (necessary to a varying degree for public procurements), a robust operations and maintenance structure, secure authenticated interface to third-party service providers, quality management, disaster recovery provisions, management information systems, and reporting. These large-scale schemes would employ many staff and a complex IT system architecture.

Services that we accept as part of the fabric of our modern life influence our expectations of what a central system for road user charging should provide. Services such as telecommunications, banking, medical care, utilities, and local government all depend on a well-managed portfolio of core services that operate within a legal framework, and are auditable, reliable, accurate, secure, and cost-efficient. The functions that comprise central systems for tolling and road user charging schemes also have the same aims.

6.2 The Role of a Central System

6.2.1 Elements

The central system, whether publicly or privately funded, plays a pivotal role in enabling an effective business operation for toll collection or road user charging. The functions that comprise a central system can be split into several areas:

- Account registration and fulfillment (e.g., meeting users' requests for OBUs);
- Account management and customer relationship management (CRM);
- Charging data capture and collection;
- Enforcement, including revenue recovery;
- Systems management and reporting;
- Payment services;
- Interfaces to other public agencies and specialist service providers;
- Provisions for data security and disaster recovery.

Each of these areas is described in the following sections.

6.2.2 Account Registration and Fulfillment

Several registration options may be available. Regular, frequent users of a road network could have a prepaid account that is debited according to the charging policy (e.g., at each toll plaza, or on each road segment, or according to distance traveled). Most plaza-based tolling operations offer cash as the primary means of payment. Although this is probably the most expensive payment option to provide, it is convenient for all road users, regular and occasional. No registration is required, and the relationship between the service provider and the road user is temporary. The road user also retains anonymity at the point of payment.

Pay-per-use accounts are mostly linked to OBU-based schemes, although prepayments were historically used for voucher systems, many of which have been replaced by ETC-based accounts. The registration process requires, as a minimum, enough information to identify and validate the charge payer's account, and then link this to a specific class of vehicle or the license plate number of a specific vehicle. The latter can provide operational efficiencies, as discussed in Section 6.6.

Registration channels should be chosen to maximize accessibility to the charging scheme, such as "tag stores" (account registration and OBU collection point), paper application forms, or a call center. The registration process enables the road user to declare eligibility for concessions or other discounts, although this may require additional proof. For example, the London Congestion Charging scheme requires an individual who claims a residency discount to prove a main or permanent home within the prescribed charging zone [1].

Declarations are confirmed by written signature, although electronic signatures could be used in the future. By accepting the contract, the road user also commits not to misuse the account, and to report any changes that may impact eligibility. Additional conditions may apply to an account that is based on an OBU. For example, the Melbourne City Link (MCLP) makes the charge payer liable for all OBU events through the registration terms and conditions. The following clause is included in MCLP's standard customer service agreement [2]: "We will debit a [charge] to your Account when we detect . . . your [Tag] . . . in a [charging] Zone."

The Light Vehicle Transponder Lease Agreements [3] for personal and business use issued by Electronic Toll Road (ETR) 407 in Canada state: "You, the Lessee, agree[s] to remain fully responsible for any and all amounts arising from the use of the [Tag] until the [Tag] is returned to 407 ETR or until you have notified 407 ETR that the [Tag] has been lost or stolen [and] to mount the [Tag] as per the [Tag] placement instructions you received from 407 ETR" [3]. Similarly, the EZ-Pass scheme does not require the OBU to be associated with a vehicle; however, "If [the charge payer] use[s] the Tag in a vehicle other than one of the class for which the Tag is designated, [the charge payer] may incur administrative fees of up to $50 per occurrence" [4] to discourage switching OBUs between vehicles of different tariff classes.

Tag distribution methods depend entirely on the OBU design, and whether or not the operator physically needs to install the OBU. For example, the charging policies of the German LKW (Heavy Truck Tolling System) require the OBU to be installed in the vehicle (i.e., a truck of over 12 tons MGW) by an authorized workshop, due to the size, power requirements, and separate GPS antenna (although many are now integrated). The contractual link between the OBU and vehicle is also matched by a physical and logical link between the OBU and vehicle. Whenever the OBU is seen at a fixed or mobile enforcement system, the electronic declaration can be matched with confidence to the vehicle. The Singapore Electronic Road Pricing scheme also requires the IVU to be installed by agents authorized by the Land Transport Authority. IVU installation is mandatory for most vehicles. The technologies that enable road usage to be measured and reported by OBUs are not currently part of the electronic systems in new vehicles [5], so a nationwide charging scheme with many millions of OBUs would require retrofitting of passenger vehicles as the primary option for 5 to 10 years following such a mandate. Retrofitting OBUs by an authorized agent may be feasible for a relatively small population of users, but for the following reasons, this approach is not feasible for the majority of the vehicle population:

- Road user charging and tolling schemes are often local and do not apply to all roads, which reduces demand and economies of scale.
- The value of a single payment event is low.
- The OBU is often associated with a road user for a period that is shorter than the life of the vehicle (i.e., the OBUs are likely to be removed by the user at least once).
- The cost of installation (opportunity cost to the user, marginal cost of installation) is relatively high.

Getting road user charging equipment fitted as standard equipment by manufacturers takes time, regulation, agreed standards, and type approval, which generally takes up to 10 years in Europe. Consequently, most projects depend on mass-market distribution of OBUs by mail, or on road users collecting the OBU from a specialist service provider or road operator. The successful business model for prepaid mobile phones in Europe was mirrored by MEL, operator of the M6Toll in the United Kingdom. MEL offers a Tag-in-a-Box to its customers of class 2 (passenger) vehicles [3].

> The box costs £35 [approximately $60], including a £5 [approximately $8.50] refundable deposit when you activate your account with us. The box comes with a Tag (that has £30 of credit already on it) a bracket, windscreen cleaning wipe for your Tag, a Tag User Guide, including details about how to fully activate your account. An account number is already allocated to you when you buy the Tag in a Box, all you need to do is fill out your details on the application form, send it us and we'll activate your full account immediately. It's just like a pay as you go phone; you tell us how much you want your account topped up with (minimum £30) and every time your credit reaches zero we automatically take the amount from your debit or credit card. It means that you don't have to wait for your Tag to be posted out to you, you just buy the Tag in a Box and away you go on the road!

The MEL Tag is sold from a motorway rest stop for a fee that includes a prepaid balance that can be used immediately. The standard 5% discount to Tag accountholders is only applicable when the account is registered to provide the user with an incentive to register the OBU. Discounts, typically around 5% to 10%, are extensively used by charging scheme operators to encourage take up of OBUs since cash handling and "stop and pay" have significant operational cost associated with them.

Interoperability is frequently mentioned as the means by which a road user can pay fees at any charging facility. Through standardization of charging technologies, common procedures, and contractual agreements between operators, interoperability can be provided. Interoperability can also encourage multiple sources of charging technologies (e.g., OBUs and roadside equipment), and potentially can simplify procurements and maintain supply chain competition for all future procurements.

Registration in an interoperable, multioperator environment means that an OBU that is seen by one operator ideally should be acceptable to another. The issuing authority can guarantee the payment of all prepaid accounts every time an

OBU is seen, regardless of whether or not the user is required to register with each operator. The third-party road operator does not need to know anything about the road user, other than the identification of the issuing authority with whom the account is held, the account number, and the declared tariff class of the vehicle, so that the appropriate charge can be made, and, if necessary, enforcement action initiated.

A road operator may need some confidence that the OBU is authentic, particularly if a third-party has issued it. Some DSRC standards describe a security scheme that allows a roadside system to confirm the authenticity based on a challenge-response mechanism. This enhances security against rogue OBUs and spoofing (fraudulent impersonation), but there is a logistical overhead cost to personalize the OBUs before securing it with access keys. For example, the four urban highway concessionaires in Santiago de Chile rely on the sharing of access keys, which then enables OBUs to be shared within the group of operators. The local Ministry of Public Works, Transport and Telecommunications (MOPTT) defined an interoperability specification [6] that describes the personalization process to be used at time of registration and the authentication process every time the OBU is subsequently accessed.

The registration and fulfillment process is therefore critical to establish contractual and logical links between OBUs and the intended vehicles. Ensuring the correct physical link with the intended vehicle could be logistically difficult to establish and maintain, although it does provide confidence that a charging event triggered by the OBU can be associated with the correct vehicle.

6.2.3 Account Management and Customer Relations Management

Account management and customer relations management (CRM) are the primary long-term means of establishing and maintaining the contractual relationship with road users. This also includes billing, account inquiries, and complaint handling.

The charging process is dedicated to converting chargeable events into transactions that are applied to a charge payer's account. The process needs to be auditable to ensure accurate financial reporting and to help respond to inquiries and disputes from charge payers. Objective criteria need to be established, so that the quality of service of the billing system can be measured, often for contractual purposes. For example, if a state DOT in the United States awards a concession to a private operator, it is in the joint interest of both to ensure that the quality of the customer services meets performance targets defined by the state's DOT to ensure statewide consistency. Incentives for concessionaires to meet these customer service targets can be contractually defined and the performance measured and reported on an ongoing basis to the state DOT or other concession-awarding authority.

Road user charging schemes can operate by prepay, postpay, or a mixture of both. In this context, the bill informs the account holder of debits that have been made from his or her account (prepay), or the charges that are levied on him or her (postpay). When the time of the vehicle's passage is an important consideration (e.g., peak hour or time-of-day charging policy), then the chargeable event should be recorded to the nearest second and traceable to a recognized time reference (terrestrial or satellite broadcast) to resist challenges to the reliability of the charge.

The desired billing accuracy also needs to be specified; however, the number of overcharging and undercharging events should be no greater than 1 in 100,000, and 1 in 10,000, respectively, based on analogous telecommunications quality of service (QoS) requirements [7] defined by key performance indicators (KPIs). The number of road users incorrectly penalized should also be very low, otherwise the credibility of the road user charging operation could suffer, and attract media interest. No operator should be made responsible for the accuracy of the output produced by another operator within an interoperable network, and this should be reflected in the terms of the contractual interoperability agreement between operators. For GNSS, the identification of a charging event is based on identification of chargeable road segments. Section 4.4.4 shows that the estimated location of the OBU is subject to error, so there is finite probability of resulting billing errors. The ERTICO-led RCI group defines the errors as probabilities of wrongly identified road segments (undercharging or overcharging) and missing road segments (undercharging) of no more than 1 in 1 million and 1 in 10,000, respectively [8]. The relationship between positioning, usage, and billing errors is shown in Figure 3.11.

The billing authority also requires a well-defined process to identify, investigate, and resolve billing complaints. The evidence that a vehicle was liable to be charged (e.g., an image of the vehicle's presence on a road network) may need to be retained for a time period long enough to give the charge payer an opportunity to challenge the charge. The alternative is to develop a profile of the charge payer and vehicle usage to enable discretion to be exercised. The financial services industry, including credit card providers, typically use this approach. The user's behavior is used to develop a user profile, such as the OBU normally used, number of daily trips, and payment history. The profile can help identify anomalies or errors, such as OBU faults, that would not be visible if only a single charging event or charging period were considered.

All communication with the charge payer forms part of relationship management. The tag or OBU may also communicate simple instructions, such as "low balance" or "contact operator." However, a charge payer who relies on the tag or OBU providing a notification that the account is low or in debt may be able to claim that the driver of the vehicle did not hear the instructions. Dependence on a simple interface within the vehicle therefore may not be enough. For example, the Stockholm Congestion Charging pilot relied on the vehicle owner being aware of the vehicle's usage to ensure that the tax on the usage is properly paid. The vehicle owner was expected to pay on time, every time, without being billed, since the charge legally had the status of a tax, making nonpayment a criminal offense. The OBU provided no audible indication to the driver; the driver was expected to be aware that a charge had been incurred. Conversely, the Singapore ERP OBU [known as an in-vehicle unit (IVU)] provides a balance to individual account holders at each charge point on strategic highways and on crossing the cordon entry to the central business district (CBD). The MCLP relies upon the audible indication to notify the driver of charging events and other simple messages relating to the status of the account and health of the OBU.

6.2.4 Charging Data Capture and Collection

The detection of chargeable events is fundamentally important to any charging scheme operator. The event will vary according to the local charging policy, and could include:

- Passing a defined location, such as a toll plaza or cordon;
- Presence within a prescribed zone;
- Presence on a defined route, road segment, or other geographical object;
- Driving a minimum cumulative distance;
- Periodic, time-based reporting of usage;
- Ad hoc streaming of usage data from vehicles.

Other conditions could affect the charges related to any of these events, such as time of day, measured congestion, vehicle occupancy, dynamic or maximum design weight, type of vehicle, purpose of trip, or benefits (e.g., discounts or exemptions) afforded to specific road users. Refer to Chapter 4 for more information on the policy/detection technology relationship.

The most robust charging schemes are able to accurately detect a chargeable event, so that it can be later audited and satisfy the conditions of financial scrutiny. An end-to-end audit trail could include the use of recognized time sources (particularly if time-of-day pricing is used), unique record identifiers, encryption of communications, and well-defined procedures for system acceptance and maintenance. The financial services industry could be used as a source of standards and good working practice for this.

Although introduced in Section 3.5.1, a charging event is a record of the time, date, location, applicable charge, identity of the organization that manages the account, and the identity of the object that caused the event (i.e., the OBU ID or the number plate). Additional information may need to be captured, such as an image of the vehicle at the time of the event, declarations made by a OBU, direction of travel, number of axles, or the classification of the vehicle declared by the OBU (or derived from the account). This information can be included on the bill or archived for a short time, so that complaints and other challenges can be effectively handled. A concession operator may also be required to provide summary reports to the local roads authority, in which case a maximum permitted retention period for data may be specified to ensure compliance with local data protection laws.

The definition of "event" will be specific to a charging scheme. A vehicle's usage may trigger several events in a trip, some of which may be chargeable and some not. Such a journey is described below as an example of a future complex charging regime based on distance traveled on all roads, with local variations to fund infrastructure development and demand management. The hypothetical process applies for every trip made by the vehicle, to ensure that the distance traveled within a charged area or along a charged corridor is accurately and repeatably recorded.

1. The vehicle's ignition is turned on, and the vehicle is initially positioned relative to a known zone or route.

2. Distance recording starts when triggered by the time of day, or confirmation that the vehicle is within or has passed onto the chargeable road network.
3. The vehicle is detected at an ORT or MLFF charge point on a road segment managed by a concessionaire. Evidence of the vehicle's presence at this location is captured and retained by each charge point.
4. The vehicle crosses a boundary between the concessionaire's road network and a congestion charging zone operated by a city transport authority.
5. The vehicle terminates its journey within the charged area.

The complex reconciliation of charges across multiple road networks managed by several service providers should not be prevented by diverse interfaces and reporting mechanisms. Standardized formats for event records, as one of the enablers for contractual interoperability, is at the core of many regional interoperability schemes, including EZ-Pass, management of EFC DSRC Interoperability in the Alpine Area (MEDIA, including Switzerland), AutoPASS (Norway) and the Open Minimum Interoperability Specification Suite (OMISS, the United Kingdom), as well as planned schemes by the South African National Road Agency (SANRA) and Transit NZ (New Zealand). The evolution of discrete charging policies to an areawide scheme for all vehicles on all roads could face institutional hurdles if the standardization of interfaces is not addressed during the development of individual schemes.

An event may be triggered under different conditions, and independently of any specific charging technology within the vehicle. For example, the video tolling option offered by Melbourne City Link is aimed at moderate users for whom the economic benefits of an OBU are not justified, or for users who do not wish to install one. The video tolling event is the detection of the vehicle's license plates at both the front and rear, which improves vehicle detection accuracy compared to single point detection. Enforcement of the London Congestion Charging scheme operated by Transport for London is currently based upon vehicle detection events from a network of ANPR cameras located on the periphery, and at locations known as "screen lines" on strategic roads within the charging zone. Finally, the German LKW truck tolling scheme uses periodic reports submitted by in-vehicle equipment over a GPRS connection that includes the distance driven on the German autobahn network, measured by a combination of GPS and on-board map-matching.

The detection event can reliably assign a chargeable event against a prepaid or postpaid account. The events may not always be reliable, and could trigger overcharging or undercharging, could be applied to the wrong account, or could incorrectly penalize a charge payer. These failures are often caused by erroneous events that were not properly trapped by subsequent business rules:

- An OBU that does not provide a periodic report when required (charges not applied to account), applicable equally to DSRC-based and GNSS-based schemes;
- License plate misreading and not being assigned to an account, possibly causing the owner of the wrong vehicle to be penalized;
- A correctly functioning OBU associated with the wrong vehicle at the point of charging, due to a positioning or database error, potentially causing the

OBU to trigger a charge and the vehicle's license plate to trigger a second charge plus an administration fee;
- Faulty OBU causing the vehicle's license plate to be used as the basis for the charge, perhaps incurring an administration charge for not (apparently) having an OBU installed.

There could be hundreds of process failure modes for any charging scheme that relies on automatic vehicle detection or event capture. The development of MLFF, ORT, and GNSS schemes needs to recognize and accommodate failure modes at the design stage, by ensuring the appropriate functionality of the underlying business processes. The objective should be to direct high confidence, high accuracy events, such as charging events, through automated business processes; and route low confidence, suspected unreliable events through manual checking processes. As ANPR detection rates reach their limit of accuracy, then the burden on manual processes used to check the license plates will fall in relative terms. The enforcement policy will still likely require all penalty notices or citations to be manually checked before being issued, to ensure that the correct vehicle and its registered owner is properly identified. Setting contractual targets for this and other measures is critical to preserving scheme credibility to external stakeholders, including bondholders, investors, the media, and the general public.

The relationship between the accuracy of the event capture technologies and the process cost of maintaining the highest levels of service quality cannot be underestimated. The detection accuracy of a mature DSRC-based scheme (99.5% to 99.995%), compared with the best single point detection ANPR systems (85% to 90%) and distance measurement equipment (95% to 98%), means that the detection method should be matched with the appropriate charging policy requirements. For example, single point detection ANPR is considered to be suitable for period-based charging schemes, such as the London Congestion Charging scheme and the CityLink Pass product offered by the Melbourne City Link, but not suitable for a mainstream pay-per-use single point detection schemes although it has been used in such configurations on toll roads such as in Bergen, Norway. GNSS is able to accurately identify mapped road segments in rural and suburban areas where visibility permits (e.g., not adversely impacted by foliage and banked curves), but needs to be augmented to maintain this detection accuracy in the urban environment (see Section 3.5.3).

All business processes and manual interventions should properly accommodate the expected detection and measurement accuracies and error rate. Although not related to road user charging, there are examples of *traject-controle* (literally, "section control," meaning the measurement of average speed over a measured distance) schemes in the Netherlands that detect speeding offenses, leading to fines issued by the Centraal Justitieel Incasso Bureau (CJIB), with 100% (claimed) accuracy to the registered owners of vehicles identified by multiple ANPR cameras deployed along the controlled road segment [9]. However, a fully automated end-to-end enforcement scheme for road user charging would need extremely high levels of violation detection, error-free business rules and, undisputed accuracy of vehicle-owner databases. The impact of failure is high: road users would be unfairly targeted, scheme credibility would suffer, and the goodwill of the road operator would be damaged. Process failures are to be expected, so a combination of automated and manual checks ensure that the number of potentially harmful errors that propagate to the end of the enforcement process do not exceed a predetermined target. Answering the following question can set the target: What is the maximum tolerable quantity of nonviolators that we could erroneously target every day? Scaling this value up over a week, a month, or a year can be sobering. The answer is more likely to be below 10 than above it, although this depends on the scale of the scheme and the public appetite for scheme failure.

Improvements in business process design include improved vehicle detection usage recording technologies to reduce manual intervention, improved manual checking for penalty candidates, regular database cleansing methods, and charge-payer/vehicle profiling to focus on likely system errors.

6.2.5 Enforcement and Revenue Recovery

As Section 4.1 describes, enforcement aims to ensure high levels of compliance with payment requirements, recover lost revenue and any associated operating costs, and minimize the temptation to evade payment for road use (i.e., the perceived risk is proportionately higher than the benefit of evasion).

A road user charging system can only be operated on a sustainable basis if the enforcement processes are efficiently and effectively operated. Evidential enforcement is not simply the capture of an image, but a process that starts with the capture of evidence. The evidence is used to apply the charge to the correct account; if an account cannot be identified at the time of the vehicle passage, the only recourse may be to identify the vehicle and its owner from its license plate to enable charges and additional costs and penalties to be recovered or other punitive measures (see Section 4.4.4).

Section 4.4.2 also describes physical (barrier-based) enforcement that enables direct, in-lane enforcement. For example, if a vehicle's measured or observed class is different from the class declared by a OBU, then a toll lane officer could resolve the difference in the toll lane. Evidential methods (described in Section 4.4.3) rely on the accuracy of the evidence captured: one or more images of the vehicle's license plate, usually coupled with a context image, showing the make, model, and color of the vehicle. Associated metadata, such as the time and date of the event, vehicle location, data from any external measurement devices, and data read from the vehicle's OBU, is also included. Information extracted from images of the vehicle, such as the license plate number, is used to check whether an account already exists, and whether the vehicle has been seen elsewhere on the chargeable road network, which would strengthen any planned enforcement actions. Mobile enforcement relies on vehicle-based equipment or temporary roadside sites to capture enough information to decide whether the road user may be violating scheme rules.

A well-designed and managed approach to enforcement is a key factor in the success of any charging scheme, and ensures fair and equal treatment for all road users. Barrier-based toll plazas typically have a 99.9% compliance rate, and the best enforcement schemes can reach 98% compliance, although this requires substantial investment in enforcement processes to filter the evidence captured or other legal constraints such as plate denial (see Section 4.4.4). It may not be possible to capture evidence that is of sufficiently high quality (e.g., the number of vehicle occupants in a HOT lane) without stopping the vehicle. The enforcement policy may therefore depend on having the appropriate powers to stop the vehicle, particularly for foreign-registered vehicles (see Section 7.3.8).

Although enforcement schemes that do not use barriers have a lower compliance (and capture) rate, the chance that a road user may not always be caught on his or her "first offense" is offset by the likelihood that he or she will reoffend. The statistical chance of repeatedly avoiding detection is low. Travel patterns of offending vehicles of interest can be recorded, and a mobile enforcement vehicle may be positioned to intercept these vehicles. Such an approach would be an important part of an authority's enforcement strategy, particularly if the vehicle has an unregistered or false license plate that cannot be traced to a valid owner. Experiences from several worldwide charging schemes suggest that police authorities are generally interested in such vehicles (and their owners), since they may also be involved in other unlawful activities.

A well-advertised escalating penalty charge regime for a single violation (or offense) is employed by Transport for London (United Kingdom) and by Stockholm City (Sweden). This has been proven to be an effective mechanism to encourage compliance and to ensure sufficient deterrence to systematic evasion. The Stockholm City escalating fee scale applied additional charges of SEK70 ($9) for payments made after the initial 5-day limit, rising to SEK500 ($65) for payments made after 28 days, and a further SEK500 for payments after 115 days, accompanied by other measures. The legal basis of enforcement will dictate whether the vehicle could ultimately be seized, or whether the vehicle owner would face criminal charges (as in Sweden). A typical escalation regime could be 5 to 10 times the nominal charge if paid within 10 days, then 15 to 20 times the charge up to 1 month, then 30 to 50 times the charge. Singapore's LTA initially demands payment of a S$10 ($6) administration charge, plus the fee. The MCLP can only apply an administration charge to reflect the true marginal cost of processing, rather than a penalty charge. Although a higher charge imposes a greater deterrent to nonpayment and encourages early payment of the penalty, the local legal framework will dictate whether administration charges calculated on a cost-recovery basis or escalating penalties can be applied.

The legal basis of enforcement also imposes additional requirements on the process. A jurisdiction may require the evidence to be submitted to an independent third-party reviewer or to a court to reach a judicial decision. The evidence, and possibly the process by which it was captured, must meet accepted requirements on integrity. The evidential requirements must be considered at the design stage, and could include the resolution of the images captured, the security of the image during transmission (e.g., its encryption), physical protection against tampering with evidence (e.g., door interlocks of enforcement site cabinets, personnel authorization checks), detection of tampering (e.g., digital watermarks), and context images showing the color and model of the vehicle. Successful prosecutions are in many countries based on case law and precedence, so it is important that no loopholes are exploited by defendants and their legal teams. If local requirements do not exist, then analogous standards can be used (e.g., data security requirements in the financial services industry) to demonstrate that accepted design standards were followed.

The increasing need to enforce vehicles from neighboring states means that cross-border enforcement cannot be ignored. The barriers to enforcement are often not technical or operational, but instead involve the differences between the definition of offenses, methods of notification of vehicle owners, and approaches to recovering lost revenue. For example, local courts in the majority of EU member states do not recognize decriminalized financial penalties imposed on their nationals by authorities in other member states. To resolve this, current EU legislation would need to add clauses dealing with mutual recognition of traffic offenses by member states. Private debt collection authorities can be used, although their success is limited by economic feasibility and availability of vehicle owner information using license plates (e.g., Norway provides public access to vehicle owner information, whereas Germany does not). Furthermore, it is difficult to identify the owner of a vehicle registered in another EU member state, since neither technical solutions nor reciprocal agreements currently exist to support cross-border sharing of such data. The quality of local vehicle registration databases reduces the effectiveness of enforcement, improvements to such databases are critical to any MLFF or ORT scheme.

The third phase of the VERA project that started in Europe in late 2005 aims to create a technical demonstrator of ENFORCE, a data exchange network for cross-border enforcement, initially piloted between four countries (France, Spain, Austria, and the Netherlands), with the United Kingdom and others as observers. The central system would be required to host an interface with foreign agencies to manage cross-border enforcement [10].

6.2.6 Systems Management and Reporting

Reporting is required for many reasons, including:

- To meet legal requirements and local accounting standards for accurate accounting of revenues;
- To interface between organizations (e.g., between a state DOT and a private concession operator, between a payment gateway provider and the road operator, or between the operator and the department for vehicle registrations);
- To provide management information (e.g., maintenance of process health, monitoring of efficiencies, and prioritization of investments for improvements);
- For fault reporting (e.g., integrity of the charging and enforcement processes, protection against security threats, and identification of breaches that may compromise financial or evidential integrity);
- To enable charge reconciliation in a multioperator environment, and to ensure a timely presentation and exchange of charge records;
- To maintain customer service levels, and handle challenges from road users;
- To provide supporting information to long-term road or transport planning;

- To enable state operations to be compared on similar bases (e.g., absolute cost per transaction, annual cost for each account, quantity of erroneous charges, payment channel mix, and total operating costs).

The toll road operating model typically assigns the risk of accurately detecting vehicles to the concession operator. Vehicles that are not detected reflect lost revenue, although the government agency that manages a concession may still expect to be paid for the vehicle passage, thus passing the financial risk to the concessionaire. Monitoring equipment that is able to independently determine the charges that should have been collected are often not accurate enough, particularly if the equipment is required to assign vehicles to each charge class, including discounted or exempt vehicles. A concessionaire could be required to make an auditable declaration of vehicle statistics to back up any claim on revenue and service quality. In this example, accurate monitoring and reporting are necessary to ensure that the contractual requirements are being met, and that the financial risk lies with the concessionaire.

The systems management function is also responsible for receiving alerts of central system components that need maintenance. This function may also act as a distributor of configuration data and parameters to other functions, including charge collection and enforcement points.

6.2.7 Payment Services

Road user charging and electronic toll collection schemes require that all vehicle passages are associated with an account or (ideally) preregistered vehicles. It is recommended that all vehicles, including exempt vehicles, are also linked to an account, merely to ensure that the minimum level of services can be offered, including market communications and updates on the contractual terms and conditions. If the vehicle cannot be associated with an account, then it is regarded as not registered and treated as an exception passage, potentially leading to enforcement action.

The traditional cash-based toll systems have not needed to develop a relationship customized to a specific road user's needs. Road users are treated as large market segments, defined only by vehicle type and exemption category. However, the introduction of ETC and RUC requires road users (i.e., charge payers) to be treated individually, even if an account holder has several vehicles associated with a single account.

The business processes for each segment of RUC that meets a specific customer need (e.g., call center, complaint handling, and payment) depend on the type of user and the congestion charging product offered. A product is defined here as a bundle of services that includes payment options, eligibility restrictions, reporting, and notification channels. The product may also require the charge payer to install a tag or OBU in the relevant vehicle to aid the account identification process, and optionally enable a vehicle-specific or user-specific declaration to be made at the point of charging.

Not all products will be offered through all channels; OBU-based accounts will apply mainly to vehicles that regularly interact with the charging scheme.

The diversity of road users may require different account types to be defined. High-usage vehicles generally should be required or encouraged to register for a type of account with the lowest operating costs (e.g., an OBU-based account). The following account types or products could be defined for a typical scheme:

- Mandatory account-based schemes for high-usage commercial operators, which are postpaid and invoiced monthly or prepaid;
- Mandatory account-based schemes for high-usage private users, which are prepaid and automatically recharged;
- Intermediate usage accounts, which are prepaid, charged on a per-passage basis, and require video recognition of license plate;
- Low-usage user accounts, which are preregistered, prepaid, valid for specified period regardless of number of trips or distance traveled, and do not require OBUs or registration of license plates;
- Exempt and discounted vehicles accounts, requiring mandatory registration and OBUs.

Table 6.1 shows examples of direct payment channels, typical means of payment, and relative operations cost for each payment channel.

Electronic charging methods generally have lower marginal costs. This remains one of the main reasons for the introduction of ETC—to absorb fluctuations and growth in demand that progressively contributes to increased congestion at toll plazas by converting cash payments into automated charging with off-line payments. A cash payment channel (e.g., a toll both or automatic coin machine) can be replaced by an ETC event and related payment. An OBU account for a regular user costs less for the operator to maintain than does the provision of cash payment facilities for the same regular user.

Table 6.1 Payment Channel Options and Relative Marginal Costs

Payment Channel	*Means of Payment*	*Relative Marginal Cost for Each Payment Transaction*
In-lane	Cash (plaza or hybrid plaza/ ORT), manual or ACM	High
Phone/call center	Credit/debit card, other payment provider	Moderate to high
Mail	Check or credit/debit card	Moderate to high
E-mail	Authorized recharging of registered account	Moderate
Retail outlets	Cash, check, or credit/debit card	Moderate
Internet	Credit/debit card other payment provider	Low
Interactive voice response	Credit/debit card other payment provider	Low
SMS or other mobile phone–based messaging	For payment notification, or payment against a preregistered account	Low

The migration of some user-initiated channels, such as interactive voice response (IVR), otherwise known as "pull" (user-demanded) services, to lower cost "push" (operator-initiated) services, such as SMS (text message) notifications, could be possible. This enables the scheme operator to choose when a payment event should be triggered (e.g., automatically requesting a payment), rather than providing payment channel capacity to manage ad hoc requests from charge payers. Convenience, accessibility, and other service differences can be used to encourage the use of lower marginal cost channels, and to accelerate the adoption of automatic prepayment accounts that require minimal manual interaction with the operator. Differentiating payment channels based on convenience may encourage adoption of these services, but if the differentiation is ineffective or inappropriate, then financial incentives can be used (e.g., reduced rates for OBU users, and higher rates for non-OBU users). For example, in March 2006, the U.S. EZ-Pass scheme offered discounts between 50 cents and 10% of the nominal tolls to encourage the establishment of OBU-based accounts. The DART-Tag ETC scheme operated at the Dartford Thurrock Crossing by Le Crossing in the United Kingdom offers a 7.5% discount to OBU accountholders.

Transport for London highlighted the migration effect in its Third Annual Report, following the commencement of its scheme on February 17, 2003 [11]: "The retail channel, which at the start of 2004 was used by 35 percent of charge payers, was by January 2005 used by only 30 percent. This decline corresponds to the growth of the web and mobile phone text message payment channels . . . At the current rate of change, web will overtake retail as the most popular channel in the second quarter of 2005."

Cash payment options should always be provided to ensure maximum accessibility to the road user charging scheme, particularly for occasional and low-usage users. This does not automatically require toll plazas with manual tollbooths. Retail outlets can be used to handle cash payments, which would depend on a high level of user awareness of retail outlet locations. The retail outlets must be conveniently placed and readily accessible. For example, the Fort Bend County Toll Road Authority in Texas provides four outlets. By comparison, the German truck tolling scheme enables the route to be predeclared via the operator's Web site, or at more than 3,500 manual registration points located across Germany and in neighboring countries. London has 5,000 retail outlets offering a cash payment option located within a 40-km radius of the congestion charging zone.

The need for accessibility drives the quantity required. All schemes must offer sufficient capacity to road users, along with other payment channel options, such as the Internet or SMS.

6.2.8 Data Security

A security strategy aims to prevent unauthorized system access (physical and logical), to allow logging of access to meet audit trail requirements, and to protect data exchanges with third parties.

The charging detection function described in Chapter 4 may be required to collect records accumulated by the OBE within the vehicle. For example, a typical GNSS-based security scheme for data collection from OBEs over a wireless connection such as GSM or CDMA could employ symmetric-key DES-protected connections with the OBEs during the collection process. Evidential records collected from roadside, mobile, or portable detection points may need to be authenticated and perhaps also encrypted following local guidelines.

Archiving and access are subject to the strictest specifications with respect to data security and data integrity. Modern security technology, such as encryption and user identification, permits controlled access, which helps ensure a secure and stable operation of the complete system.

All schemes could benefit from a security policy, ideally based on a recognized approach. For example, the United Kingdom's e-Government Interoperability Framework [12] "defines the technical policies and specifications governing information flows across government and the public sector . . . [and covers] . . . interconnectivity, data integration, e-services access and content management." External interfaces could typically use a secure socket layer (SSL) to comply, and all messages could be tagged with a unique identifier to ensure auditability and integrity. Message authentication codes or digital signatures could provide a means for authenticating the source of messages before they are processed.

Privately hosted managed services can support interfaces to fixed locations, such as charge points, enforcement points, and OBE installation workshops. Data flows would be protected by alternative routings if there is a fault on the network, and additional bandwidth could be provided on demand. Procuring a managed service can often provide a cost-effective, highly scalable solution, although some countries may lack a competitive market for this level of service.

6.2.9 Disaster Recovery

The purpose of the charging scheme dictates the relative importance of its facilities. Loss of a payment channel or a temporary reduction in call center capacity may not be considered as of great importance. The complete loss of the means of gathering charging reports can also be sustained for a short period. However, the loss of one site that is critical to central system operations through intentional or natural causes could initiate a collapse of many of the functions that comprise the central system, if a comprehensive disaster recovery and business continuity plan were not in place.

Disaster recovery may amount to no more than preserving a solitary server that constantly replicates account updates and MIS reports. In the event of a disaster or evacuation at an operational center, central service operations could temporarily relocate to a second secure location. A disaster and business continuity strategy could be based on main and backup sites, which are physically separated but synchronized, with both running the same application software. If a disaster happens at the main location, then all functional operations can be readily transferred to the backup system. Although temporarily, central services are then being delivered from a backup site that has no protection of its own. Two recovery procedures could be started in parallel: first, to commence the rebuilding of the main site back to an operational status, and second, to activate a shared disaster recovery site in case a second disaster occurs.

6.3 The Operations Life Cycle

The operations life cycle of the central services scheme is inextricably tied to the life cycle of the charging scheme. The following sections outline the progression from conception through the development of requirements, the benefits of trials, and approach to procurement, implementation, and operations.

6.3.1 Development of Requirements

Development of charging policies can have two possible starting points: *greenfield* (where no charging policy previously existed), and *brownfield* (where some form of direct charging had already been applied). Greenfield projects have several hurdles to overcome: public acceptability, development of acceptable legal framework, development of an attractive competitive procurement, development of requirements, and a well-managed implementation. On the other hand, brownfield sites have already developed public awareness, enabling legislation probably exists, and standard working practices have already been proven and are being slowly refined.

The main option for charging for specific road segments is either to use a toll plaza or employ MLFF/ORT, and both methods can be introduced on greenfield sites. The scope of the central services is defined by this strategic choice. If a barrier-controlled, plaza-based toll collection is employed, then a small core team can manage several tens of thousands of ETC accounts. Enforcement is immediate, and exceptions are handled in the toll lanes. Removing the barrier requires that evidential enforcement and new functions must be accommodated by the central system. A vehicle owner can only be traced via an image of the number plate, so an interface to the state vehicle registration authority would be required. The introduction of a project that only employs MLFF/ORT means that direct cash payment is not an option. Many of the payment channels described in Section 6.2.7 may need to be provided, and cash would be payable only through authorized outlets. A progression from cash payment systems to sole dependency on MLFF/ORT systems substantially impacts the scope of the central system. The Florida Department of Transportation (FDOT) and some of the EZ-Pass operators in the Northeastern Unites States are currently planning these changes, along with the necessary changes in legislation, customer awareness, charging technology capability, image-based enforcement systems, and assumption of enforcement risk.

The requirements development process is often based on precedence and preferences for competitive or negotiated tendering. A concessionaire will often procure a charging scheme that is part of an infrastructure development concession. Instead, it would be the responsibility of a local government agency to procure an ETC scheme on a state-owned highway. The evolution from separate, isolated charging schemes to a network of interconnected chargeable routes means that the development of requirements also needs to consider any regional integrated transport strategy. A concession contract should ideally include any locally accepted precedence or preference for standardized interfaces, interoperability, evidential strategy, charging technology choice, and periodic technology updates to ensure that road users are not disadvantaged by regional charging schemes having progressively different approaches to charging.

The requirements may not need to be detailed. For example, MOPTT developed a regional interoperability strategy and specification [6] to ensure that road users would not need multiple incompatible OBUs to travel within Santiago de Chile. Other initiatives define a top-down prescriptive model to enable, but not require, operational convergence. The prescriptive requirements often relate to the following factors: external interfaces (e.g., vehicle registration authority, format of statutory and performance reports, OBUs); service quality targets; standards (e.g., security, evidential quality); legislation (e.g., health and safety, worker protection, data protection); and quality.

6.3.2 Pilot Deployment

Pilots and trials have been used worldwide since electronic toll collection was first used commercially in 1987. The reasons for the trials are often to meet a range of technical, operational, and political reasons:

- To raise public awareness of new approaches to charging for road usage;
- To help develop competitive procurement strategies, and promote interest from potential system integrators, service providers, and technology vendors;
- To help develop requirements and reduce procurement strategy risks;
- To confirm operational viability, scheme effectiveness, and to assess public acceptance;
- To select between possible solutions as part of a procurement.

More than 50 large-scale trials have been conducted since 1987, including: the Inter Agency Group technology trials in New York State (1991/1993); Hong Kong Transport Department ERP trials (1983 to 1986, and 1997 to 2000); FDOT and Florida's Turnpike (1993); ADEPT (1990/1995); German A555 trials (1994/1995); Ministerie van Verkeer en Waterstaat in the Netherlands (1996/1997); Singapore LTA (1999); Taiwan Area National Freeway Bureau off-road trials (2003/2004); Transport for London technology trials (2004/2006); and Stockholm pilot (2005/2006).

Pilot programs can offer significant internal and external benefits, particularly where local requirements are not easily accommodated by existing standards or interoperability specifications. Pilots also raise the level of knowledge and debate regarding policy and technology, both among professionals and the general public.

The development of the TIS initiative in France, initially based on prestandard CEN DSRC interoperability specifications, relied almost entirely on multiphase trials on the host operator's networks, supported by multiple vendors. Trials can also provide system integrators and technology vendors with a useful insight into the standard operating practices of the group of operators or highways agencies, and can help develop the final test and interoperability specifications. Similarly, ASECAP states that the Common EFC System for ASECAP Road Tolling European System (CESARE) project, which was also based on a multiple operator/multiple vendor trials, had the objective of "specifying, designing, developing, promoting and implementing a common interoperable Electronic Fee Collection System (EFC) on European toll roads." Phase 3 of the CESARE project commenced in April 2005 to refine earlier draft specifications, enabling contractual interoperability between operators in several EU member states [13].

6.3.3 Procurement Strategy

The procurement strategy can be based on sorting tenders received (e.g., scoring against targets, including price), achieved trial system performance, and final negotiations. The purchase can be limited to the provision of equipment, an integrated system, facilities management, or central services.

Services can be internally developed and delivered, or externally procured from other road user charging service providers or third-party providers. Contracted services often result in lower setup costs, but reduced flexibility and/or increased the management costs. Nevertheless, a state transport authority often has no choice but to procure bundles of services and technologies. A risk assessment and check on internal expertise may highlight the need to contract out for system integration rather than develop internal expertise. Other services, such as feasibility studies into developing new system architectures, could be contracted out to specialized consultancies, with the aim of ensuring that the road operator owns the resulting new intellectual property. Some services, such as retail outlets to distribute OBUs, or specialized workshops to install more complex on-board measurement equipment for heavy trucks, would be contracted out, with an agreement to meet well-defined, measurable quality standards.

Lack of expertise may prohibit a local authority or private toll road operator from developing a procurement specification for central services. The competence requirement typically includes the definition of operational resource requirements, development of an overall scheme design, and selection of service providers and retail outlets. A private toll road operator may have the benefit of a standardized set of procurement documents and a good understanding of electronic tolling principles gained from earlier projects. The operator may also have a preferred list of subcontractors, and may already be familiar with international technology standards.

Procurements will typically require the systems or services provider to assign a dedicated quality assurance resource, which would ensure that the agreed project requirements are fully satisfied through systematic detailed quality planning defined in a quality assurance plan (QAP). The program director will review and issue the QAP, which will include: the development of an approved audit program, initiation of a project start-up audit to ensure all systems have been put in place; assessment of subcontractors' quality management systems; liaison with customer quality assurance representatives; analysis of quality data/audit results; and the use of quality inspectors to assure the quality of the installation or services provided meets agreed requirements and statutory obligations. The vendor should be required to apply a documented quality management system (QMS) for quality assurance in design, development, production, installation, and servicing to ISO9001.

6.3.4 Supply Chain Structure

The needs of a new highway concessionaire requiring MLFF in Taiwan, for example, should be similar to a concessionaire requiring MLFF in Israel, Canada, Australia, or anywhere else. Assuming that the charging and enforcement infrastructure comply with local interoperability requirements (if any), the most significant differences usually relate to the scale and structure of the central system and the range of user payment options.

The components required for a road user charging solution are listed below. The end client for system integrators and technology vendors may well be a highway operator, private concessionaire, local authority, or national government agency. In many cases the end client may need to also act as an end-to-end systems integrator and project manager. Moreover, a supplier may be able to deliver only part of the bundles listed, perhaps some combination of the following:

1. Central system (core services): payment and violation event consolidation/archiving, enforcement services, security and risk management, projects management, contracts management, legal support, internal accounting and reporting, and so forth;
2. Central systems (peripheral services), customer relations management, retail outlet management, and so forth;
3. WAN communications, regional or national, for any distributed fixed infrastructure;
4. Charging infrastructure/equipment (e.g., GNSS requires in-vehicle equipment and interfaces to WAN vehicle-to-roadside communication gateways), or toll plaza equipment and related integration;
5. Enforcement infrastructure and related services (e.g., fixed and mobile infrastructure), requiring vehicle-to-roadside communications and vehicle measurement equipment (e.g., ANPR and classification);
6. OBU development, personalization, distribution, and installation;
7. Retail outlets;
8. Maintenance services;
9. Marketing and public communications;
10. Interface to third parties, such as vehicle registration agencies and other scheme operators, or to other external functions, such as payment services providers, clearing, security key managers, and so forth.

6.3.5 Managing the Start-Up Demand

Launching a new scheme that relies on evidential enforcement (regardless of the method of charging) creates the potential for short-term increases in demand on back and front office operations, including the manual image-handling processes and customer enquiry channels. This short-term increase in demand is known as a "bow wave," which denotes its shape, as shown in Figure 6.1.

The bow wave reflects a short-term demand on back office and front office processes, which leads to a temporary increase in nonpayment rates and complex customer inquiries (by call center or by mail). The bow wave can impact all

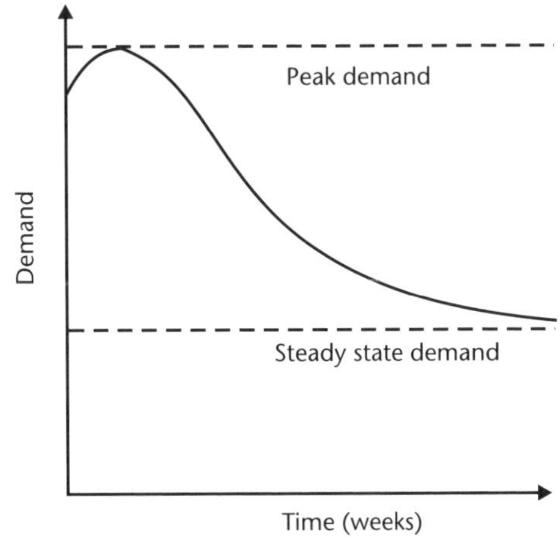

Figure 6.1 Start-up demand: the "bow wave" effect.

peripheral and core functions that comprise the central system, although not all symptoms caused by temporary overload would be externally visible to account holders or other external third parties. In the worst case, a short-term increase in demand could lead to a temporary reduction of quality of service experienced by charge payers. Other symptoms could be longer waiting times to talk with a call center operator or delays in issuing violation notices. This could reduce the effectiveness of the system and in the worst case trigger a negative media campaign. Chapter 9 deals with this in more details, using worked examples.

Establishment of a population of OBU-based accounts before commencement of a new scheme (e.g., 40% of potential road users) would increase the proportion of transactions that is automatically handled on day one. Users in some schemes would have fitted their vehicles with OBUs, and incurred dummy transactions in advance of the full scheme rollout and commencement of charges. The following steps could be taken to encourage the early adoption of tags and OBUs:

- Mandating the concession operator to achieve a minimum level of adoption before commencement of charging or tolling services (e.g., Toll Collect, Germany);
- Providing a financial incentive (e.g., waiving a registration fee before the scheme starts, or providing free trips);
- Withdrawing existing token or voucher schemes, and replacing with OBU-based accounts;
- Cross-marketing with other services, such as prepay mobile phones (e.g., Costanera Norte, Santiago de Chile).

In general, the cost of provisioning additional, short-term capacity to accommodate a bow wave would be reduced approximately in proportion to the number of OBU-based accounts and quantity of vehicles equipped with OBUs when the scheme goes live. Section 7.7.4 covers this in more detail for a hypothetical case study. The magnitude of the bow wave suppression will also depend on other effects, including the potential number of vehicles and road users that are eligible to register for OBU-based accounts. Ensuring high levels of awareness of payment options and providing easy access to acquire an OBU can help ensure a high initial penetration. For example, a new road could be operated without charge for a few weeks to market the benefits of the new road link. The German truck tolling scheme initiated an intense marketing campaign to heavy goods vehicle companies and their drivers within Germany and all neighboring countries before the scheme started. OBUs were only able to be installed by authorized workshops, so the workshops were trained and equipped to install the OBUs 6 months before the start of the charging scheme. Difficulties in setting up this supply chain partially arose from delays in OBU availability, and from attempts at consistent levels of quality across all workshops despite the broad variety of vehicles, with a significant minority over 10 years old. "The average age of vehicles seems to depend on taxation group [and] has been increasing throughout the last decade . . . [ranging] from 5.5 years to more than 10" [14]. This diversity of vehicle types and ages will impact all national road user charging schemes that require mandatory installation of OBUs. For example, in 1997, over 36% of vehicles were more than 10 years old in the European Union [15].

6.3.6 Operations and Maintenance

Individual processes and systems can be monitored to ensure that the performance meets agreed contractual requirements. These systems typically include performance management processes; production of periodic management reports (including summaries, unmet performance targets, and other issues); capacity planning databases; resource scheduling; systems monitoring; network status monitoring; and other custom tools. One of the most widely accepted approaches to the management of IT services provision is the IT infrastructure library (ITIL), developed by the Office of Government Commerce (OGC), an office of the U.K. Treasury [16–18].

All of the critical central services should also be subject to planned maintenance and performance monitoring on a continual basis. The facilities management company or service provider should be expected to have a maintenance resource management scheme that provides fault alerts, and warns when service quality degrades below a predetermined level. Specialist expertise may not be locally available, so remote access granted on request by a local shift supervisor can help meet first-response time targets. The maintenance operation will also define the strategy for stocking spare parts to meet service availability commitments.

6.4 Scalability

Scalability of a central system is likely to be required, in order to accommodate the following external changes:

- Increased demand on any of the central services functions, including payment, enforcement, and customer account management;
- Incorporation of other existing charging schemes;

- Establishment of new geographically independent charging schemes;
- Enhancement of the charging policy to accommodate new types of account and discount classes;
- New procurement strategies relating to some of the central services functional blocks.

Section 6.4.1 discusses the most common requirement to scale central services: geographic expansion to include new roads. It may be possible for a road operator to procure services (or alternatively offer services) at or near their marginal cost. Section 6.5 discusses the opportunities to exploit the economies of scale from combining similar processes within a specific back office function.

6.4.1 New Road Segments

The most prominent road user charging schemes worldwide have all passed through several stages of evolution. The most visible extensions have been geographic:

- EZ-Pass (Northeast United States): expansion from 7 to 22 tolling agencies since 1993, now including 7 million accountholders, and more than 12 million OBUs in 11 states;
- Singapore LTA: from the CBD in 1998, to six additional roads and major expressways, including the Pan Island Expressway (PIE) and the Ayer Rajah Expressway;
- Autopass (Hong Kong): from the Aberdeen Tunnel in 1994, to each of the 10 remaining tolled tunnels and bridges in Hong Kong and New Territories;
- Taiwan Area National Freeway Bureau (TANFB): planning to award a 20-year operating concession for an ETC scheme on all national freeways in 2006;
- Transport for London (United Kingdom): expansion to include an area west of the central charging area—the Western Extension Zone (WEZ);
- Autopass (Norway): from Oslo and Trondheim, to more than 22 city and road operators throughout Norway, for a total of 45 toll schemes;
- Santiago Urban Concessions (SUC): from Costanera Norte, to an additional three urban concessions.

The scalability objective and its impact on central services vary between each of the examples. Of the examples above, only the Singapore, London, and Hong Kong schemes were scaled without significant (if any) modification of the ownership and management structure. The Taiwan national scheme is expected to follow the same strategy. Conversely, the four Santiago Urban Concession (SUC) operators in Chile, the EZ-Pass operators in the United States, and Autopass operators in Norway are joined only through contractual relationships. Scalability is based on the inclusion of independent, vertically integrated schemes into a single operation that meet the minimum requirements for contractual interoperability.

6.4.2 Interoperability

The benefits of interoperability are often treated as purely technical, encouraging Europe, the United States, Japan, and other markets to achieve a regionally standardized roadside-to-vehicle interface (see Chapter 4). However, the commercial benefits are equally as significant and are detailed in Section 3.6.3.

The central system can also enable interoperability, particularly when the means of charging differs between regions. For example, a road user may be regarded as occasional in one city, and a regular in another. One region may operate an area pricing scheme, but a second region may only charge on entry. The approaches to vehicle detection and charging may be different, so it may not always be possible to depend on a vehicle being equipped with the same OBU. Despite the apparent benefits of OBU-based interoperability, the contractual interface can offer the road user the ability to use both schemes by registering with only one. The area scheme that depends on prepayment and ANPR for enforcement can refer to a locally updated database of license plate numbers that is provided by the authority at which the charge payer registered the vehicle. An OBU may also enable a valid charge event on entry to the second city. Note that the charge payer would be required to register the vehicle's license plate, and have a valid OBU, since this meets the minimum charging policy requirement for both cities. This form of interoperability through central services not only encourages regional convergence, but, given the number of disparate schemes, it may actually require the convergence as a precursor to end-to-end interoperability.

The choice of communications technology within the EU as a component in road user charging is subject to the Interoperability Directive [19]. Interoperability generally implies transparency throughout different systems with respect to the range of services offered, and enables the OBU to be used with a single contract in road user charging schemes operated by other EU member states, independent of country of origin, service provider, or OBU manufacturer. Contractual interoperability allows a scheme operator to recognize, authenticate, and interact with an OBU issued by another contract provider. Similarly, the contractual relationship states that OBUs issued by the scheme operator would be accepted by other operators, such as other on-road service providers (ORSPs). Each operator would expect to have a guarantee of payment for every vehicle that is equipped with an OBU issued by a third-party contract provider. The mechanisms for this payment will vary between regions, but all of the charging and enforcement services will need to employ business rules that allow enforcement to be deferred until the payment guarantee can be acquired. If not, then the vehicle's passage would be subject to enforcement, which would require cross-border or interstate message exchange with vehicle registration authorities to establish the identity of the owner to begin recovery processes.

Interoperability is therefore not limited to a paper document focused on the vehicle-to-roadside communications, but on a series of procedures, contracts, and standardized message exchanges, which collectively enable a vehicle to roam between charging policy zones while complying with all charging policy requirements in each zone.

6.5 System Architectures

6.5.1 Open Minimum Interoperability Specification Suite (United Kingdom)

The U.K. Department for Transport (DfT) initiated a project in 1999 to develop the Open Minimum Interoperability Specification Suite (OMISS). An associated trial program known as Demonstration of Interoperable Road-User End-to-End Charging and Telematics Systems (DIRECTS), established in 2001, included off-road controlled testing, followed by on-road trials in Leeds (Northern England) from 2004 to 2006. The trials researched the feasibility of electronic charging for road use, and studied the implementation of local road user charging schemes within a national context. DIRECTS provided the platform for the validation of many of the end-to-end requirements within OMISS, from the roadside-to-vehicle interface, to the reconciliation of charging transactions within an external clearinghouse. The trials included two approaches to determining road usage (DSRC and GNSS), and an evidential approach to enforcement (digital image capture by fixed roadside infrastructure). The DIRECTS trial is described in more detail in Section 8.7.5.

OMISS outlines roles for all levels of the charging and enforcement service delivery chain, including:

- Roadside system (RSS);
- On-road service provider (ORSP);
- Payment services provider (PSP);
- Data clearing operator (DCO).

Other entities are defined for enforcement process management. OMISS is defined in three volumes: Volume 1 (Functions and Performance), Volume 2 (Interfacing Between System Entities), and Volume 3 (On-Board Unit to Roadside Equipment Communications).

Most toll road operators embody the functions as a vertically integrated structure within the existing organization. Programs such as OMISS highlight the role of the central system within a regional interoperable structure. Any national model must accept that charging polices may vary between road types and between cities. Tariffs may also vary for the same vehicle categories in different locations when other possible variations are taken into account, such as time-of-day pricing.

According to OMISS, a toll road operator or city authority would be described as an ORSP. The road usage of each vehicle generates events that are collected by an ORSP, either at DSRC-only charge points, or remotely via GNSS. The DCO needs only limited information to process the event, including knowing the PSP that manages the reported account. Information on the road user is not necessary to process the event. Depending on the local charging policy, fixed or mobile enforcement operations provided by the ORSP (or an agent employed by the ORSP) would capture vehicle-specific and account-specific information to decide whether or not a violation has been committed. OMISS shows that an organization that is contemplating the introduction of tolls or road charging does not need to establish all central services functions in-house.

Large service providers providing banking, telecommunications, and utility services, for example, benefit from economies of scale in providing central services. Consolidating similar specialized functions, such as billing or fulfillment, generally delivers a lower cost for each transaction. This suggests that national interoperability architectures offered by initiatives such as OMISS, Autopass, or EZ-Pass would only require a toll road or congestion charging operator to enable external trusted parties to perform some or all of the central system functions. The cost per transaction generally varies from approximately 10% for a toll road scheme to more than 30% for some RUC schemes. These figures must be treated with caution, since the charge per passage is often not linked to the actual cost of charging for the passage. Comparisons are usually only effective when made within a specific category of operation, and if absolute costs per passage, per vehicle, per account, or per unit of distance traveled (for distance-related charging schemes), are also considered.

Section 6.6 elaborates on the potential for operational efficiency improvements, using the economies of scale that are available through coordinated planning with other operators within a regional or national framework supported by regional or national DOTs. These benefits of economies of scale arise through scale and repetition, but do not necessarily drive growth. Unusual or specialized tasks may counteract the economies of scale.

The functional partitions introduced by OMISS also aim to reflect modules that can be separately delivered by a competitive supply network.

6.5.2 EZ-Pass (United States)

The EZ-Pass scheme in the Northeastern United States probably represents one of the best examples of regional cooperation to establish central service operation for 22 toll agencies, at the time of this writing.

The Interagency Group (IAG), formed in 1991, developed an electronic toll collection system at that stage only embracing seven independent toll agencies: the Pennsylvania Turnpike Commission (PTC), the Port Authority of New York and New Jersey (PANYNJ), the New Jersey Turnpike Authority, the New Jersey Highway Authority, the New York Metropolitan Transportation Authority (NY MTA), the New York State Thruway Authority (NYSTA), and the South Jersey Transportation Authority (SJTA). A technology trial included three types of charging technologies from international and U.S. vendors. The New York State Thruway first employed the newly branded EZ-Pass in 1993, beginning with the Spring Valley toll plaza (trial location), and extending to the full length of the highway over the next 4 years. By 2006, the groups' combined facilities included 20 transportation agencies, more than 200 toll plazas, and 2,500 km of highways, tunnels, and bridges. This area generates $2 billion of the $3 billion total toll revenue collected annually in the United States, and approximately 40% of the total number of toll transactions nationwide. A single proprietary charging technology was adopted under a multiyear supply agreement to provide the basis for a common charging platform. Interoperability was achieved solely through central system functionality.

A customer service center (CSC) is associated with regional grouping of operators. For example, the New York Service Center handles MTA bridges and tunnels, NYSTA, NYSBA, and PANYNJ customers. Each CSC is logically connected to every other CSC by a reciprocity network to exchange charging transactions between operator pairs. Most of the accounts are prepaid, and users have the option to permit the account to be automatically refreshed when the account passes below a preset low-balance level. Postpaid accounts are offered by some agencies, although charge payers are required to place a security deposit with the operator that manages the account.

The EZ-Pass scheme has been extended through contractual agreements with several regional airports under the EZ-Pass Plus brand, enabling parking fees to be charged to the road user's EZ-Pass account, or directly to a credit card, depending on the transaction value. Technical interoperability was not considered to be feasible, so rather than depend on OBU-based transactions, the paper entry ticket and EZ-Pass account information is used when exiting the parking facility to apply the charge to the correct account [20].

6.6 Economies of Scale

The business drivers for consolidation in many service industries, particularly mobile telephony in the United States and Europe, have been focused on specialized activities that are similar. Economies of scale occur when the average transaction cost is reduced as the transaction volume increases. These savings can occur within an organization (i.e., internal), or across a group of organizations (i.e., external). The source of internal economies of scale for central services within road user charging includes:

- Natural growth in demand (e.g., more chargeable transactions);
- Shift in demand between functions (e.g., Internet payments of road user charging accounts increasing relative to cash payments);
- Offering spare capacity to other road operators (e.g., issuing tags or OBUs).

External economies of scale occur when reduced costs accrue across similar organizations. Examples include:

- Local technology standards (e.g., data transmission);
- Local operations standards (e.g., agreed and tested evidential strategy, common mobile enforcement policies, and common payment channel providers);
- Locally developed procurement policies, and coordinated procurements of some elements of the central services;
- Common services branding, which improves road user awareness and presents opportunities for joint marketing.

Available funds and the business case for direct charging for road use often limit the cost of setting up a scheme. For example, 90% of the setup costs of the London Congestion Charging scheme related to customer services and central facilities, and only 10% to the roadside infrastructure. By comparison, a new toll road with electronic tolling capability typically spends 95% on construction, and less than 5% on telematics and operations. Most of that 5% goes to central services and general administration, leaving less than 1% of the total cost to be spent on roadside telematics infrastructure, such as automatic incident detection and variable message signs.

The absolute costs of a scheme may be prohibitive, particularly when there are competing sources for public funds. A potential bidder to build and operate a new road will need to demonstrate to shareholders a commercially viable venture that meets the project needs established by the appropriate transport authorities. Lower operating costs improve commercial viability and release funds for use elsewhere. Regardless of the objectives of the project, the economic performance is prioritized. The operational cost is typically compared with the revenue collected, regardless of any significant difference between the fees set to manage demand compared with a toll rate to fund construction costs, operating costs and return on investment (ROI) criteria over a long concession period. Distance-related charging schemes obviously generate revenue in proportion to the quantity of vehicle kilometers traveled on the prescribed road network. The German LKW scheme currently charges an average of €0.124 (15¢) per kilometer. If a fee that partially replaces fuel duty supplemented this charge, then the fee would need to be above €0.15 (18¢) per kilometer to approach fiscal neutrality. This approach may provide a smooth policy transition to ensure that the road user does not pay extra charges, but, by definition, this does not cover the additional operating costs. Being able to reconcile the operating costs with alternative measures is therefore critical to sound policy development of road user charging.

The charge required to induce a change in driver behavior depends on the local elasticity of demand. In some cities, a lower charge may achieve the same demand reduction targets and forecasted public benefits as would a higher fee elsewhere. However, a lower charge regime places increased pressure on achieving a lower cost operating model. Recognizing the cost drivers and potential savings through outsourcing or selling spare operating capacity at or near marginal cost may make a congestion charging scheme economically viable. Independent development of customer interfaces and other central services may not be possible, leaving outsourcing to another operator as a potential option.

There is no single source of demand on a toll road or road user charging scheme. Schemes are compared by the quantity of road users equipped with OBUs or more complex on-board units. The cost drivers for a scheme include:

- Quantity of road users (more road users increases demand on all services and channels);
- Quantity of accounts (10% to 20% of accounts typically represent 70% to 80% of the registered vehicles);
- Mix of account types (some accounts are more costly to operate than others, such as occasional users);
- Quantity of chargeable events (e.g., an ETC transaction at a toll plaza, or distance reports transmitted from an OBU), or charge data records (GNSS);
- Quantity of payment events and mix of channels (e.g., cash is the most expensive payment channel, Web-based payment is one of the cheapest);

- Demand on customer service channels [nonroutine customer contacts via a call center are more expensive than standardized inquiries that can be handled by IVR or frequently asked questions (FAQs) on a Web site];
- Complexity of tariff table (e.g., mixing time of day, vehicle classification, and road types increases the probability of customer confusion and misunderstanding, resulting in higher cost of compliance);
- Size of catchment area (larger areas incur larger marketing costs and the need to ensure a larger geographic access for some customer channels);
- Evidential strategy and violation rates (a barrier-controlled system should have a violation rate of less than 0.1%, compared to a range between 1% and 10% for an image-based open highway system).

Little public evidence of operating costs and resourcing exists [21–23], but a core staff of 10 dedicated to noncash ETC should be able to handle 50,000 accounts for a physical (barrier-controlled) scheme. Conversely, a staff of 200 would be required for a barrierless road user charging scheme, handling approximately 1 million chargeable events per day with a 5% violation rate. Both could be economically viable, though.

A simple focus on cost saving will miss many of the benefits of seeking economies of scale. Access to proven routines and procedures, procurement policies, standards, and ad hoc expertise can ensure that economic viability will lessen the risk of scheme implementation by reducing the need for the learning and developing untried processes. Leveraging assets that have been already developed by other operators or regional transport authorities can also reduce the complexity of operational arrangements and standardize central services. This would ensure consistency in the application of charges, enforcement, and the treatment of road users in general. As Chapter 5 explains, the principles of evidential enforcement require the vehicle and the owner to be identified from a database of vehicle registrations. This process can be complex and expensive for a small ORT operator, particularly since the violation rate should be kept acceptably low by ensuring high levels of deterrence. A low-volume operator may not be able to achieve critical mass of compliance while maintaining an economically viable charging scheme. This may result in the continued use of conventional toll plazas, rather than a migration to ORT. The procurement of enforcement services from an external source, based on a per-transaction basis at or near the marginal cost of processing, could enable an economically viable migration to ORT.

Hence, there is an alternative to self-contained, vertically integrated operations. These benefits could be enabled if the assets can be leveraged by other existing (or new) scheme operators by:

- The creation of standard operating procedures;
- Definition of common evidential strategies (including image management policies);
- Development of a common approach to procurement and services validation (standardized procurement model, definition of services, and a menu approach to services procurement);

- Specification of quality metrics that can be incorporated into procurement specifications (e.g., transaction error rates, measures of congestion and throughput, data transmission latency);
- Standardized definition of external interfaces (e.g., department of motor vehicle registrations, credit card companies, banks, and courts).

The inevitability of direct charging for road use as one of the principal instruments for demand management, fuel tax replacement, and transportation financing requires policy changes. Contracts (or bond agreements) that define the purpose of the charging operations often cause resistance to change.

Being able to outsource services to an external specialized organization may need a revision to these principles. The benefit usually outweighs the difficulty: realizing the benefits of economies of scale can improve operational efficiency and reduce costs. Examples of economies of scale applied to electronic tolling in particular are visible worldwide:

- The Télépeage Inter Société (TIS) in France uses a single roadside system validation, a common vendor list, and common procurement definitions and transaction clearing, with the participation of nearly all French operators.
- Autopass in Norway uses common specifications for charging and procurement, enabling interoperability between 22 local charging schemes. The national Autopass Common Service specification requires one contract for all schemes, although separate registration with each scheme is required. Autopass includes the specifications for the OBUs, roadside equipment, central systems, interfaces between the system elements, AutoPASS logo and trademark, the AutoPASS contractual framework, and the AutoPASS security architecture.
- The Singapore Electronic Road Pricing scheme extends central services to enable the incorporation of pricing on local highways operated by the Land Transport Authority (LTA).
- EZ-Pass in the United States uses a single roadside system validation, common vendor list, common procurement definitions, and transaction clearing, led by the IAG.
- The Department for Transport in the United Kingdom is developing the OMISS system architecture to enable local authority procurement.
- Santiago de Chile uses a common transaction definition for the four current urban concessions and common transaction validation procedures [6].
- In South Africa, credit card transactions are routinely accepted to enable convenient payment on South Africa's toll road network.
- Sydney, Australia, uses a common roadside system transaction definition, and common OBU issuers.
- In Japan, 100% of freeways in the national scheme are designated toll roads, and all have ETC capability.
- In New Zealand, an initiative led by Transit NZ, the government agency responsible for all state roads, created a "nationally integrated toll processing/management system, based on international best practice models" for all charged roads [24].

There are many more examples, including schemes in Portugal (Via Verde), Spain, Florida (SunPass), Hong Kong (AutoPASS), Taiwan (currently being deployed), and Malaysia.

Overall, a common approach to system design, development and use of standards, and service provisions will enable the future policy integration of road user charging and a consistent approach that can help develop public acceptability. Enabling a phased approach to implementation to support multiple approaches to charging and enforcement will also enable interoperability with other schemes, as described in Section 6.4.

6.7 Summary

As road operators shift from 100% wholly owned and operated, vertically integrated models, through buying services to reduce the scope of operations as members of a broader network of service providers, the need for management across the interfaces becomes critical to ensure that the operator can still deliver a consistent quality of service and meet its contractual obligations. The pressure to outsource some services may also be driven by the need to exploit economies of scale. However, the natural inertia imposed by long-term concessions (e.g., from 5 to 99 years) may create some resistance to enabling efficient and effective regionwide delivery of road user charging services. The business models employed by a state DOT that focuses on regional transportation will differ from the model employed by a highly focused road operator (either public or private). The need for accurate charging, and a robust enforcement regime, impacts all road operations.

Offering contractual interoperability through central services to enable a road user to comply with all regional charging schemes also crosses these policy boundaries. Often a road user does not care who operates the road network. Current mobile phone technologies already shown the expectation for mobility, and a road user could reasonably expect to have a single bill for all road usage. Enforcement policies may also need to be harmonized, to improve the efficiency with which the local vehicle registration authority provides owner information, and to enable increasingly automated methods of handling evidence.

Road users may find themselves paying different rates according to vehicle types, emissions class, number of occupants, and time of day. A single journey may include routes and zones that charge on different bases, to reflect local environmental sensitivity, to manage demand, or to pay for the cost of construction and operations. The road user may not be aware of the transition between the policy areas, since a single account for all charges is enabled through effective integration between central services, even if no single operator owns, or controls, all functions.

The distribution of services between external organizations and the road operator will be impacted by any contractual interoperability framework that is created, such as the EZ-Pass IAG architecture (Northeast United States) or the proposed MEDIA model (Switzerland) for a number of European countries. The local decision on contracting for services may range from facilities management to outsourcing of noncore functions. Legal and data protection requirements also restrict information sharing. Initiatives, such as the U.K. OMISS, recognize that a road operator need not know the identity of road users, provided they comply with local charging policies. Similarly, the service provider that manages the road user's contract need not have any knowledge of the journeys that have been made. A simple reference to the road operator that generated the charges may be enough, which is analogous to the process by which credit card transaction reports are reconciled.

Any road user charging scheme will need to respond to evolutions in charging technologies, enforcement regimes, legislation, transport policy, integration of urban and interurban schemes, new approaches to fuel duty taxation, and changing political priorities, such as the EU Directives in Europe.

References

[1] Transport for London, *Registering for a Resident's Discount*, 2006, http://www.cclondon.com/downloads/ResidentsLiving.pdf.

[2] Transurban, *Everyday Account: Customer Service Agreement, CityLink*, consolidated for all amendments on July 1, 2005.

[3] 407 International Inc., *ETR 407 Sale Agreement Schedule 23 Toll Collection and Enforcement Procedures*.

[4] New York State Thruway, *EZ-Pass Customer Agreement Terms & Conditions*, May 2005.

[5] Frost and Sullivan, *Strategic Analysis of the North American In-Car Wireless Network Technologies and Protocols*, publication #F681, December 2005.

[6] Public Works, Transport and Telecommunications Ministry (MOPTT), *Electronic Fee Collection and Other Applications, Specification for Interoperability in the Beacon—Transponder Transaction*, Chile, Vol. 1.35, January 10, 2005.

[7] Office of Telecommunications (United Kingdom), Office of Telecommunications Standard for Telecommunications Metering Systems and Billing Systems, 2001, ITR003:2001.

[8] ERTICO, *RCI Supplier Workshop—Presentation*, February 16, 2006.

[9] Bureau Verkeershandhaving Openbaar Ministerie (the Netherlands), Section Control, Verkeershandhaving Dossiers, 2006.

[10] VERA Project Team, VERA 2 (Video Enforcement for Road Authorities), *Final Report*, September 2004, http://www.veraprojects.org.

[11] Transport for London, *Impacts Monitoring—Third Annual Report*, April 2005, pp. 140–142, http://www.tfl.gov.uk/tfl/cclondon/pdfs/ThirdAnnualReportFinal.pdf.

[12] U.K. Cabinet Office, *e-Government Interoperability Framework*, Version 6.1, 2005, http://www.govtalk.gov.uk/schemasstandards/egif_document.asp?docnum=949.

[13] ASECAP, *The CESARE Project*, 2005, http://www.asecap.com/pdf_files/The%20CESARE%20Project%20-EN.pdf.

[14] European Commission—DG Taxation and Customs Union, *Study on Vehicle Taxation in the Member States of the European Union*, Final Report, Table 9, January 2002, p. 19.

[15] Motorist's Forum, *SMMT Report, Commission for Integrated Transport*, Tables 1 and 2, March 2001, http://www.cfit.gov.uk/mf/reports/laq/index.htm.

[16] U.K. Office of Government Commerce, *ICT Infrastructure Management*, October 2002.

[17] U.K. Office of Government Commerce, *Service Support*, June 2000.

[18] U.K. Office of Government Commerce, *Service Delivery*, April 2001.

[19] European Parliament, *Directive 2004/52/EC of the European Parliament and of the Council of 29 April 2004 on the Interoperability of Electronic Road Toll Systems in the Community*, 2004.

[20] Port Authority of New York and New Jersey, *E-Z Pass[SM] Plus Is the Easiest Way to Pay for Airport Ticketing,* http://www.panynj.gov/ezpass.html.

[21] Federal Highway Administration, *Highway Statistics 2004—Disbursements of State-Administered Toll Road and Crossing Facilities*, 2004, section by section copy at http://www.fhwa.dot.gov/policy/ohim/hs04/index.htm on March 26, 2006.

[22] IBTTA, *IBTTA Data Warehouse*, 2006, http://www.ibtta.org/Information/?navItemNumber=847.

[23] Transport for London, *The Greater London (Central Zone) Congestion Charging Order 2001: Report to the Mayor—The Procurement, Financial and Cost-Benefit Implications of the Scheme Proposals*, Ch. 14, February 2002, http://www.tfl.gov.uk/pdfdocs/cc/14_social_cost_of_recommendations.pdf.

[24] Transit New Zealand, *Developing a Nationally Integrated Toll System for New Zealand's Toll Roads*, 2004.

Selected Bibliography

Department for Transport, *DIRECTS Road Charging Research*, 2002, http://www.dft.gov.uk.

Department for Transport (United Kingdom), *OMISS Volume 1 (Functions & Performance)*, November 2005.

Department for Transport (United Kingdom), *OMISS Volume 2 (Interfacing Between System Entities)*, June 2005.

Department for Transport (United Kingdom), *OMISS Volume 3 (On-board Unit to Roadside Equipment Communications)*, November 2005.

Interagency Group, home page http://www.e-zpassiag.com/.

Midland Expressway Limited, *About Tags, Get a Tag Save Money*, 2006, https://secure.m6toll.co.uk/account/tags.asp.

CHAPTER 7

Assembling the Pieces

7.1 Background

This chapter aims to walk through the end-to-end process from conceptualization to operations of a fictitious charging scheme, fully reflecting the relevant technical, economic, and political contexts that surround high profile RUC projects. Different perspectives are adopted, including from that of the highway authority, a potential operating concessionaire, and a technology vendor.

The chapter emphasizes the planning and development of a procurement specification that meets local, regional, technical, and political requirements, and relates this progression to the conclusions made in earlier chapters. For this reason, readers should familiarize themselves with at least the constraints on the selection of technologies for charging, enforcement and vehicle classification in Chapters 3, 4, and 5. This chapter extends the context within which RUC schemes develop, and makes specific recommendations based on the context. However, it is not the recommendations that are important here, but the process to derive the recommendations. Every project is different, but the general approach to weighing up strategic options is similar and often bears little relationship to the size of the scheme.

The chapter discusses the development of the procurement strategy, including the development of requirements and ensuring adequate competition, but does not present details on the routine of preparing the tender documents and managing the selection process. The process, scoring of responses, and development of an acceptable contractual relationship with technology suppliers and system integrators is often very specific to the procurement authority.

7.2 The Story So Far

A hypothetical local government, through its department of transportation, is considering the development of a multilane free-flow, privately operated charging system. The primary method of charging will be DSRC, although the procurement specifications will be technology-independent and describe the intended output as far as possible. The government is also considering a distance-based truck tolling system, although no date has been set for this, and will require all road operators to provide the necessary charging and enforcement support.

Long-term policy objectives are still being considered, including:

- The adoption of road user charging as a policy instrument for demand management and congestion reduction, along with other options;

- Making the current state-owned highway operators responsible for developing transport corridor strategies;
- Possible migration to an all-roads/all-vehicles charging policy to offset declining revenues from fuel taxation;
- Consideration of demand management issues by local authorities;
- Transit charges that reflect local externalities, varied for some road segments by time of day, including use by heavy goods vehicles.

This chapter does not relate to any specific country, province, state, or region, but instead aims to reflect current thinking relating to the use and application of road user charging:

- A possible approach for charging for existing infrastructure;
- Hybrid MLFF and plaza-based tolling (a natural extension of the U.S. ORT technique);
- Technology choice, influenced by local preferences and experience;
- Technology choice from an output-based (requirements-led) approach;
- Business case for different methods of charging all vehicles on a specific road;
- Provision of social and environmental benefits;
- A developed region view (economic, demographic, land use, technological);
- A developing region view (accessibility, nondiversion of local traffic);
- Implications of congestion charging in the context of other approaches to demand management;
- Ensuring policy flexibility to migrate to more sophisticated forms of charging, such as charging all vehicles on all roads according to distance driven, with local variations;
- Consideration of users' concerns and priorities.

Each of these will guide this story from start to finish.

7.3 Context

7.3.1 Global

The political, economic, environmental, and technological context of charging for road use has become more complex over the last 40 years. Since the idea of electronic toll collection was first proposed in 1973 by the operator of the Golden Gate Bridge, nations have become increasingly interdependent; two medium Earth orbit (MEO) navigation satellite constellations (GPS and GLONASS) have been launched, with a third (Galileo) at a pilot stage; the commitment to reduce emissions agreements is gaining momentum; and defense spending cuts have pushed defense technology into civilian markets. Technology is becoming cheaper, has increased functionality, and, as Moore's Law predicts, squeezes greater processing capability into ever smaller packages. In the last 10 years, increased integration between the member states of the European Union has led to increasing centralization of directives, aimed at harmonizing practices and reducing barriers to trade within an expanding economic region.

Most countries do not have toll roads, but where they exist, the private sector has been largely responsible for funding their construction and development, most recently in South America and Southeast Asia. The United States, Japan, France, Italy, and Spain have a public sector–dominated road ownership policy [1]. Government policy has assisted with the development of new infrastructure, including provision of enabling legislation, planning, traffic studies, public relations, and the development of related infrastructure. New roads in developing countries aim to stimulate economic development, provide access to employment, and reduce the barriers to trade. China, India, Southern Africa, and Central and Eastern Europe (CEE) provide high profile examples of these programs.

The traditional view of tolling still applies, but the shortfall in public infrastructure funding, growth in car use and congestion, and reduction in fuel tax revenues in the United States and the European Union has encouraged the debate of direct charging for road use. The expansion of the European Union, and the infrastructure developments required in road and rail to connect previously separated economies to new markets, drives innovation in funding and public-private partnerships.

In very sophisticated toll road markets, such as Australia, where MLFF and customer service–driven tolling are very advanced, authorities increasingly demand proposals that go far beyond the single road/bridge/tunnel development. The proposed corridor management schemes integrate road, rail, and sometimes sea transport into multimodal developments. Similar trends are emerging elsewhere, with major infrastructure schemes in South Africa, multimodal corridors in the United States, and integrated transport schemes in Europe. The major differences relate to the degree of partitioning between public and private sector financing.

7.3.2 Regional

The deployment of intelligent transport systems has reached many road users in developing nations in the last 10 years, although has not always met stated political ambitions. Vehicle use has generally increased proportional to economic growth rather than population growth, and vehicles are increasingly dependent on electronic systems for safety, fuel efficiency, and driver assistance. Tolls are becoming increasingly recognized as an acceptable method of funding infrastructure developments, and innovative debt/equity funding plans have increased the role of the private sector in many road building programs in the United States, Australia, South America, and Europe.

New private financing models in the United States offer the prospect of significant cash injections, although potentially with some loss of transport policy flexibility over concession periods that extend as long as 99 years. Singapore introduced an electronic scheme in 1998, London launched a congestion charging scheme in February 2003, and some progress has been made towards creating positive political support for charging in other cities.

Modern road user charging principles offer a potentially effective integrated transport policy instrument while complementing public transit and land use planning strategies.

7.3.3 Local

This chapter introduces a step-by-step analysis of a hypothetical program to develop and deploy an open road, interurban electronic toll collection system. The road exists in an important transportation corridor that includes several communities, and extends from a high population metropolitan city, through mixed-use suburban districts, to low-density, outlying residential areas. The route already exists as road fragments of variable quality, developed incrementally over the last 50 years. The civil works program aims to upgrade existing segments, and construct new interchanges to strategic commercial and residential areas. The road extends into a metropolitan area and is subject to localized congestion at various points, since the road has traditionally attracted road users hopping between intersections. This behavior was not expected when the route was established, but nevertheless serves to keep traffic out of suburban and residential areas. Passenger car use continues to grow, even with higher quality public transport provisions.

7.3.4 Technological

No decisions have been made on the method of charging or enforcement. It is expected that new approaches would be considered to ensure that the true cost of road usage is (or could be) reflected in the charges, rather than only attempting to cover the cost of upgrading and maintenance. It might be politically more acceptable in some cases to initially charge only to cover upgrading and maintenance, to which road users would be less likely to object.

The overall charging strategy aims to introduce a closer dependency between distance traveled and the charge for all vehicles, with some limited exceptions for physically challenged road users who have limited modal choice, and for emergency services. Financial incentives are already available to owners of hybrid vehicles, and the charging strategy for the road is expected to focus these incentives on meeting accepted definitions for low emissions and low fuel consumption. There already is a heavy goods vehicle tolling scheme that charges heavy truck operators for the use of the roads in the region. Many European and U.S. states are subject to transit traffic, and in both regions, electronic vignettes [2, 3] log usage and calculate charges. New Zealand has a similar system [4, 5]. Allocation of charges at a state and county level has been possible by a recording of the route taken by the vehicle, on which a charging policy for trucks already exists.

The existing policy has focused on heavy goods vehicles, instead of the remaining 90% of vehicles (i.e., passenger and light goods vehicles). The new toll route needs a policy for all vehicles. A partial reconciliation mechanism will apply to heavy goods vehicles that are already subject to the electronic vignette policy, to ensure that there is no policy conflict with the new planned road user charging scheme so that these users are not charged twice.

Several charging methods are being successfully employed worldwide, based on detecting the vehicle's passage at discrete points on every road segment. The approach has traditionally depended on DSRC (see Chapter 3) to determine the account to be charged, and enforcement has depended on either barriers or video enforcement (see Chapter 4). New methods of charging that have emerged in the United States and Europe based on satellite positioning methods have challenged the business case for infrastructure-based solutions for large road networks. Knowing the road segment on which the vehicle is traveling enables the distance traveled on mapped road networks to be measured and the charge calculated. This approach reduces the overall infrastructure costs, but increases the cost and complexity of the on-board charging unit and related supply, installation, and maintenance logistics.

There are claims and counterclaims from the proponents of each method of charging. The only deployment of GNSS in Europe is in Germany, and for specific road network types (high-quality interurban roads in this case), this method can require less roadside infrastructure, although enforcement infrastructure is still required. The charging policy and road network topology contribute to different regional preferences, as shown by neighboring Germany (GNSS/DSRC hybrid), Switzerland (odometer/DSRC, GPS backup), and Austria (DSRC-based). More complex in-vehicle units are also currently required, often forgotten in marketing messages, although this reflects the demands of the heavy goods vehicles market.

Many back office processes are independent of the method of measuring road usage, so they could accept charging events from electronic toll collection systems at toll plazas or MLFF/ORT schemes. Multiple charge point MLFF back office systems employ segment detection and exception-handling methods that differ from GNSS-generated solutions, where technology supplier preferences dominate. For example, GNSS OBU reporting demands differ between thin and intelligent client solutions (see Section 3.5.3.3) that aim to optimize road detection data capture, transfer, and storage, in an effort to allocate different elements of the tolling technical value chain functions between the OBU and the back office. The central system for the example route will need to accept specific requirements of MLFF charging and the evolution to GNSS-based charging mechanisms, if an all-roads/all-vehicles policy is to be adopted in the future. This challenge will belong to the selected concessionaire, and the decision on whether the system architecture would be able to evolve to a new form of charging, despite the policy uncertainty.

The main disadvantages of GNSS-based schemes are the poor positioning accuracy in the urban environment [6, 7], and the relatively high cost of on-board units if vehicle downtime is included. Nevertheless, the use of GNSS as the primary method of charging has been shown to be feasible on a large scale in Germany, with heavy goods vehicles as part of a highway-only charging policy [8]. The gap between charging policy and technology capability is expected to significantly narrow in the future, which would enable broad use of hybrid technology solutions that include a satellite positioning element, each adapted to specific road user segments.

The procurement process aims to keep open all long-term options, and to recognize that the forces that enable advanced forms of road user charging will probably emerge from outside the road operator's area of expertise and influence.

The operating concession in this example will be defined by a well-defined financial and risk structure, although it is expected that the future operational strategy of the road network will be subject to legislative and technological influences that will enable more complex road user charging policies to form part of a regionally integrated transportation strategy. The method of charging initially will be based on a conservative approach suitable for all vehicles, and based on detecting the vehicle's presence at well-defined points along the route. The greater the number of points at which a vehicle is detected, the greater the charge will be, up to a maximum.

Occasional users will be encouraged to preregister the vehicle's license plate before using the road to obtain a license valid for a specific number of days. Postregistration will be accepted up to 2 days after the trip. The number of journeys within the period of the license will be unlimited. The charge for the license will be proportionately higher than a prepaid OBU-based account, with a limit on the number of licenses that can be purchased in any one year. The use of the period license may reduce revenues compared with a pay-per-trip scheme for the same number of journeys, although this depends on the relative proportion of types of charging products. The occasional user product relies on automatic license plate detection, so the cost of processing payments, manual confirmation of the unreadable or erroneous images, and the account setup means that occasional user trips proportionately cost more than OBU-based trips. The occasional user scheme could be unwieldy and inconvenient from the user's perspective if there is more than one charged road network, each of which requires local registration.

The passage of each vehicle on the road at defined payment points must be associated with a local prepaid account, so account identification based on DSRC, complemented by video tolling and ANPR for occasional users, will be initially employed. The possible long-term transition to regionwide pricing will be complemented by an alternative means of measuring and reporting road usage, together with in-vehicle equipment that can measure distance traveled and location relative to geographic tariff boundaries. The support of automotive manufacturers, as original equipment manufacturers (OEMs), is regarded as critical to facilitate installation of distance/position recording equipment in new vehicles, and of after-market upgrades in existing vehicles. CN/GNSS is expected to be a viable solution for an all-roads/all-vehicles scheme within 10 years, although this will require declining per-vehicle costs, emerging acceptance through proven use of GNSS in other worldwide projects, and developing charging policies that are enabled by it.

The necessary business process and operations regime should allow interoperability with other regional road network operators, which are enabled by contractual agreements between the operators or their intermediaries. This interoperability will also be required to support parallel charging policies for some road users (e.g., heavy trucks paying based on distance traveled). The procurement specifications will therefore define the functional requirements, but will prescribe where necessary, including standards, performance, and external interfaces. Since a long-term migration strategy for all vehicles on all roads is being considered, potential operating concessionaires will be asked to comment on how to achieve scalability of each of the main elements of the proposed solution. While they might be able to suggest how their particular systems would scale and how they might become interoperable, concessionaires will not be able to solve this problem on their own. The authorities will require interoperability in their procurement policies, and actively provide for it. This could be achieved by offering central, accessible lists of license plate numbers, or access to vehicle registration databases. In Australia, where interoperability is well advanced, the regulator is able to recoup much of the cost of maintaining the central list by charging a fee each time an operator checks a license plate number.

The importance of vehicle registration authorities, transactions clearing, payment service providers, and other roles need to be properly defined as more charging schemes are planned.

7.3.5 Policy and Politics

All private and public road operators will be required to support any future charging and enforcement regime. Many of the existing operators are bound by operating regulations that were written in an earlier political and economic environment. The cost of including new means of charging and containing traffic growth through demand management would have previously been in conflict with the operators' financial expectations and business models that were based on conventional tolls. The lack of statistics on operations efficiency from other regional road operating authorities has made it difficult to identify the best examples.

The cost to the state of changing the existing regulations will probably require some compensation mechanism to existing operators, which will be factored into the overall setup cost of any future area pricing scheme. It will be easier and less expensive to ensure alignment for future private operating concessions rather than for existing operations, which are bound by inflexible financing assumptions and operating models. Thus, all future procurements for operations, whether or not tied to new infrastructure, will also require long-term transport policy flexibility. Operators will have to accept the likelihood of new taxation policies and regional road user charging, and the need for long-term operating models and underlying operating contracts.

The effective lifetime of future operating concessions will be as long as the regional transport policy planning horizon allows. All new operating concessions will be required to follow minimum reporting standards to the department of transport, for the purposes of auditing performance and long-term transport corridor planning. The redistribution of freight services between rail networks and public transit presents challenges, frustrated by various definitions of the term "load." For example, it would appear that railways in Europe carry 18% of freight traffic, when the official units of "ton-kilometer" are used. However, "ton-kilometer" is not correlated with economic value. Using economic value, only approximately 3% of freight goes by rail; using "vehicle-kilometer," the percentage drops to less than 1% [9].

Table 7.1 gives the relationship between government objectives and the resulting actions, once the decision is made to charge users for the use of the infrastructure.

The policy objectives of the example project were aimed at encouraging private sector participation, enhancing regional equity, and introducing a user-pays

Table 7.1 Meeting Policy Objectives

	Actions Available to Government to Meet Objectives					
Objective	*Involve Private Sector*	*Provide On/ Off Ramps and Access and Feeder Roads*	*Specify Land Clearance and Resettlement Arrangements*	*Specify Lower Tolls for Local Residents, Buses, and So Forth*	*Provide Financial Support*	*Specify Lower Tolls by Time of Day, Level of Congestion, and So Forth*
User pays/ internalize externalities	No	No	Yes	Yes	No	Yes
Regional equity	No	Yes	Yes	Yes	Yes	No
New, stable, and dedicated funds	Yes	No	No	No	Yes	No
Private sector participation	Yes	No	No	No	Yes	No

Source: [10].

approach to road usage, to ensure consistency with the long-term plan to introduce regionwide road pricing.

The use of tolling implies that the route can be commercially operated. Where the economic justification is weak or nonexistent (e.g., requires expensive specialized tunneling through an environmentally sensitive region), then alternative sources of funding or subsidization would be needed. If charging as a demand management tool is needed, then the broader policy will often include enhancements to other transport modes (funded by road user charges), which aims to establish them as viable complements.

In our example region, the long-term policy objectives are still being considered, including:

- Expansion in phases of the mandate of current state-owned highway operators, making them responsible for supporting the development and delivery of future transport corridor strategies;
- Reinforcement of the move (through legislation, pilot schemes, development of private sector interest) toward an all-roads/all-vehicles charging policy, based on distance traveled, to offset declining revenues from fuel taxation;
- Consideration of the benefit of demand management mechanisms, especially road user charging, by local authorities and state-owned highway operators, using budgetary incentives;
- Establishment of charges for transit traffic (including heavy good vehicles) for using local roads (e.g., by charging a pro-rata fee based on distance traveled);
- Adoption of broad road user charging and complementary measures as part of an integrated transport strategy;
- The refocus of transport and infrastructure planning around people and goods rather than vehicles.

Charging for the use of the upgraded infrastructure had met some political resistance, although it was felt that charges could be justified, would intensify the public "user-pays" debate, and would ensure more effective long-term use of the capacity.

In our example, the growth in demand for personal transport shows no sign of decline, although users will be faced with road usage charges that reflect the consequences of travel on the route. The former U.K. Department of Transport lists a number of components in environmental assessment: air quality, cultural heritage, disruption due to construction, ecology and nature conservation, landscape effects, vehicle travelers, water quality and drainage, geology and soils, and policy and plans. Its guide "sets out steps that need to be taken to encourage the systematic consideration of environmental costs and benefits alongside other factors within the overall appraisal process" [11].

It is difficult to consider all of these factors, so a subset is often used, including: the costs of accidents; road construction, maintenance and services; and the environmental externalities, such as air pollution and noise [12]. The economic assessment and political bargaining will determine the actual fees to be charged, and the fees will be balanced with the equivalent cost of public transport for the same journey. The roads authority that grants the concession makes the pricing decisions, but, in practice, pricing is often far more complex and economically rational. Risk, cost, margin, and expected returns will drive potential concessionaires over extremely long concession periods. They will bid using prices that very precisely balance demand and usage indicators, and very often these prices can be wildly different from those that would result by applying the view described above. A concessionaire uses the level of charges (adjusted by peak period) to fundamentally assess of attractiveness of a project. In the case where the authority sets the charges, the figures will be scrutinized alongside expected traffic demand at each price level, time of day, and background growth in traffic.

An initial assessment of traffic and vehicle-type mix showed that the route can be commercially operated even when a demand pricing element is added to maintain quality of service. In this case, quality of service is provisionally defined as a variation in travel time between peak and off-peak periods, rather than simply measuring throughput at a specific point. This looks at demand management from the authority's perspective. A potential concessionaire is likely to have a different view. The ability to accurately predict traffic mix and demand will dictate the importance of meeting operational cost and margin. Pricing is a mechanism to control demand, so accurate traffic modeling attempts to predict the point at which a specific price either increases demand to the extent that servicing the additional customers becomes uneconomic, or reduces demand to the extent that the extra charges collected fail to make up for the reduced number of individual trips. This is a very important element of the price analysis conducted by potential bidders.

It is clear that the strategic decision made by the authority to set charges (including the peak period levels) will influence the commercial attractiveness to bidders. The authority should therefore assess the attractiveness from the bidders' perspective, to show that the objectives of the upgraded route can also be met, despite the differing objectives of authority and concessionaire.

7.3.6 Regulatory Environment

The regulations for an open highway charging system must permit an effective enforcement regime to be established and operated. The following regulations, standards, and specifications are required for the tolled route.

- Legislation must permit evidence of a vehicle's presence on the charged route to be accepted as the basis for an enforcement action.
- Specifications for evidential capture must be developed, which can be regionally adopted to ensure that all future schemes can benefit from a proven approach. These specifications would describe the method of capture, methods to secure the evidence, and management of its use and distribution. The application of this to image-based evidence would apply, but would not preclude other future technologies, such as electronic registration identification (ERI).
- The use of revenues collected from tolls must be dedicated to specific purposes, partially to fund the operations and partially to invest in improved traffic management and (currently unspecified) enhancements to other modes of travel.
- The interface to the motor vehicle licensing agency must be standardized, along with an investment program that increases the quality of the data held by the agency, a program that defines a standardized form of electronic inquiry to vehicle registration databases held by neighboring states, and cooperation agreements with neighboring jurisdictions on mutual recognition of nonpayment offenses.
- The liability for payment of tolls and penalties must rest with the owner of the vehicle, by agreement, to reduce the burden of establishing the identity of the driver at the time of the offense. The prevalence of barrier-based toll schemes reduces the importance of this issue, although the use of MLFF/ORT requires the liability for payment to be fully defined. Liability for taxis, for-hire vehicles, and fleets also needs to be clarified, since the driver is often not the owner of the vehicle.
- The charges must be defined and allocated according to category of user, purpose of travel, and vehicle type, with allowance for local variations, including zone of travel, vehicle occupancy, and time of day. The future imposition of a taxation (rather than a charge) element would be permitted, to accommodate national legislation that progressively replaces the fuel tax by a usage-based tax. Future flexibility shown by any scheme operator would therefore be desirable.
- Procedures for notification of an offense, the imposition of penalties, the right of the road user or vehicle owner to challenge the fees, and related escalation of penalties must be established. The procedures will permit evidence to be submitted electronically to the courts or other accepted authorities, and will provide some certainty to the operator that the enforcement claim will be predictably and efficiently scheduled.
- The documents that define the road upgrades and charging and enforcement solutions must be established. These documents will include project time-

table, payment terms based on meeting service quality targets, definitions for minimum quality of service, operations review processes, change control, reporting requirements, and provisions for early contract termination.

Other regulations and instruments will be required to ensure that the use of private finance is encouraged in this case to cover some of the shortfall of public funds available for transport infrastructure enhancements. Regional legislators are presently facing competing demands from other potential investment opportunities in operations of regional transport. The upgraded route, with the introduction of charges, is only one of several legislative initiatives aimed at meeting future transportation needs identified over the next 50 years, which are expected to be driven by changing work patterns, increasing economic dependency with other regions, and changing demographics.

The upgraded route cannot be seen in isolation. Recent statistics have shown that nationwide vehicle miles traveled (VMT) has increased (see Figure 7.1), while vehicle ownership per capita has slowed (and has declined in some regions). Simply building infrastructure reduces travel times and improves the consistency of travel time (both benefits), but may also promote population dispersal and thereby increase VMT. Over the period from 1982 to 1997, the U.S. Federal Highway Administration (FHWA) highway performance monitoring system (HPMS) database reported in the Washington, D.C.–Maryland–Virginia region a 28% increase in population, an 82% increase in daily VMT, and a 107% increase in freeway VMT [13]. The VMT increased only due to population growth, but also due to the size of the local workforce and increasing distances between home and work.

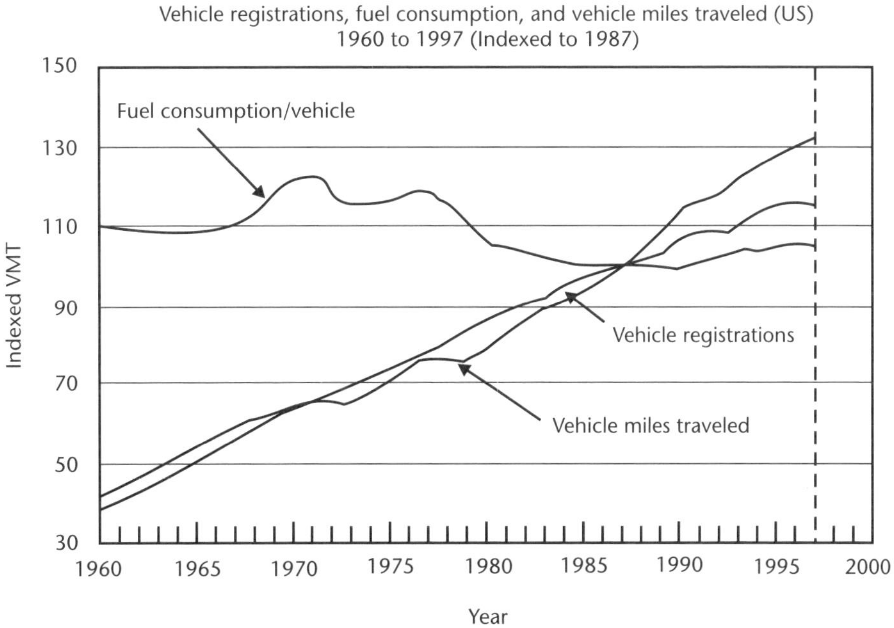

Figure 7.1 Vehicle registrations, fuel consumption, and vehicle miles traveled (United States), 1960 to 1997 (indexed to 1997). (*After:* [13].)

In addition, "transportation system demand and land use patterns are linked and influence each other" [14].

The pressures on transport infrastructure in countries such as India, China, and Korea have increased as the gross domestic product (GDP) per capita has increased. China was the world's fourth-largest producer of passenger vehicles in 2003 [15], and the third-largest passenger vehicle market in 2006 [16]. Further growth in ownership is generally seen to be beneficial. It is thought that increases in disposable income, rather than economic growth, create the spending power to drive sales of passenger vehicles. The average price per vehicle in China, the United Kingdom, and the United States is similar, yet an average passenger car costs 20 times the annual average salary in China, compared with 0.7 and 0.6 for the United Kingdom and the United States, respectively. In 2001, China had only 1.5 vehicles per 100 households, versus 170 vehicles per 100 households in the United States [17]. Other demand suppression mechanisms affect vehicle ownership, such as import duties, initial registration taxes, and the certificate of entitlement (COE), as used in Singapore [18].

The example region applies an initial registration tax and annual reregistration fee. Import duties are applied by the national customs and excise authority, and fuel duty is centrally collected, independent of funds distribution to the regions. No overall pricing mechanisms exist to suppress latent growth in vehicle ownership, which would also be difficult to apply regionally, although recent fuel price increases were believed to have had some effect on vehicle usage. A policy transition from dependency on fuel duty to direct user-pays charging has been debated, although this is considered to be outside the scope of the example project.

Legislation and guidelines for the capture and management of digital images to be used in criminal prosecutions exist in the example region. Enforcement based on digital image evidence that shows a vehicle at a specific time and location has been used successfully in enforcing electronic toll collection, and this precedence will most likely be directly applicable to the example project. The registered owner of the vehicle and not the driver will be held liable for violations. Administrative processes that identify the driver would have been difficult (i.e., expensive), and would have required the cooperation of the vehicle owner. Systems of driver "nomination" are in place in many charging schemes and road law enforcement schemes throughout the world. These systems do not need to be too burdensome or expensive, if properly legislated. An image of the license plate will be regarded as acceptable evidence, but the image must not include the driver or any passengers (e.g., in the Stockholm congestion charging pilot).

The level of compliance could theoretically be increased by imposing an escalating penalty regime, as used by Transport for London [19, 20]. The operator of Melbourne City Link (Australia) issues an infringement notice, requesting payment of the outstanding toll plus an administration charge. If payment is not forthcoming, the Victoria State Government then imposes a penalty and pursues violators through the courts. Nonpayment is an infringement, but it does not attract demerit points on a driver's license [20]. Enabling regulations for the example route must consider the possible levels of compliance, and whether an escalating penalty charge regime can be used (e.g., driving license demerit points, or prevention of vehicle reregistration). U.S. states may prevent the renewal of the license plate until traffic offenses are settled, which is a useful mechanism to ensure compliance and payment of outstanding tolls. The level of compliance is a measure of the proportion of vehicle passages that can be attributed to the receipt of correct fees. With the deterrent of not being able to reregister a vehicle, a compliance rate of up to 98% is achieved in some states in the United States. In other circumstances an efficient enforcement regime should be able to reach at least 95%. 407 ETR in Canada operates a similar plate denial scheme, where the "... Registrar of Motor Vehicles ... shall, at the next opportunity, refuse to validate the vehicle permit issued to the person who received the notice of failure to pay ... and refuse to issue a vehicle permit to that person" [21].

7.3.7 Local Precedence

A provisional business case (PBC) for the route must be defined, to cover upgrade works and operations. Some understanding of the effect of price changes on the demand also needs to be performed, based on stated preference surveys and local tolling precedence. In common with many other regions [1], heavy goods vehicles would be expected to pay at least twice the rate paid by passenger cars, with a variation for the number of axles and tires, although these parameters could not be reliably measured on an open road for enforcement purposes.

Infrastructure programs in the example region, aimed to provide access between economic zones, so far have been exclusively funded from public sources. The source of funds had been partially dependent on taxes collected from fuel purchases and other general taxation. Increased congestion had traditionally been met by increased infrastructure to maintain vehicle flow rates and travel times. The example route will use private funds to fill the public funding shortfall, and will focus on demand management, rather than on maximizing throughput.

There is a moderate understanding of ETC within the region, although cash collection at toll booths remains the dominant form of payment made by road users. Users need a more comprehensive understanding of forms of payment in order to implement MLFF, since cash cannot be paid at the point of charging. Users must be encouraged to migrate to more efficient and lower-cost payment methods, such as using a credit card via the Web. Cash payments will still be accepted, but only at retail outlets that offer accredited cash deposit services.

A public communications program must be launched to overcome local skepticism, and legislation must ensure that images will be accepted as primary evidence for enforcement. The relevance of existing tariff tables is questionable, since they had been developed for toll plazas where measurement of selected vehicle parameters in a constrained space (the toll lane) was easier. The development of the procurement documents will consider whether a new classification scheme, based on existing vehicle categories, should be developed to aid measurability and enforceability (see Chapter 5).

Public acceptability for the upgraded route has already been secured, although there had been many objections to charging for road use, which were mostly lessened when additional commitments had been made, including more consistent journey times, higher quality transit (subsidized by public funds), and discounts for local residents. These concessions may reduce commercial viability, but they secure public acceptability. The political future of the route, funded by tolls with premium pricing for peak periods, had thus been secured. The overall transportation corridor, in which this route plays a major role, had always been aimed at benefiting business and individuals, which was part of the communications strategy throughout the planning phase. The planning phase included the long-term objective of migrating to advanced road user charging policies, although in the absence of any fixed plans (they would be impacted by long-term national plans for RUC anyway), it was considered critical to ensure long-term contractual flexibility. Communications with potential construction and operating companies had to emphasize that regional road user charging policies would be implemented throughout the lifetime of the tolling operating concession.

7.3.8 Cross-Border Issues

Regional economies do not exist in isolation. Increasing economic interdependency means that transport infrastructure will enhance the connectivity and economic linkages between regions. In addition to the obvious ethnic and cultural differences, a border usually represents a discontinuity in many areas [21], including:

- Transport planning policies;
- General taxation policies;
- Currency control and restrictions;
- Recognition of the benefits of, and orientation towards, transport telematics;
- Applicable standards for wireless communication and electronic data interchange (EDI);
- Approaches to traffic management;
- Vehicle registration procedures, and access to and quality of vehicle registration databases;
- Privacy and data protection laws;
- Definition and prosecution of traffic offenses;
- Evidential quality requirements;
- Approaches to benefit cost analyses;
- Public procurement policies;
- Radio frequency spectrum management policies and licensing rules;
- Value added tax and sales tax levels;
- Language.

All of these represent potentially significant hurdles to achieving an integrated transport strategy, and incorporation of road user charging across borders, which would benefit the exchange of goods, services, and labor. The alleviation of nonphysical barriers to cross-border commerce would increase efficiency, reduce costs, and maximize economic benefits. The reduction of administrative and logistical barriers can transform cross-border transport corridors into economic corridors, where infrastructure developments are directly linked to production, trade, and investment potential.

The European Interoperability Directive, published in 2004, focuses on road user charging at a pan-EU level, and states that [22]: "artificial barriers to the operation of the internal market should be removed, while still allowing the Member States and the Community to implement a variety of road charging policies for all types of vehicles at local, national or international level. The equipment installed in vehicles should allow such road-charging policies to be implemented in accordance with the principles of non-discrimination between the citizens of all Member States. The interoperability of electronic toll systems at community level therefore needs to be ensured as soon as possible." However, the Directive makes no reference to cross-border enforcement.

Differences in taxation policy and treatment of value added taxes and general sales taxes means that cross-border reconciliation of road user charges between scheme operators is complex, although the success of the credit card and mobile telephony industries shows that mass-market transaction clearing of relatively small values (micropayments) across borders is already pervasive. Differences in vehicle registration policies, privacy and data protection laws, definitions and prosecutions of traffic offenses, and evidential quality requirements complicate cross-border enforcement of road user charges and make them expensive to operate. For example, the procedures to identify the party responsible for a vehicle and to serve a notice requesting payment differ widely between EU member states. Section 4.7 elaborates further on this problem, and introduces two initiatives centered on Europe, but which have worldwide relevance:

- VERA, which aims to develop the necessary tools and relationships to enable the effective identification of violators, notification, and recovery of revenue;
- CAPTIVE, which aims to examine the current legislation and suggest improvements for member states.

CAPTIVE and VERA are described further in Section 4.7.

The legislative planning for the upgraded route included a budget for cross-border consultation to identify the expected shortcomings of laws to request personal data on vehicle owners (violators) and the procedures for serving notices for payment of outstanding tolls. The legislation will help develop an enforcement strategy, and, if needed, manual on-road enforcement, although its usefulness is arguable. Manual enforcement may be limited to checking for lack of a functioning OBU, which is very difficult at highway speeds, and could be more easily and efficiently performed by image capture. Unless immediate fines for violation were allowed, the enforcement authority would still retain the problems of the subsequent collection of fines/fees, including from vehicles registered outside the region.

Contractual interoperability between operators would enable cross-border charging. Any future operator of the example route will be required to accept declarations made by OBUs issued by other regional operators. As the use of ETC expands regionally, the operator will be required to accept other OBUs, provided that they are issued by an authorized organization in accordance with accepted procedures, are technically interoperable, meet regional specifications, and have an acceptable guarantee of payment from the contract issues. These policies and guidelines will need to be developed as the regional use of ETC expands.

The same outcomes could also be achieved using an industry-based (but authority-supervised) interoperability agreement, to which all scheme operators would need to subscribe. This would not necessarily require all scheme operators to accept OBUs issued by other organizations (whether operators or not), nor to accept OBUs from another authority. They would have to accept the declaration by operators that they have detected a passage on their roads by a customer of another operator, and the acceptance of the relevant toll charge and an appropriate handling fee. This approach would maintain interoperability as a business process, rather than requiring technical interoperability as described above. This approach would apply where OBU supply, retrieval, maintenance, and operations logistics is important to operators as a strong source of competitive advantage (assuming a competitive cluster of road operators). The interoperability strategy chosen by an authority can inadvertently affect efficiency savings through economies of scale. A priced road network in which the authority defines charges could restrain competition. This reflects the difference between tolling and demand management road user charging, and the dilemma facing authorities who wish to enable a commercially viable operation with an element of demand management, that is, the blurring of tolling and charging identified in Chapter 1.

7.4 Timetable

7.4.1 Project Timetable

The political timetable can constrain projects, but strategic investments in transport infrastructure and land use planning have a time horizon that can extend 10, 20, or even 50 years into the future.

International approaches to the upgrade of infrastructure and construction of new infrastructure vary widely. Several U.S. states have already sold the right to collect tolls for periods up to 99 years: equivalent to about 25 changes of government, 10 charging technology product life cycles, and 5 human generations. Forecasting traffic demand over this period is difficult, but nevertheless, the attraction of the upfront infusion of funds from the private sector to deal with short-term transportation measures could prove to be attractive [23]. One alternative approach defines a shorter operating concession, either fixed by time (e.g., 25 years), or automatically terminated upon achievement of an agreed-upon ROI target. A third option procures the construction and operations funding separately, to enable long-term transport and charging policy flexibility. The operating concession could be as short as 5 to 7 years to enable a complete reexamination of charging and enforcement solutions, and to ensure that they remain aligned with the regional approach to road user charging.

In this example, the attractiveness of an upfront payment did not offset the loss of flexibility that the region required. The existing route, although requiring significant upgrade, was regarded as economically critical to the future development of the region and offered no additional room for vehicle capacity expansion. However, dedicated lanes were provided to improve transit time, while sharing with registered carpool vehicles. The additional complexity, and uncertainty of the benefits of mixed-use charging in an environment that had not previously been charged, reduced its practical value. Potential experienced operators may have found the task of modeling revenues and devising charging and enforcement regimes too daunting, and the best and most qualified operators might be reluctant to bid. Dedicated lanes were considered as a future policy option, assuming that there was sufficient flexibility in the concession agreement to permit this change. The PBC of the project, even if peak period pricing was applied, was regarded as being sufficiently profitable to attract international bidders for the road upgrade and operations concession.

The procurement would be based on a design, finance, build, and operate (DBFO) concession over 25 years, to ensure that the project was sufficiently attractive to bidders and within the time horizon acceptable to investors (typically 40 years). The construction period was estimated at 4 years (incremental upgrade, plus one new segment), leaving about 21 years for operations, consisting of regular reviews every 7 years to ensure that the enforcement and charging solutions and its central system were technically and operationally aligned with other emerging and future road user charging projects in the region.

7.4.2 Pilot Deployment

Section 6.3.2 describes a pilot system that could be deployed to meet a range of technical, operational, and political requirements. Depending on the point at which a pilot solution is introduced into the procurement process, the additional investment can also raise public awareness, inform and reduce the risk of competitive procurement strategies, help develop requirements, and choose among solutions as part of a procurement.

The options for pilot and trial schemes generally are:

- Funding for a comparative multistage pilot procurement [e.g., NYSTA (EZ-Pass trials), Rekeningrijden (the Netherlands), and Transport for London (DSRC urban road user charging minizone trial)];
- Funding for a competitive multistage procurement to select the most appropriate technical solution as part of an overall assessment [e.g., Land Transport Authority (Singapore ERP procurement)];
- Competitive procurement of a single pilot solution (e.g., Stockholm City/SNRA Congestion Charging pilot);
- Deployment and operations without a pilot, with scheduled detailed design reviews and reliance on concessionaire-led acceptance testing (e.g., Melbourne City Link, Cross Israel Highway, Santiago Urban Concessions).

For the example route, electronic tolling solutions were relatively mature (although MLFF had not been used regionally); public acceptance had already been assured (although only marginally); and outline performance requirements, consistent with international deployments, had been developed during a preliminary study for the project as part of the economic analysis. It was further expected that each consortium bidding for the project would include a civil works contractor, teamed with a systems integrator and an MLFF charging/enforcement systems vendor. It was also anticipated that a single company would provide route operations and would be responsible for the development and deployment of the end-to-end charging and the enforcement solution.

It was agreed that a pilot would not be necessary, but the successful concessionaire would be required to implement a multistage design, systems development, and trials program prior to operations. The operations risk would rest with the concessionaire as main contractor, and the regional transport authority would specify the outline solution that was known to be technically deliverable. The concessionaire would also be required to participate in a public awareness program to local stakeholder groups, ensuring an early marketing of the scheme, its payment channels, and the approaches to enforcement.

7.5 Procurement

7.5.1 General

New technologies and evolutions of existing well-known charging technologies are being continuously introduced, and highway operators are now more intensely engaged in establishing contractual interoperability between systems that were initially purchased separately. The ongoing desire to introduce competitive procurement specifications, while reducing the procurement risk to manageable levels, is the subject of simultaneous efforts in many regions, including the United States and Europe.

The example route will be the first regional scheme that will implement road user charging for demand management on a specific corridor as part of a regional or national strategy. This section aims to provide some broad guidelines on developing the approach to procurement for this project.

7.5.2 Procurement Strategy

The procurement could be structured in many ways. Each of the main works elements (e.g., civil works upgrade, charging and enforcement solutions, central system, operations, and maintenance) could be procured separately. The functional split can reflect natural industry boundaries, but would require that the procurement authority act as the program manager and systems integrator. The alternative is to subcontract the end-to-end solution that includes operations as a service, funded by road user charges. Part of the overall project risk can be transferred, but at the cost of having less visibility and control of the design and development of the individual functions.

The commercial viability of the route for a tolling scheme will be dictated by construction costs and forecast demand. If the cost of construction is high (e.g., requiring tunneling) or if demand is low, then it is likely that the project will be split into separately paid subcontracts. This includes construction of the road link, and integration and operation of the charging/enforcement systems. The limited examples of road user charging, tolling, and access control systems on existing roads (e.g., Singapore, Oslo, London, Stockholm, and Rome) limits the procurement only to the supply and integration. Operations may remain with the urban transport authority, with limited operations control handed over to the private sector perhaps through outsourcing. Figure 7.2 shows a typical structure of organizations at each stage of the development and operations of civil works contract, including charging and enforcement operations.

The procurement process itself can include several phases: invitation of expressions of interest, invitation to prequalify, invitation to tender (invitation to negotiate in the United Kingdom), and contract award. The selection process can be limited to a simple comparison of tender documents, or could include competitive on-road trials to reduce the number of potential suppliers. The final phase would then include technical and commercial negotiations to select one candidate for contract award. Notably, EZ-Pass (Interagency Group), Singapore's ERP system (Land Transport Authority), Taiwan Area National Freeway Bureau, and the Western Extension Zone (Transport for London) used this systematic approach.

The emerging model in the United States emphasizes the submission of unsolicited bids from interested parties (usually construction or finance provider–led consortia). There could be any number of unsolicited proposals over many years. Once one particular project is deemed "feasible," then the proponent of that bid will be invited to negotiate the financial details on an exclusive basis, using environmental impact studies, traffic and revenue modeling, community consultants, and so forth.

Tenderers must be prepared to take the risk of entering a multistage project without any certainty that it will lead to awarding of the contract. Projects with a high civil works contract procured under a BOT or DBFO style contract often require that the private sector operating company initiate the design and acquisition of charging and enforcement systems. This approach applied to the M6Toll (United Kingdom) and Melbourne City Link (Australia).

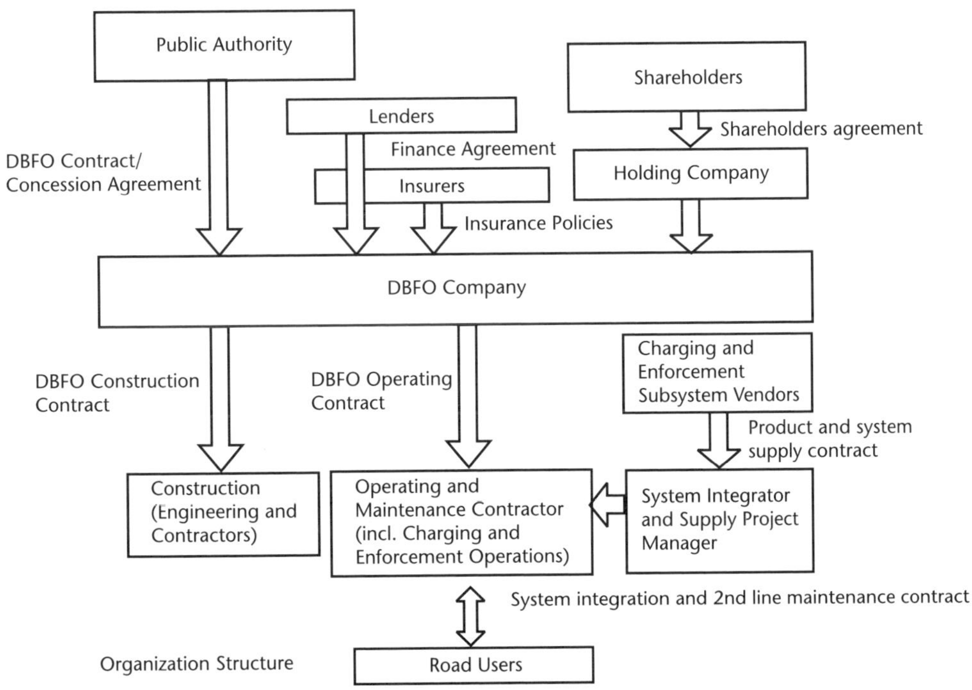

Figure 7.2 Organization structure.

7.5.3 Developing the Requirements

7.5.3.1 The Context

The requirements for a road user charging system depend on whether it will be implemented on a toll plaza or on an open highway, whether it will enable migration from single-lane tolling to open highway tolling [ORT (MLFF)], or whether both approaches will operate simultaneously.

Procurement processes vary widely between countries. Charging and enforcement technologies could represent at least 50% of the value of procurements for new combined manual, automatic, and electronic toll collection projects. If the project includes the design and construction of a new highway, then the value of the integrated charging and enforcement solution would fall to typically 2% to 5% of the overall project value. Consequently, if the project is dominated by charging and enforcement technologies, then it would naturally be regarded as an IT procurement following IT procurement rules. If the civil works value dominates, then the project would be regarded as a civil works project, following civil works procurement traditions and attracting a different mix of tenderers. The largest projects would include civil works contractors, financiers, specialized professional engineering companies, traffic engineers, construction support services, operations and maintenance, and program management. A system integrator could deliver the smallest projects, based on internally developed central systems, and charging and enforcement technologies.

Examples of the various types of projects could include the following:

- A new electronic toll collection system for a small toll plaza;
- A World Bank–subsidized infrastructure project to link economic hubs in developing countries;
- Private finance initiative (PFI) projects, based on new infrastructure and long-term operating concessions in developed countries;
- Incremental expansion of electronic toll collection in a region, based on a well-defined and proven set of requirements;
- New multimodal infrastructure to upgrade an existing transport and economic corridor;
- An area or cordon pricing scheme within a major metropolitian area.

Existing systems may require the development of a migration strategy that would completely replace earlier solutions or parallel operations, either temporarily or indefinitely, if it can be shown that the benefits can be economically and socially justified.

7.5.3.2 Development of Requirements

Although the requirements for a specific deployment vary, some general pointers can be used to guide the development of a toll collection development strategy, approach to technology selection, development of a procurement model, and support for a business case to ensure economic and social benefits.

The requirements for a new charging and enforcement system must enable tenderers to be innovative and unconstrained in approach, yet these requirements must be detailed enough to ensure that the upgraded route fits into the charging policy and local transport strategy for the region. Regulatory requirements would include data communication standards, data protection laws, aesthetic guidelines for roadside infrastructure, mitigation of environmental impact, technical and contractual interoperability, consistency with a future distance-based truck tolling system, and future regional road pricing needs.

The development of requirements must also consider their deliverability. For example, the development of the London Congestion Charging Scheme was underpinned by recommendations from earlier studies that "considered the feasibility, acceptability and transport impacts of potential charging options [for London]" [21]. The use of ANPR to enforce an area pricing scheme was initially proposed, since ANPR was considered to be mature, had a known performance in the urban environment, and was deployable within a single mayoral term within budget constraints. It was felt that more sophisticated technologies did not meet these criteria at that time, although the 2005–2006 DSRC and GNSS trial program showed that these technologies were being considered as potential candidates for London alongside ANPR for occasional users and for enforcement. Similarly, the perception that EZ-Pass was easily rolled out across 20 agencies in the Northeast United States ignores the 3 years of initial trials and the development of a scaleable package, which included charging and enforcement strategies that could be used by other agencies with minimal local variation, to maximize the customer benefit of a single tag solution.

A time-sequenced, informed development of requirements is essential to the overall procurement of any charging and enforcement scheme to ensure that the local context is properly captured. MLFF charging schemes have been deployed in many countries (e.g., Canada, Singapore, Australia, Sweden, and Israel), and the regulatory context and approach to enforcement vary between these countries. Reducing the procurement risk by ensuring that the requirements can be met, in some cases by contacting potential tenderers or agencies that had already employed MLFF charging schemes, is critical to ensure ongoing development of positive public and political acceptability. The ongoing regional assessment of distance-related charging for trucks and the long-term migration to an all-road, all-vehicle scheme must consider the charging and enforcement technology, along with the necessary enabling legislation.

7.5.3.3 Categories of Requirements—Charging Technologies

A short list of questions that a local or national highway authority should ask before developing a charging strategy are given under five headings: procurement, performance, technology characteristics, enabling the business case, and evolution.

Procurement

- Compliance with standards: What public domain standards exist and what level of vendor support can be identified for the standard(s)?

- Existence of standards and interoperability specifications: Are there existing specifications that can be used to reduce implementation risk? If not, then does the procurement schedule allow time to generate minimum specifications to ensure the interoperability of separately procured systems?
- Completeness of standards and specifications: Have the standards and specifications been fully debugged, or, if not, can a subset be extracted and used for procurement without introducing unnecessary additional risk or delays to the project?
- Multivendor support: Would it be possible to procure the OBUs separately from the roadside infrastructure? If so, how many vendors would be prepared to compete for a procurement of OBUs, both now and in the future? If, as in the example project, the concessionaire provides the charging systems, then this benefit would only be realized by the concessionaire, although road users may benefit from greater choices if additional applications are offered in the future.

Performance

- Demonstrable performance: Has the road charging technology category been sufficiently proven to warrant deployment from a limited scale to regional and national use? If not, then how can the risk be mitigated without compromising the long-term operational objectives?
- Communications capability: Does the standard support central accounts and on-board accounts, strong communication link security, OBU authentication, battery level monitoring, and simple extension of security schemes to accept OBUs that are encoded with other operator's security keys?
- Security: What techniques are provided to resist spoofing, replay attacks, or unauthorized access of data carried by an OBU? What control does the operator have to ensure that the OBU data is not provided to third parties? If none, can the risk be contained?
- OBU basic capability: Can the OBU be used at mainline speeds in all of its modes of operation, and, if so, what communication margins are provided to enable the transaction to be completed successfully every time?
- OBU (future): If a distance-related charging scheme were introduced in the future, could the charging reports from all OBUs be accommodated? If not, what changes would be required?

Technology Characteristics

- Spectrum allocation: Does the RUC system operate in frequency bands that are permitted, or preferably encouraged, by local spectrum management authorities? For example, there are no RUC systems in Europe operating in the 902- to 928-MHz band, which is dominated by GSM. There are no systems in the United States operating in the 5.8-GHz band, although the 5.9-GHz band is a safe haven for RUC systems in the United States [24], and the focus of a related interoperability initiative [25].

- Emitted power: Does the OBU or roadside reader generate power levels that are greater than locally permitted levels? If so, could an exemption be granted without compromising the safety of road users, local staff, and service technicians?
- Integration with enforcement: Can the RUC technology be integrated with an enforcement subsystem wherever the charging policy is to be employed, either in a toll lane and/or on the open highway? How is the OBU localized in the toll lane, or, if open highway use is envisaged, can the communication link help localize the OBU to help match OBUs with the correct vehicles?

Enabling the Business Case

- OBU unit cost: Does the unit cost have the potential to enable the business case for mass-market OBU deployment, or, if not, can the cost be factored through other service providers to ensure low-cost access to all potential road users without impacting the level of fees collected?
- Roadside system cost: Does the life cycle cost (including acquisition, operations, and maintenance) enable the business case for wide-scale regional or national deployment?
- Policy flexibility: What charging policies are enabled by the technology options? Is the capture accuracy sufficiently high to enable toll collection by commercial operators, and can the technology be used and enforced in both single-lane and open highway environments?

Evolution

- Along which dimensions does the RUC technology have the potential to evolve? Directions for migration can include policy, geographic (incremental, acquisitive, and contractual), and application (capability to support added value services). Section 7.9 further elaborates on each of these topics.

7.5.3.6 The Example Route

The requirements for the example route were split into several categories, including:

- Charging system, including tariff table, means of charging, accuracy, payment options, complaint handling, and redress;
- Enforcement, including penalty levels, image evidential quality requirements, and accuracy;
- Operations and maintenance, ensuring route availability;
- Reporting;
- Construction timetable, including key contract milestones;
- Trials system requirements and acceptance criteria;
- Interfaces to third parties, including departments for motor vehicle registration, police, and management agencies (e.g., for revenue and traffic reporting, local traffic management, and incident management);
- Regulations, including data protection, quality targets, health, and safety;

- Service levels targets and project milestones;
- Scalability;
- Public relations.

The metrics for the operator's quality of service must be defined. For example, typical measures include the vehicle throughput per hour (typically used in interurban, capacity-enhancing measures. and at toll plazas), or the differences in time to travel a prescribed distance at congested and uncongested times of the day (typically used in urban congestion charging schemes). The latter method was adopted for the upgraded route. The concession operator would be required to report whether or not agreed performance levels were being met, and would be financially incentivized (through penalties) for not achieving the targets, especially those targets that are linked most closely to social benefits, such as ensuring accessibility to the upgraded road. Ensuring that the performance criteria were generally decoupled from factors outside of the control of the concessionaire was necessary but presented a challenge.

Eighteen months was allocated to the development of complete requirements for charging and enforcement, including contact with other agencies. This reflected the necessary learning curve of planning and operating an MLFF charging scheme, and time to rapidly implement the necessary legislative changes, although in the worst case this could have taken a further 12 months. Potential bidders needed the certainty to adequately quantify risks important to their stakeholders.

7.5.4 Local Expertise and Global Sourcing

The specific mix required for any project is unique, even with well-known and well-understood policy, technology, and regulatory components. The attractiveness of this project to potential tenderers will be assessed alongside other competing projects. The supply chain, from the licensees of core processors for the OBUs to the auditors in local government, all face competing time and interest pressures. Civil engineering labor rates vary due to local supply and demand effects, but specialized expertise in data encryption for OBU data protection, for example, face more global pressures. The objective is to capture sufficient experience at each stage of the project to meet the stated objectives and community expectations. Universities can also provide expertise in applying ITS technologies in a policy context, with prominent examples in Germany, China, the United States, Singapore, and the United Kingdom.

Road user charging projects generally cross the boundaries between politics, law, and technology. The program management needs the breadth and diversity to easily cross these boundaries, and recognize when a decision made by one discipline could impact a decision or plans already made by another. For example, it is necessary to identify and communicate risks early and indicate when regulations are lacking. Similarly, the development of new regulations must be matched by an ability to comply with these regulations. The procurement strategy also should consider the supply side perspective, along with the demand side.

Public and media relations are an essential component of the local expertise within the procurement team. Most of the media's attention has been on the charging technology, which makes the project more tangible and newsworthy. No attention has been given to the long-term policy plans to make better use of existing infrastructure, including any necessary extensions and upgrades. The media interest has been reasonably positive, and any mistakes have been swiftly corrected by the public relations executive. Public presentations have been more frequent, to raise awareness of the implementation phase, the methods for payments by road users, and the consequences for not paying.

7.5.5 Technology Options

Technology options for charging and enforcement are described in Chapters 3 to 5.

The technology options were evaluated to enable a low-risk, predictable deployment. Section-based charging for the road (which has 15 "at grade" intersections) had been chosen, and this suggested a DSRC solution, enforced by camera images, showing the license plate and a color image of the vehicle in context, but excluding the driver and passengers. This solution was thought to be viable at least until the next technology review in 5 to 7 years. Over that period, mass-market solutions that included positioning technologies such as GPS or Galileo were expected to be generally available and competitively priced, reflecting greater accumulated volumes shipped worldwide and acceptance for RUC.

Other competing solutions that combine communications methods for safety and driver information, such as ISO TC204-CALM and WAVE, will also be more prominent by 2010. The acceptance of these technologies will depend on the region (e.g., WAVE in the United States), acceptance by vehicle manufacturers, availability for retrofit, initial cost to the road user, and services mix available (e.g., open or walled garden). This is discussed further in Sections 9.2.1 and 9.2.5.

Vehicle class definitions will not be modified for this project, since it was assumed that, for locally registered vehicles, the department for motor vehicle registration would only be able to confirm vehicle class information as accurately as their database would allow. Nonlocally registered vehicles (approximately 20% of the total traffic volume) still present a measurability problem. Unless they were regular visitors, only a small proportion of users would open a tag-based account with the required down payment. The procurement team had also met with regional vehicle registration authorities and had agreed to a policy that would allow efficient and rapid access to vehicle registration details. Versions of the most common vehicle taxation classes were redrafted, partially based on externally measurable parameters. Migration of the taxation classes to include externally measurable parameters was regarded as the optimal policy from a road operator's perspective, although this was put into the long-term action plan.

Occasional users will be offered all payment channel options, including cash. Signs will be located on the approaches to and from the upgraded road, showing where and how to pay. All road users, including those from other jurisdictions, will be required to register, ideally prior to driving on the road. Signs explaining the registration process will complement the multimedia-marketing program. A short period of grace (maximum 2 days) will be permitted for users who do not preregister. The alternative approach of not requiring vehicles to register (e.g.,

407 ETR) was considered to be operationally expensive, since this approach would shift the burden and cost of identifying all vehicles to the road operator. This project includes peak period pricing for demand management element, and indicating the costs of road usage requires users to participate more fully than they would in a pure tolling scheme.

7.5.6 The Case for Standards

The issues that must be addressed by a regional or national administration include: development of a technical specification for the project; background information sufficient to ensure that all tenderers make similar assumptions; statements of minimum requirements, particularly if interoperability is a strategic objective; and the decision criteria. The tendering phases may include an initial on-road test or pilot evaluation to help develop tendering requirements, and provide visibility of technology variants that meet a provisionally agreed core set of requirements. The regulatory basis needs an early assessment.

The availability of specifications and standards for charging technologies effectively means that viable commercial-grade technologies already exist for procurement, and often suggests that there are multiple suppliers willing to initially compete, and that there is an ongoing basis for system expansions or additional batches of OBUs. The geographic expansion strategy could be based on separate procurements of RUC systems (e.g., plaza, ORT, or MLFF) within a regional interoperability framework that is not part of the example project. This would allow separate competition for technology suppliers and system integrators, without compromising the objective to develop a regionally integrated road user charging scheme that is also envisaged for the long term. In the absence of any technology strategy or objective, isolated islands of charging and enforcement technology may create a case for a de facto regional standard based on a single technology precedence. If several uncoordinated technology procurements occur in the region, then additional hurdles would be presented to road users who wish to travel freely without additional registration at other regional schemes.

Traveling between roads with separate technology specifications will most likely prevent users from roaming between operator areas without having separate contractual relations (and possibly different OBUs) with each operator. An OBU offers positive identification, but if an OBU were not recognized, then the vehicle's license plate could provide a reasonably accurate identification of the vehicle for enforcement purposes. If a common technical standard is not agreed upon, then a user who registers on a home road network could be simultaneously registered on other chargeable road networks in the same region, although the user may have to pay a rate appropriate for non-OBU customers of the other networks. If this registration option were not offered, then the road user would need to register with the operators of all the toll roads that they are likely to use. This would prove onerous, discourage use, and could reduce compliance.

National policy that gives the private sector a greater role in infrastructure development could lead to privately managed RUC system procurements that could be less than optimal nationwide. Setting early expectations with private concession operators could greatly ease the adoption of any technical standards and interoperability specifications, which suggests that communication of the intended approach to interoperability is a critical part of the development of the specifications.

The development of an interoperability specification provides some reassurance that a nationally integrated network of toll operators would be technically feasible, assuming compliance with the specification. The central system itself can act as a transitional measure for interoperability for some road users. The EZ-Pass solution provided similar benefits to road users with a sole source technology and transaction clearing between schemes. In the case of EZ-Pass, although none of the operators "own the customer," they simply report usage to one operator that issues charge and infringement (citation) notices under its name and authority. This is not optimal, because it could precludes the ability to leverage the benefits of increased efficiencies in customer service processes that competition engenders among individual operators.

7.5.7 High Occupancy and Toll

The complete route is chargeable, and differentiating charges based on vehicle occupancy are being considered. Chapter 4 highlighted the dependency between measurability and enforceability. Vehicle occupancy has so far proven difficult to measure automatically, particularly on the open road. Trials have provided limited success of automatic occupancy counting [26], and integration with an enforcement regime had not been attempted. Traffic officers typically manually enforce HOV and HOT lanes. The practicality of manually capturing image-based evidence of occupancy from mobile enforcement vehicles was regarded as unreliable. The current practice is to physically apprehend vehicles suspected of violating high-occupancy regulations.

HOV lanes have been provided alongside general purpose uncharged lanes, usually by constructing two new lanes (one in each direction) within the median, or by providing an additional lane as part of the original construction. Single occupancy vehicles (SOVs) can use high occupancy and toll (HOT) lanes by using a tag located in the vehicle. Thus, HOT lane enforcement would also need to include tag-reading equipment. The deterrent to non-HOVs using the high occupancy lanes depends on the level of the penalty and the perceived probability of being detected, as compared to the toll saving. The visible deterrent offered by traffic enforcement and incident response teams can be further enhanced through physical lane separation and clear signage on the approach to the HOV lane entry points.

A future development target for automated occupancy counting would be to provide evidence that unambiguously shows the true vehicle occupancy, based on evidence that would be acceptable in court. In the meantime, the approach to HOV/HOT lane enforcement will generally be based on a visible deterrent (fines, plus visible presence of traffic officers), a public awareness program explaining the purpose of HOV/HOT, and continued applications research into automated occupancy counting.

HOT is not applicable for the example route, since the whole road is subject to charges. However, it was also decided that the introduction of discounts for HOVs would be included as an option that could be exercised at the request of the procurement agency at a later date. HOV discounts would be offered when the enforceability had been perfected, including appropriate regulations and feasible working practices. For the example route, the local regulations dictate that HOV lane offenses would need to be witnessed by a traffic officer, so the introduction of automated capturing of image-based evidence would depend on new regulations and sufficiently accurate detection and recording equipment.

7.5.8 Support for Truck Tolling

Tolls are normally used to pay back the cost of developing and operating the road infrastructure. The term "truck tolls" is a misnomer, since the fees collected are not normally applied for the purpose of tolling (i.e., infrastructure development and operations), but instead are applied as a charge for several reasons, including:

- A transit tax or charge on the owners or operators of trucks for that part of the vehicle's journey within the region (territoriality);
- A fee that more closely reflects the external impact caused by the vehicle [27].

In the example region, the vignette system captures electronic declarations from trucks equipped with ETC-type tags. All vehicles with an MGW of 38 tons and above are expected to be equipped with these tags, and enforcement currently uses traffic officers and fixed identification points on strategic routes as a deterrent. The upgraded road will need to accommodate this existing policy, and ensure that any roadside infrastructure can be equipped with truck tolling identification equipment provided by the truck tolling scheme operator. The existing truck tolling system was developed for a specific purpose, and is not currently technically inter operable with any commercial MLFF charging solution. This technical difference will require trucks to have both tags in the short term, which is technically inconvenient for road users, but impacts neither scheme (upgraded road and truck tolling). It is unlikely that this approach would be sustainable, given the need to secure broad stakeholder interest in road user charging and scheme interoperability.

Trucks are generally treated differently from other vehicles, in areas such as taxation levels, supply chains, regulations, and safety requirements. The paper-based period vignette has been replaced by electronic vignettes in the European countries of Switzerland, Austria, and Germany, with the Czech Republic, Slovenia, and Sweden following close behind at the time of this writing. This is providing an early experience of road user charging that is potentially applicable to all vehicles in the future. The policy migration, from period-based charging to distance-based charging for heavy goods vehicles, has proven to be less of an institutional hurdle than has an all-vehicle scheme that is presently based on an annual tax and fuel tax, both poorly linked to road usage.

The long-term aim is to ensure that road users can easily access and understand road user charging. Policy overlap in this case will mean that truck owners and operators initially will need to see the schemes as separate. Future technical innovations may mean less equipment in the vehicle [28]. Policy alignment, contractual relationships between scheme operators, and an increasingly integrated approach to road user charging will ultimately lead to a more sustainable solution.

7.6 Perspectives

7.6.1 The Procurement Team

The procurement team that was being assembled for the route contained civil engineers, structural engineers, traffic engineers, telematics consultants, requirement capture analysts, program managers, a public relations executive, legal advisors, and financial advisors. The mix included expertise from the demand side and from the supply side, obtaining a balanced view of risk and responsibility. Communication with road users, motoring organizations, and potential bidders all require different perspectives, so this diverse expertise was regarded as beneficial.

Although a formal market testing program was not attempted, a presentation was prepared by one of the program managers to summarize the project, which showed the attractiveness of the project to potential tenderers (see Table 7.2).

The expected competition for the project includes a mix of regional, national, and global players. Local expertise is specifically relevant to civil and structural works. Several local engineering companies have been involved in traffic engineering and other highway instrumentation projects, and it is expected they would provide some useful local knowledge and context for the project, effectively reducing the risk from bids from teams that are led from companies based outside the region.

Many system integrators, which are responsible for the assembly of subsystems from many other suppliers, have attempted to contact the procurement team to find out more about the project. The procurement team accepts that the charging and enforcement technology is only an enabler, even if this is the only focus of some companies. The winning bid will need to demonstrate that the solution can be delivered, include service commitments, and demonstrate operational flexibility during the concession period. Price and quality will remain the two areas on which the competing team's proposals will be ranked.

Bidders will be informed of the criteria on which they will be judged and the relative importance of some of the requirements, to test the willingness to bid. The quality of bids is expected to be high, so constraints will be imposed on the procurement, including limiting the bidders from a minimum of five to a maximum of seven, limiting the proposal to a fixed number of pages, and requiring auditable references from the response to the initial requirements. Electronic submission of bids will be encouraged, although not mandated (a hard copy will be required anyway). Computer-assisted auditing of bids is being considered, at least in collating all the text that is specific to each requirement and assembling notes made by the reviewers.

As a rule, the bidder who fulfills all the terms and conditions of the requirement, and offers the lowest price, is generally selected. If only one bidder meets the requirements, then the procurement will likely be abandoned, although the risk of this is thought to be low. The procurement team routinely maintains and internally distributes a risk register, alerting to changes in risk profile and responsibility for mitigation.

7.6.2 The Integrator

Potential tenderers would be expected to see the project as being in line with corporate objectives. The structure of a bidding consortium aims to allocate respon-

Table 7.2 Perspective: The Procurement Team

The charging solution can be performed with existing technologies and systems.
Vehicle class definitions will be difficult to measure. To reduce the quantity of road users incorrectly targeted, it will be necessary to cross-check every violation candidate with its vehicle registration record held by the department of motor vehicle registrations.
We control the level of tolls, with a minimum and maximum, indexed to inflation, since one of the objectives is to impose peak period pricing similar to demand management, and since it was established that in this specific case, the road could be operated commercially. Potential tenderers will want to understand the relationship between price and demand if they are to believe the traffic figures that they have been given. The alternative, where concessionaires set the charges based on their estimate of traffic, was rejected. However, experience suggests that since estimated patronage is arguably the single biggest revenue risk over the life of the concession, tenderers would definitely not accept traffic risk on the basis of figures they had not compiled (or validated) themselves.
The amount of specific development of an MLFF solution should be minimal. If the project does not align with current demand, then the tenderers will assess whether the requirements reflect future mainstream demand. If so, the tenderers may see this project as providing them with a strategic advantage in being able to offer this specific solution elsewhere.
We cannot ignore the emergence of distance-related charging and related technologies. It has already been confirmed that the region envisages an "all-vehicles, all-roads" charging policy, so we must ensure that the cost and risk of this goal is not increased by the award of the upgraded roads project in the short term.
We need to understand more about the risks that the tenderers will face, including commercial viability in a demand managed area, revenue (variation in charges over time), enforcement (ability to ensure compliance and recover revenue), and regulations (acceptance of the timetable to develop and secure enabling legislation and regulation).
Public support is moderate but fluctuating. Compliance can only be assured through high levels of awareness, and we would need to ensure that positive support is further developed during the development phase. As the contracting authority, we need to preserve positive public support. Project delays, poor safety, and early problems could be detrimental to the short- and medium-term health of the project.
Tenderers will wish to market themselves at all levels of government, since many regard this as critical to international success. This policy will continue to extend from awarding of the contract, to the road upgrade, systems integration, and operations phase.
Good marketing of the facility to road users, with visible signs and other information, is paramount to deter nonpayment and ensure that users understand how to pay the charges. We need to ensure effective management of charges by the road operator.
Occasional users will not be required to have an OBU. Preregistration of the vehicle's number plate will be sufficient, although the lower accuracy of license plate capture means that the occasional user scheme will offer a temporary license, with a minimum duration of one day.
A good bid from tenderers should be straightforward, and we expect that four or five consortia will prequalify. However, a winning bid will only emerge from a team that is clearly differentiated from competing teams, offers good value, has a track record, accepts responsibilities within the capacity of each member of the team, shows an understanding of the risks, and works in cooperation with the client. Being able to separate the order winners from the order qualifiers [29] will be critical to winning this project.
We will encourage tenderers to submit high-quality bids, and will provide equal guidance to ensure that our requirements are properly prioritized and understood. Bidders will be reimbursed a proportion of costs for unsuccessful bids, to encourage high-quality responses.

sibilities and risks where they can be best managed within the operational capacity of the organization. The risks of a project change throughout its lifetime, from tendering, design and systems integration, deployment, and operations. This often suggests that a separate organization, with a common core of members and a program management board, will be required for each phase of the project. If a special purpose vehicle (SPV) is used, then equity stakes can be varied throughout each phase to reflect the different responsibilities. A small equity stake, even as low as 2.5%, may permit a global company to join the team, offer its expertise at a commercial rate throughout the project development period, and appear within the team assembled for the bid.

The procurement team will be concerned about ensuring sufficient competition, so the project needs to be perceived as being attractive to potential bidders. The operations and design innovation rests with the tenderers, but the main aim of bidders will be to ensure that the initial cost, followed by a net positive contribution (marginal revenue less marginal costs), meets corporate objectives over the lifetime of the project. The innovation risk therefore rests with the bidders, but commercial investors (i.e., the consortium members) will need some general reassurance that the requirements can be met without taking risks that cannot be quantified and mitigated. The procurement team that is defining this project will reflect some policy innovations, including the use of peak period pricing. This must be taken into account to assess the expected traffic volumes, without ignoring the emerging long-term reality that unbounded traffic growth is unsustainable.

The existence of the project had been well promoted through presentations at international trade and investment fairs. Some of the main professional construction and transport telematics journals had contained editorials and short articles on the project opportunity. The expected global interest means that the project would compete alongside other similar opportunities in other countries. Potential bidders would consider the risk profile of this project and other projects when deciding whether or not to bid, complemented by its strategic importance, available resources, and potential teaming opportunities.

A competent, intelligent bidder known as Integrator ABC was busy preparing potential partnering profiles and an initial project risk register. This standard approach split the project into its constituent phases, from construction (road upgrade) and systems deployment, to operations. Each risk was scored from 1 to 5 according to the its probability of occurrence, its impact on cost and time, potential mitigation measures, and resulting changes on impact. Strategically, this helped in the search for potential team members, regardless of how they would ultimately be contractually linked. The risk register was sorted, and the top tier risks were used to compare this opportunity with other opportunities worldwide available to the bidder. Ensuring accuracy of the risk assessment was critical, so any potential uncertainties were validated through a local team member who was very familiar with the project from its inception. The allocation of enforcement risk had not yet been defined (or at least Integrator ABC did not know it).

Private investors need reassurance that the level of risks and its profile throughout the life of the operating period are acceptable. Table 7.3 highlights some of the questions asked within corporate management of Integrator ABC.

The FHWA outlined some of the new methods in procurement at a project level that included "the evolving relationships among public agencies, contractors, and private engineering firms, which are transforming risk allocation processes, quality control/quality assurance, and general contract administration procedures. Emerging delivery methods include the use of non-traditional procedures such as design-build contracts, public-private arrangements, maintenance and warranty

Table 7.3 Perspective: The Integrator

Is there a political risk? Could a change of government force a complete rethinking of traffic demand management policy that would impact the commercial viability of the project?
Are the taxation and value added tax (VAT) laws clearly defined, particularly for foreign companies?
Is it strategically important to have a local partner(s), and what should the profile of the partner(s) be?
Do we need a joint venture or equity-holding-plus-knowledge-transfer agreement as part of our market entry strategy?
Are there any long-term plans for transport infrastructure that may weaken the long-term viability of the upgraded route?
How are foreign consortiums regarded, and what level of local content would be desirable?
What is the quality and cost of obtaining regional resources, particularly civil works and structural engineering? Are there any projects competing for the same resources at the same time?
How much risk allocation is the procurement authority willing to bear?
What are the quality of the local vehicle registration databases, and are they centralized or fragmented? Are they adequate to correctly identify suspected violators?
What standards will be applied to key interfaces?
Can the charging policy be delivered, and will it be expensive to deliver?
Are there any local technology preferences for charging and enforcement? Is there a strong interoperability requirement that would dictate specific technology, and will this meet our operational requirements on accuracy and lifetime cost?
What partners would be needed to help set up the payment channels, particularly cash payment?
What legal constraints could we face if we require prepayments? Are their any banking regulations of which we should be aware?
What evidential strategies would be acceptable, and is there any relevant local precedence?
We need to market OBUs only to regular users. In this way, we recover the cost of the tag by maximizing the savings in transaction costs. Occasional users will be offered temporary licenses of one day, one week, or longer.
What is the penetration of customers with bank accounts, or is the local preference to deal with cash?
What other projects should we be considering, which would be more commercially attractive for us or our partners?
If we decide to bid for the charging and enforcement part of the works, will we form a system design group to develop the most feasible operating solution? If so, then our specially selected Red Team [30] will try to tear open the solution to expose its weaknesses so that we can improve it further before bidding.

requirements, and use of third-party consultants to perform contract management . . ." [31].

The development and deployment of the charging and enforcement infrastructure was only one part of the project, and bidders were reminded through their research that the aim was to enable high-quality accessibility between communities along the upgraded road, under demand-managed conditions. Charging and enforcement technologies are regarded as enablers of this greater objective.

7.7 Delivery and Operations

7.7.1 Countdown: From Integration to Launch

Section 7.5.3 outlined the relationship between the procurement route and the scale of the project. The development, introduction, and final preparations for launch will also depend on the scale of the project. A new road funded by tolls paid at barrier-controlled toll plazas can be completed in isolation of surrounding communities and businesses. If the plaza capacity assumes that ETC will be the primary means of charging at peak hours, and if the penetration of OBU accounts is not reached in time for the opening of the road, then the toll plazas will suffer from congestion and the benefits of ETC will be lost. An urban congestion charging system applied on existing roads is deployed in the full glare of the public and media spotlight. The design and implementation will pass through the same phases, but the failure to achieve many of the targets could result in loss of public confidence, poor quality of service on customer service channels operating at capacity, and overall poor compliance of road users.

Earlier sections outlined the earlier phases: agreeing on the objectives of the scheme; securing any necessary legislation; developing a procurement strategy; assembling a procurement team; defining the procurement and deployment project plan; raising external awareness of the project; developing requirements; creating tender documents; prequalifying tenders; issuing tender documents; and, through a mix of trials and negotiation, finally selecting the supplier. If the project has a high civil works content, then the selection of tenderers will focus on their ability to complete the civil works construction, and proportionately less attention will be paid to the systems and services relating to charging and enforcement.

Ignoring the civil works component of the example route, the development and introduction of the charging and enforcement solution will include the following activities:

- System and business process designs, with periodic client reviews;
- System architecture design;
- Traffic surveys to establish benchmarks;
- Critical subsystem proving (e.g., classification);
- Use case designs and test script development;
- Acquisition of permissions and site designs;
- Factory acceptance tests;
- Site acceptance tests (see Figure 7.3);
- High-volume performance testing;
- Failure mode and anomalous behavior tests;
- Central system functional testing;
- Payment channels and third-party interface tests;
- Introductory, silent running, and system readiness tests;
- Commence operations and maintenance;
- Launch.

These activities can be regarded as being part of the technical roll-out. The technical and administrative interfaces to all third-party suppliers, including payment service providers and external enforcement bodies, will be simultaneously tested to ensure that capacity is adequate and that process failures can be properly identified and trapped. A postlaunch "revenue cycle" test will be performed to

Figure 7.3 Off-road tests (Hong Kong, 1998). (Courtesy of Ian Catling Consulting.)

establish the effective operation of the end-to-end revenue collection and reporting regime, including interfaces to external agencies.

Schemes that are fully in the public domain, such as charging on urban highways or existing roads, will also require information and marketing programs directed towards the media and potential road users (see Figure 7.4). The geographic scope of the marketing will depend on the strategic position of the route in the regional road network. A local route has a smaller marketing area than does an interurban or economic corridor. Chapter 4 emphasized that a high-level awareness of payment options contributes to a high level of compliance. If the primary method of charging is with OBU-based accounts, then the penetration of OBUs among the target (regular) users is critical to achieving high levels of compliance. Note that if the expected proportion of regular users is low (e.g., less than 30%), then the penetration of OBUs among this segment will also be proportionately low (e.g., 70% of all regular users). Introducing the OBUs well before the opening of the road can mitigate this with associated strategies. The aim is to drive up the adoption of OBU-based charging accounts, as well as assist in the delivery of the marketing message. Achieving high levels of compliance will therefore depend on targeting the occasional users and ensuring high levels of occasional user preregistration before launch.

The German LKW truck tolling scheme systematically targeted foreign-registered truckers, either directly at service stations, or by direct mail to the larger companies, targeted advertising in the trade press, presentations at local trade associations, and stimulated awareness-building media coverage in each of the surrounding countries. A handful of researchers, each equipped with a PDA, can conduct a roadside survey of commercial operators using the existing route. Internet research can be used to provide contact details for each company, from which a targeting plan for the top 100 fleet operators can then be developed.

Figure 7.4 Raising the awareness of travel options.

The operational objectives of the example route included:

- Securing and equipping sites with a charging and enforcement system (Section 7.7.2);
- Developing all back office functions and external interfaces (Section 7.7.3);
- Ensuring high levels of compliance, sufficient capacity at scheme launch, and high levels of user awareness (Section 7.7.4);
- Establishing efficient ongoing operations (Section 7.7.5).

This then is the ecosystem of the example MLFF road user charging scheme. Barrier-controlled toll plazas are not an option, and, unless a national program of road pricing exists, each scheme will need to develop a local system meeting political, public, and economic targets.

The successful development of a road user charging scheme does not stop when it is launched. It merely passes into a new phase, in which policies can be refined, specific charging products for new user groups can be developed and tested, a regular monitoring program commences, and technology development monitoring and system scaling options are identified. The technical operations phase is described in Section 7.7.5. A private operator will also have developed expertise that can provide the basis for securing other road user charging projects and supporting the developing of road user charging policies elsewhere as part of its long-term strategic development. A public operator will also have developed expertise that can further support other local authorities to build local solutions.

7.7.2 Site Selection and Infrastructure

The desire to collect road user charges without toll plazas led to the emergence of multilane free flow charging and enforcement systems. The use of DSRC was complemented by geometrically similar vehicle detection, classification, and image-based enforcement systems. Chapter 3 demonstrated that this method of charging requires physical infrastructure, typically portal-style gantries across the full width of the road, for interurban highways to achieve maximum capture accuracy. New developments specifically dedicated to the urban environment have demonstrated that full road coverage can be achieved with roadside-mounted cantilever gantries (e.g., London), or even pole-mounted systems (e.g., Rome [32]) for narrow streets. The adoption of GNSS, which is not considered in this project, means that infrastructure would only be required for fixed enforcement sites and for isolated beacons to provide the OBU with position information where satellite information is poor, or where chargeable and nonchargeable roads pass close to each other.

The example project comprises highway grade road segments (i.e., not minor or secondary roads) between intersections. The charging policy has not been defined, but will be based on a fee for each segment, proportional to its length, and including a cap (capping is a typical pricing strategy) that limits the total for any one journey to approximately the charge for the equivalent of one-half the length of the road. Each charging location, referred to as a charge point (CP), will be independently specified, but, as far as practical, based on common modules, including the gantries themselves. Local highway and traffic engineering policies dictate and constrain the outline design, intended lifetime, crash-loading requirements, paint protection methods, access requirements, civil works requirements, and installation process.

The route of the road includes existing road segments (see Figure 7.5). Site selection needs to consider environmental impact, aesthetics (particularly for the suburban sections close to residential areas), availability of power and communications, road closure options, access and parking arrangements for service staff, ground conditions (elasticity, drainage, stability), and wind patterns.

The road will not be closed except for scheduled overnight maintenance. Alternate routes will be introduced for some of the maintenance periods, and a 15-minute rolling road closure will ensure 15 minutes of clear working time to lift, position, and secure each gantry (e.g., the method used for the Austrian LKW truck tolling system gantry installation).

The main cost factors of roadside infrastructure are typically:

- Number of sites (e.g., site-specific design, impact statements, permissions, civil works, and utilities);
- Width of road (e.g., quantity of gantry equipment);
- Availability of a secure (physically protected and reliable) power feed;
- Availability of secure, fixed-line communications (images can occupy 95% of available transmission capacity, depending on workload, split between roadside infrastructure and the central system);
- Accessibility to service personnel.

Figure 7.5 Charge/enforcement point location.

Road width is often a weak cost driver when compared with the overhead cost of civil works for each gantry. The above-ground cost will typically be 50% of the civil works costs, depending on permissions and accessibility to power and communications.

Bidders would expect to be granted a site visit prior to bidding. This ensures that any risks can be identified early in the tendering phase, and may expose some potential weaknesses in the procurement team's assessment of site feasibility. This is not unusual; some projects, such as Costanera Norte (Chile), M6Toll (United Kingdom), and Cross-Israel Highway (Israel), included civil engineering work through sensitive geological areas.

7.7.3 Back-Office Operations and Customer-Facing Processes

Chapter 6 describes the scope and operation of the central system, payment channels, and interfaces to external third-party stakeholders.

Bidders will define the detailed functional architecture of the central system, but it will contain interfaces to all components, including road users, the charging system, the evidential enforcement system, vehicle registration authorities, cash payment outlets, banks, credit card payment providers, payment clearing operators, and other regional charging schemes, as shown in Figure 7.6.

A central system core could be set up at relatively low risk to manage frequent events from distributed asynchronous sources, including fixed charge points, all

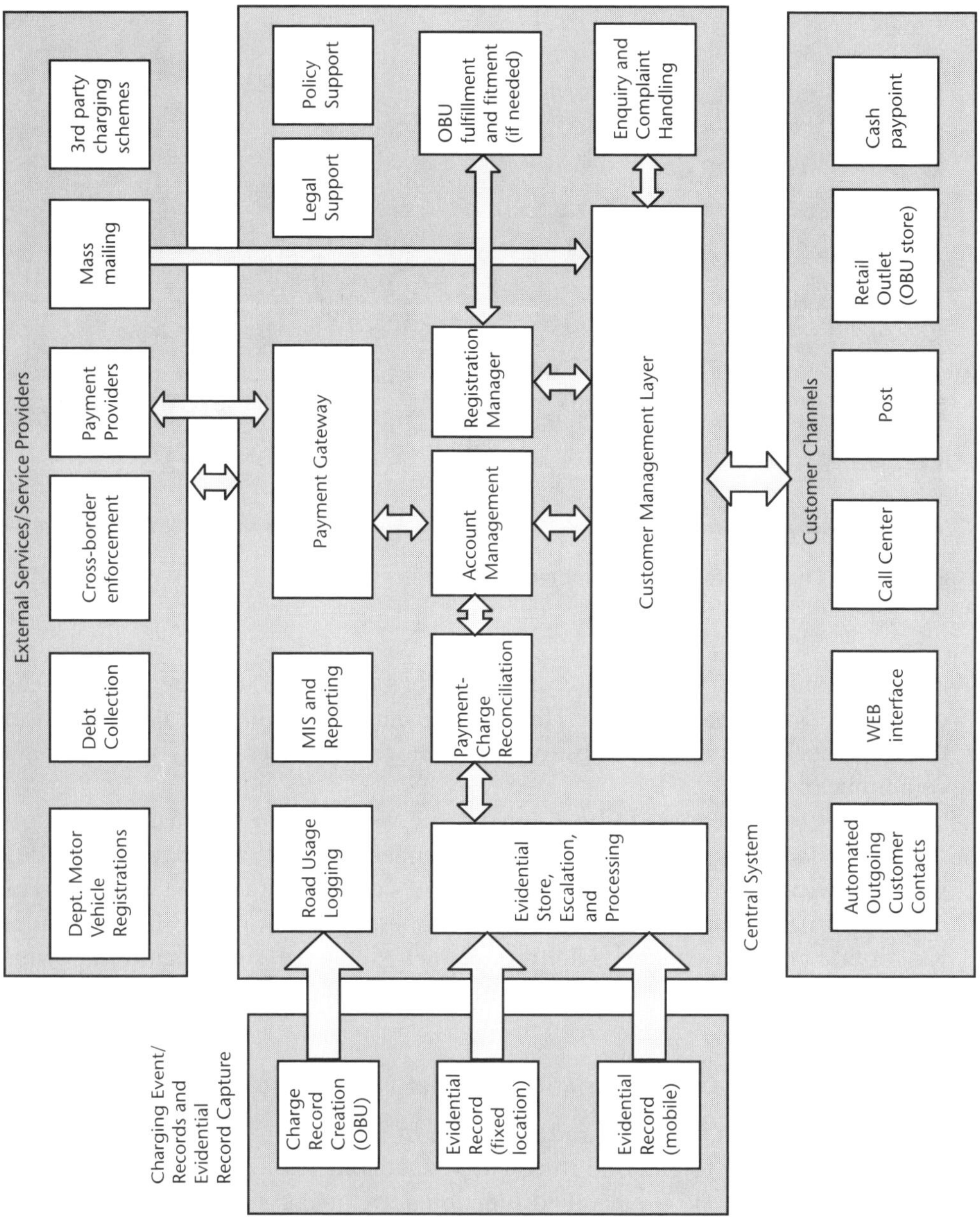

Figure 7.6 Interfaces to stakeholders and external systems.

enforcement points, payment service providers, departments of motor vehicle registrations, and other scheme operators. Local utility companies and mobile network operators (MNOs) have already shown this to work in a complex, multioperator mass-market consumer environment for high volumes of low-value transactions.

Cash payment channels can be rather expensive to operate, so operating concessionaires would be expected to capitalize on existing cash payment operators. Accepting cash as a prepayment in some countries (e.g., South Africa) would require a road network operator to register and operate as a bank, and therefore comply with strict banking codes.

Other service providers, such as fuel card operators, used by nearly all of the truck operators, could offer prepayment and postpayment. Although technical synergies are obvious, it would also be necessary to confirm that account holders of these existing service providers would enable their data to be used for the payment of road user charges on the upgraded route. Data protection provisions may constrain data sharing.

7.7.4 Fulfillment and Managing Start-Up Demand

7.7.4.1 Account Types

Users (i.e., their vehicles) vary according to the frequency by which they use the example route. There will also be a separate distribution for the number of road segments against frequency of use. Chapter 3 showed that it was not economically efficient to aim for 100% penetration of tag-based accounts for two reasons:

- The marketing cost of reaching infrequent travelers compared to regular users is expensive.
- The total savings from lower OBU-based transaction costs is not economically justified from the lower transaction rate for occasional users. An image-based account is cheaper for the operator to maintain for occasional users over the same period.

Reference [33] presents the 80/20 rule for one MLFF facility in Australia and suggests that OBU-based accounts can be operated more cost-effectively than can image-based accounts, such as those that depend on preregistered vehicle number plates. OBU transactions can be accepted by the scheme's central system with virtually no error. Chapter 6 describes the charging data capture and collection function. Image-based accounts may have been set up using more expensive registration channels (e.g., IVR and call center), and each registration is for a limited number of vehicle trips on the road network, since it is marketed towards ad hoc and occasional users. The revenue loss associated with each type of transaction is also different, as Table 7.4 shows.

Fulfillment (distribution and management of OBUs) is aimed at meeting OBU users' requirements, rapidly and at low cost. The type of scheme will dictate the fulfillment strategy. The Singapore ERP, Swiss LSVA, and German LKW truck tolling schemes require vehicles to be equipped with OBUs by an agent of the scheme operator. For commercial vehicles, it is possible to calculate an opportunity cost (i.e., the loss of productive time while the vehicle is temporarily off the road),

Table 7.4 The 80/20 Rule (Melbourne City Link)

Account Type	*OBU-Based*	*Image-Based*
Operational costs	20%	80%
Traffic	87%	13%
Transaction-related revenue loss	0.5%	99.5%

Source: [33].

plus the cost of the installation work. The benefit of controlled installation is that the OBU can be confidently associated with the vehicle, and, for truck tolling schemes, reflects the high charge level per vehicle trip.

Many ETC schemes rely on the user installation of the OBU in the vehicle. This may be prone to mistakes, such as the user not installing the tag, or installing it in the wrong vehicle. Failure rates, discussed in Chapter 3, may be as high as 0.3% of all transactions, although the effects of random errors may be reduced through repairing lost transactions (one of the failure modes) if a journey contains several consecutive charged road segments.

The fulfillment strategy should aim to establish a substantial proportion (at least 40%) of OBU-based user accounts, and distribute OBUs to users before charging starts. The higher the starting value, the lower the potential volume of enforceable events when charging does start, due to users not registering an OBU. If the charging policy requires an initial account payment, then a proportion of this could be waived if registration is made a few weeks before charging starts. User segments can be targeted to ease the load on the registration and fulfillment channels.

Typically, the encoding of the OBU exists in two stages: initialization (by the manufacturer), and personalization (by the scheme operator). The manufacturer-specific information could include a unique OBU ID, battery initialization information, and any branding and packaging ordered by the scheme operator. The scheme operator has the option of proceeding with generic personalization, such as account ID, date, and vehicle class type, even before the OBU is allocated to any specific user account. Other vehicle-specific information can be added, including its license plate number [34] and any other declared vehicle characteristics. All vehicle-specific information that is encoded increases the logistical overhead to ensure that the OBU (including its vehicle-specific data) is applied to the intended vehicle, although this allows the match to be confirmed independently of the host charging system, including third-party charging schemes.

7.7.4.2 Managing Start-Up Demand

MLFF schemes are designed to handle expected demand at all external interfaces: charging events, violation events, payment events, customer contacts and inquiries, enforcement actions, and revenue recovery. This demand is usually calculated under steady-state conditions, or at least under predictable growth conditions. Capacity planning reduces the impact of the initially high demand (referred to as a bow wave in Section 6.3.5) that could swamp the back office, payment channels, and enforcement channels.

The estimated split of account types needs to be matched by the capacity of back office customer services, to manage initial demand, or at least to ensure that it does not exceed available capacity. Effective prelaunch marketing can help stimulate early demand. Examples of such marketing include: imposing registration deadlines for residents (Transport for London); encouraging OBU acceptance by bundling with mobile phone contracts (Costanera Norte, Chile); scheduling vehicles for installation (Land Transport Authority, Singapore, and LKW Germany); and opening the route early (Melbourne City Link). Fulfillment must populate the OBU delivery channels if retailers are used, although this will depend on the OBU personalization strategy. If users need to apply for exemptions, then the application process would need to be based on trusted agents with an incentive to ensure that OBUs are properly personalized, and, if needed, installed on the appropriate vehicles (e.g., the Singapore ERP scheme and the German LKW truck tolling program). Web-based registration, OBU shipments by mail, and user installation of OBUs serves most road users in many schemes. Mandatory national schemes based on distance traveled will require a different fulfillment strategy that may also include vehicle manufacturers and accredited installers.

In the example project, the procurement team was particularly interested in whether or not the proposed operation could provide additional capacity at short notice for any change in the project, including initial launch, up to the long-term change of charging policy. If road users genuinely do not understand payment options, then this will increase the volume of enforceable events. This could increase the volume of complaints, drive up the costs (to the operator) of inquiry channels, generate poor press coverage, and could trigger further attempts at violation. The procurement team has an active interest in ensuring that road users have a high level of awareness of payment options, and the probability of being detected if evasion is attempted. Techniques used to measure the level of awareness include market surveys, questionnaires, and general monitoring of the media.

7.7.5 Operations

The operations of a road user charging scheme is usually divided along functional lines: maintenance of road-based equipment, partner management, enforcement operations, payment channel management, revenue collection and billing, marketing, reporting, ongoing project management, legal, media relations and marketing, core IT facilities, process and property security, and administration. A small scheme (e.g., less than 100 people) would rely on some individuals having two or more roles, including procurement of system upgrades. Large schemes (e.g., more than 300 people) would depend on well-defined groups, each dedicated to a specific function.

A management information system (MIS) would monitor system health and the achievement of key performance indicators (KPIs) that would be routinely reporting to the managing road authority. At the time of system design, the frequency of reporting and number of reported parameters would be defined. These parameters may include: traffic statistics (split by vehicle classification); journey times at hourly intervals; revenue collection and distribution between channels; volume of violations detected and resulting actions taken; cross-border actions taken and results; planned and unexpected system downtimes; account status by volume and type; results of disaster recovery sites drills; planned system upgrades and timing; traffic accidents; status of route monitoring equipment (accident detection, surveillance); and adjacent route traffic demands.

A national road user charging scheme such as LKW Germany or LSVA Switzerland requires significant resources to maintain the road-based equipment, regardless of charging technology. Maintenance is preventative and remedial, and contributes to maximizing service availability measured at each location, either separately or as an average across the network of locations. A combined charging and enforcement point, or, for GNSS, an enforcement point, would typically be continually monitored, with inspections made at least every 2 weeks for a visual check of integrity. A regional maintenance operation would typically include a full management structure, supported by management tools, asset management programs, corrective maintenance infrastructures, planned and proactive maintenance programs, asset replacement and updating programs, spare parts management and logistics using central and localized warehouses, trained systems and infrastructure field engineers, remote system monitoring and diagnostics, asset condition monitoring, reporting, review, and training infrastructure. Resources would be configured to meet defined response time targets. Monthly and semiannual preventative maintenance cycles would focus on equipment security, visual cable checks, equipment alignment checks, local diagnostic checks, air filter replacement, UPS battery integrity checks, and any local firmware upgrades. Mobile enforcement vehicles would receive maintenance checks based on records of engine running times or distance traveled.

The external infrastructure, OBU operation, and media reports represent the contacts of the scheme with road users. Any of these could create favorable or adverse impressions. Poor site planning (particularly near residential areas) of any roadside infrastructure can lead to criticism of the scheme. For a variety of reasons, new roads suffer less specific criticism of roadside infrastructure (compared to adding new street furniture to existing roads), although environmental impact statements would still be required. Placing infrastructure in the urban environment as part of the initial deployment or scheme extension places stress on public relations with residents and local landowners. The operator of the example route does not expect any additional sites, and restricts maintenance to short visits on a monthly basis to minimize disturbance.

7.8 Scaling

There are at least five directions of scaling:

- Policy (e.g., more complex charging rules, new user categories);
- Geographic, incremental (e.g., growth to include linear extensions to the charged route, adjacent primary routes or areas);
- Geographic, acquisitive (e.g., incorporation of remote roads or regions, each with its own distinct operations policy, charging policy, and branding);

- Geographic, contractual (e.g., contractual interfaces with other schemes);
- Application to support value-added services.

The operating company may be set up for a specific purpose (e.g., an SPV) and would therefore be strategically constrained to provide services to meet contractual commitments over the concession period. It is therefore likely that the company would not have the resources or the mission to focus on projects along any of the five directions of growth.

A change of policy towards charging all vehicles on all roads according to distance traveled will affect the operations of the existing road. Strategically, the operator may be able to change the operations to accept different forms of charging data from OBUs that declare charges according to specific road segments or provide a simple measure of location. Other scaling events may occur in the lifetime of the concession:

- Tenders for operating other existing toll roads may be launched. The operating capacity of the central system for the example route may be scaleable to incorporate the new route and benefit from the economies of scale that have been developed. How would an additional road be added? Is there sufficient technical capacity to provide back office services to a second road?
- Trials for new forms of charging technologies may be offered by the regional government, and early experience of this may help bid proposals for similar future projects.
- Similar projects may arise internationally, and the expertise developed to integrate and operate the example road could be provided to bid for these projects. It may be possible to reuse some of the key back office function designs, since continuity of procurement from one scheme to the next reduces the time of assembling a procurement team.
- Contractual interfaces with other scheme operators could be required. Interoperability will be offered to other road users registered with other road operators.
- Discounts on road user charges may be offered for low-emissions vehicles. Although this could be seen as contrary to demand management principles, this depends on whether the policy aims to reduce emissions or manage demand. The policy may be regionwide, so it will impact the example route and reduce the revenue from these vehicles.
- New charging technology may be developed that complements the delivery of other road services relating to safety, fleet management, and navigation. It may be possible to create a commercially viable bundle of services, although future business models for commercial services would need to be defined first.
- A regional transaction clearing facility may be established to reconcile chargeable events (or charging data from future GNSS schemes), or to route this information to the service provider with whom the account is held. It will be necessary to determine the most effective form of data clearing between the various schemes across and within regional borders.

- The volume of road users from other regions may grow. It will be necessary to deliver effective cross-border enforcement perhaps through harmonizing the form of evidence, while accommodating the constraints of different national legal systems.
- A national road user charging architecture may be developed. This could introduce specialized service providers, such as OBU issuers and organizations that focus entirely on processing evidential enforcement information. It may be necessary to establish agreements to support business, data processing, billing and payment interoperability between all schemes.

Regional growth in road user charging may require operators (public and private) to assess their own organizational and operational flexibility in being able to participate in any of these opportunities. The political uncertainty of some operators will make it difficult to anticipate the order in which the events will occur (if at all) and the impact on the example route.

Other cost drivers would be visible when scaling an operation, including road infrastructure maintenance, traffic incident detection and response, and value-added services (VAS). All these would typically require an operator to have expertise, a suitably equipped and staffed control room, field equipment and staff, and well-defined strategies and tactics to ensure efficient operations. They would contribute to operational cost, but would also help to maximize acceptance of the road and the charging regime.

7.9 The Future

The example route was conceived as a commercially viable toll road with the addition of peak period pricing to manage demand. Revenues from the road were aimed at funding the cost of the road upgrade and ensuring predictable, high-quality travel. Maximizing throughput was not the objective, and the business model would have taken this into account. The land use development strategies for the region are being continually developed to ensure that they complement roads and transport development policies.

The difference between different charging policies may be subtle (see Chapter 2). Terminology includes electronic fee collection, congestion charging, road user charging, road pricing, and electronic toll collection. The most obvious difference is usually geographic or organizational. The objectives usually reveal the strategic differences, and these objectives include revenue collection, funding of infrastructure or operations, pricing to manage demand, and usage-based charging as part of direct taxation policy. In principle, any of these forms of road user charging may be implemented regionally, so the operator of the example route needs to decide whether or not to participate. Strategically and operationally, the differences may be so large that there would be no efficiency gains and therefore no economic advantage in participating. However, if there is alignment of the charging and enforcement systems due to a similar charging policy and enforcement regime, then it would make strategic sense to compete for any contracts that may be awarded (unless there are SPV constraints). Offering a lower-cost means of processing charging data, or enabling any of the back office functions to be reused, may enable a local scheme that may otherwise be too expensive to operate (see Chapter 6).

An all-vehicle, all-roads policy is considered to be long term, and depends on political acceptability, technology capability, and public acceptability. Its implementation will be regionwide, although most likely not for at least 10 years. Forecasting the evolution of local, regional, and national policy could be challenging. The development of new technologies for charging and enforcement over the lifetime of the concession would also be fraught with uncertainty, although a watching brief could be maintained through local ITS societies and regional road research associations, such as the Transport Research Board (United States); ERTICO (Brussels, Belgium); International Bridge Tunnel and Turnpike Association (IBTTA, Washington); ASECAP (Paris); and other local associations in the Middle East, South America, South Asia, and Southeast Asia.

The impact and relevance of the future therefore depends on the strategic objective of the operator of the example route. The length of the operating period will necessarily require the technologies for charging and enforcement to be reviewed and refreshed. At any time, any of the scaling events listed in Section 7.8 may be relevant, so some operational and organizational flexibility must be shown in the long term. This may require a change in the purpose of the operating company.

It is up to the operator to monitor changes that may have a strategic impact (positive or negative) on the organization, and as Gibson [35] remarked that ". . . the future is already here. It's just not very evenly distributed. . . ." This is a cautionary note for any operator who thinks that external changes will always impact others first.

7.10 Summary

The story in this chapter focused on the local procurement of a road upgrade, funded by tolls and varied by peak period pricing. The time the project was conceived until its final delivery represented less than 20% of the overall duration of the contract. The procurement team grappled with procurement in a policy context that was due to change over the lifetime of the concession, including the possible introduction of an areawide user-pays scheme.

A shortfall of public funds for roads, and recognition that the route could be operated commercially by a private operator, led to the decision to develop a project that would be attractive to international tenderers when compared with other global investment opportunities. Moderate public and political support and confirmed operations feasibility was sufficient to launch the procurement.

The different aims of the regional transport authority and private tenderers were highlighted. The procurement specification for the charging and enforcement solution was sufficiently descriptive to ensure service delivery quality, and included requirements on the technologies for charging and enforcement, underpinned by the appropriate legislation, particularly relating to the use of image-based evidential enforcement to help identify vehicles and the party liable for payment. A weakness here could have postponed the project by at least 12 months. Furthermore, the charges were differentiated by vehicle class and were sufficiently (but not perfectly) measurable, so manual checking was needed for some class discrepancies.

The procurement did not include a pilot, since the operational feasibility had already been shown in other similar projects. The means of charging was based on an approximation to distance traveled, although a maximum charge was applied to ensure public acceptability and consistency with the cost of public transport for trips of similar length. DSRC was used for account identification and charging, but it was expected that GNSS would be feasible during the lifetime of the concession. If it made commercial sense in a changing policy context, a technology review strategy would consider this alternative approach.

Requiring adequate capacity for central system functions was regarded as critical to preserve service quality towards road users. However, the MLFF approach to charging means that direct cash payment is not an option. The payment and fulfillment strategy instead depends on increasing the awareness of all possible payment methods, backed up by marketing to ensure that road users know that they must preregister vehicles to use the route. Misunderstanding only leads to violations being incurred, and the possible loss of public confidence in the scheme.

It could reasonably be expected that over the life of the concession, there would be pressures on the operator to provide expertise to support other tenders, and to offer the benefit of the economies of scale developed on the example route to other similar schemes. Regardless, policy shifts may require the operator to change its strategy, at least to maintain the commercial viability of the operation.

References

[1] The World Bank Group, *Toll Roads and Concessions*, http://www.worldbank.org/transport/roads/toll_rds.htm.

[2] Underhill, P., "Road User Charging and Trucks—An Update on U.S. Programs," *Proc. IBTTA Spring Technology Workshop*, Edinburgh, U.K., June 14, 2005.

[3] Schelin, E., I. Gustafsson, and P. Blythe, "European Road User Charging for Heavy Goods Vehicles—An Overview," *Proc. 12th World Congress on Intelligent Transport Systems*, San Francisco, CA, November 6–10, 2005.

[4] Land Transport New Zealand, *Road User Charges and Light Diesel Vehicles—Factsheet 38*, June 2005.

[5] Government of New Zealand, *Road User Charges Act 1977 and Its Amendments*, 1977.

[6] Patchett, N., et al., "Assessing the Use of GPS for Congestion Charging in London," *Traffic Engineering & Control*, Vol. 46, No. 3, March/April 2005.

[7] European Transport Pricing Initiative Newsletter, *GPS on Trial in Copenhagen*, No. 4, September 2002.

[8] Jung, S., "The German Heavy Vehicle Tolling System," *Proc. IEE Road Transport Symposium*, December 5–6, 2005.

[9] Gerondeau, C., *Transport in Europe*, Norwood, MA: Artech House, 1997, p. xl.

[10] The World Bank Group, *Toll Roads and Concessions*, http://www.worldbank.org/transport/roads/toll_rds.htm.

[11] Department of the Environment Transport and Regions, *Environmental Appraisal of Development Plans—A Good Practice Guide*, 1993, Documents Div:UK/P93/213/5.

[12] Small, K. A., *Urban Transportation Economics*, Luxembourg: Harwood Academic Publishers, 1992.

[13] Federal Highway Administration, Office of Policy Information, *Highway Statistics*, 1997, http://www.fhwa.dot.gov/ohim/hs97/hs97page.htm.

[14] Texas Transportation Institute, *2005 Urban Mobility Study*, 2005, http://mobility.tamu.edu/ums/report/.

[15] Xinhua News Agency, "China's Private Car Ownership Tops 10 Million," June 14, 2003.

[16] BBC News Online, "China Introduces Chopstick Tax," March 23, 2006.

[17] Goldman Sachs, "Global Automobiles—The Chinese Auto Industry," February 21, 2003.

[18] Singapore Land Transport Authority, *Vehicle Ownership, Vehicle Policies and Schemes, Vehicle Quota System*, 2006, http://www.lta.gov.sg/motoring_matters/motoring_vo_policynschemes_quota.htm.

[19] Transport for London, *Congestion Charging—Penalties*, 2006, http://www.cclondon.com/penalties.shtml.

[20] PATAS, *Regulations for Congestion Charging (Charges and Penalty Charges) London Regulations—Consolidated Order*, 2006, http://www.parkingandtrafficappeals.gov.uk/regulationsCongEnforce.htm.

[21] Government of Ontario, Canada, Highway 407 East Completion Act 2001, c. 23, Schedule B, S.20(4).

[22] Official Journal of the European Union, *Directive 2004/52/EC of the European Parliament and of the Council of 29 April 2004 on the Interoperability of Electronic Road Toll Systems in the Community*, L166, of April 30, 2004.

[23] "Northern Indiana Leaders: Show Us the Money, More of It," *The Times of North West Indiana*, January 24, 2006.

[24] Federal Communications Commission, "FCC Allocates Spectrum in 5.9 GHz Range for Intelligent Transportation Systems Uses—Action Will Improve the Efficiency of the Nation's Transportation Infrastructure," Press Release, October 21, 1999, ref. report #ET99-5.

[25] OmniAir, *OmniAir—Ride the WAVE (Mission Statement)*, 2006, http://www.omniair.org/mission.html.

[26] Pavlidis, I., et al., *Automatic Passenger Counting in the HOV Lane*, Minnesota Department of Transport, 1999.

[27] Directorate General for Transport (European Commission), *Towards Fair and Efficient Pricing in Transport—Policy Options for Internalising the External Cost of Transport in the European Union*, 1995.

[28] MEDIA Final Report, *Management of Electronic Fee Collection DSRC Interoperability in Alpine Region (MEDIA)*, March 2005, pp. 52–53.

[29] Hill, T., *Manufacturing Strategy—The Strategic Management of the Manufacturing Function*, 1994, pp. 59–104.

[30] U.S. Department of Defense, *DoD Information Operations Red Teaming*, draft DoD Directive 3600.3.

[31] U.S. Department of Transport, Federal Highway Administration, *Common Ground: Construction Management Practices in Canada and Europe*, Summer 2005, doc ref #FHWA-IF-05-029, http://www.fhwa.dot.gov/construction/scan05.cfm.

[32] Forestini, F., and M. Tomassini, "Access Control in Rome," *Traffic Engineering and Control*, July/August 1999.

[33] Daley, K., "Open Road Tolling, An Australian Perspective—The Melbourne City Link," *Proc. IBTTA Spring Technology Workshop*, Miami, FL, June 2004, Slide 16 of 18.

[34] Department for Transport (United Kingdom), *OMISS Volume 3 (On-Board Unit to Roadside Equipment Communications)*, November 2005.

[35] National Public Radio, Talk of the Nation, interview with William Gibson, author of *Neuromancer, Mona Lisa Overdrive*, and *Virtual Light*, November 30, 1999.

CHAPTER 8

Case Studies

8.1 Introduction

Throughout the previous chapters of this book, a number of past and present examples of road user charging and electronic tolling have been used to illustrate where experiments or implementations of new forms of charging technology and innovative demand management techniques have been undertaken.

It is not possible to review and describe every single significant trial of charging technologies, since the profusion of such systems in the past decade would run into hundreds of examples. However, the authors have selected a range of case study examples to illustrate the breadth of system types and the technologies deployed. These are divided into the following categories:

- Urban demand management;
- Small-scale toll schemes;
- Regional and interoperable tolling schemes;
- Heavy goods vehicle (HGV) charging schemes;
- HOT and HOV lanes;
- Significant trials and pilots.

8.2 Urban Demand Management

8.2.1 Singapore

When considering urban demand management, Singapore is often (and rightly) cited as the pioneer of successfully implemented innovative demand management schemes. The Singapore government's history with urban demand management goes back more than three decades, prior to the introduction of the current ERP scheme in 1998, with the highly effective area licensing scheme (ALS), first introduced in 1975 [1].

Prior to the introduction of the ALS, the Singapore government had recognized that economic growth was beginning to accelerate, and the rate of growth of car ownership was predicted to reach 10% per annum by the mid-1970s. Singapore was experiencing rapid economic growth, along with significant increases in car ownership and car use, as was Hong Kong (as described in Section 8.7.1). Singapore explored in the early 1970s fiscal regulations, such as making the first registration tax a sizeable proportion of the value of the vehicle, and increasing the annual license fee (ALF) and fuel duty. Each of these measures was criticized by the public, since they were seen as instruments that penalized the less well-off car owners, who were less likely to use their vehicles to travel into the central business district (CBD), which was where the key problems with congestion occurred [2]. The policy in Singapore gradually moved away from a strategy that tried to manage demand by emphasizing vehicle ownership costs, to a policy that was a better balance with costs of usage.

The Singapore government introduced the world's first area licensing scheme in 1975, to begin the process of bringing car usage into the demand management equation. The ALS initially covered the most congested road networks leading into the CBD, initially including 22 entry points. As illustrated in Figure 2.6, each entry location was marked by an overhead gantry that clearly indicated an entrance point to the ALS and its operating hours. Red lights were illuminated on the entry gantries when the scheme was in operation [3, 4]. By the time the ALS scheme was replaced by an electronic road pricing scheme in 1998, the physical area of the CBD covered by the ALS scheme had grown by 25%, and the number of entry points had increased from 22 to 31.

The ALS enabled drivers to enter the restricted zone (RZ) by the purchase and display of a paper license. Daily or monthly licenses could be purchased, with the price of a monthly permit being equivalent to 20 one-day passes. Daily passes could be purchased at roadside sales booths, while monthly passes could be purchased from post offices.

The licenses had various shapes to indicate the class of vehicle, and the color varied from month to month to indicate the validity of the pass. The passes covered all classes of motor vehicle, including motorbikes and scooters, and were displayed on the windshields of vehicles or on the handlebars of motorbikes and scooters. The color coding helped enforcement officers to perform manual checks of vehicles as they crossed the gantry point into the restricted zone. These checks were generally made by observing moving vehicles passing into the zone. If a vehicle was detected with an invalid permit, or was not displaying a permit at all, then the enforcement officer recorded the vehicle's license plate, and a summons was issued. Offending vehicles were not stopped at these entry points, since this would disrupt traffic flow. This process of detection and recording obviously was open to human error.

The ALS operated for a period from 7:30 a.m. to 9:30 a.m. daily, except on Sundays and public holidays, and the cost of entering the RZ was initially \$2 for taxis and \$3 for cars, in Singapore dollars. Initial peak period traffic volumes fell by 75%, and traffic before 7:30 a.m. increased by 23%, with a lesser increase after 9:30 a.m. After three weeks of operation, the weekday operating period was extended from 7:30 a.m. to 10:15 a.m., to deal with the increase in traffic after the end of the RZ operation. Plans to offer free access to the zone for high-occupancy vehicles were abandoned. It was found that teenagers were renting themselves as bodies to enable a vehicle to reach high-occupancy status, boarding vehicles just before entering the RZ and then alighting soon after. This caused traffic chaos, and was deemed a possible danger to those offering themselves for rent. Once the scheme settled down, the reduction in traffic entering the restricted zone was approximately 44%.

The reduction in traffic had lessened to approximately 31% by 1988, so in 1989, the ALS was fundamentally revised. The RZ was now operational not just during the morning peak hours, but also in the evening between 4:30 p.m. and 6:30 p.m. The cost of permits was also reduced to $3 for a full day and $2 for a part day (in Singapore dollars), since the daily pass had been increased to $5 in Singapore dollars in prior years. Under this revised scheme, the morning peak traffic increased in volume by just over 10%, and the evening peak was reduced in volume by 56% [3].

The effectiveness of the ALS in bringing actual vehicle usage into the demand management equation encouraged the Singapore government to explore the possibility of introducing a more flexible and automated road user charging system through electronic road pricing. The government specified requirements for an ERP scheme in 1993, and in 1994, three consortia were asked to set up demonstration systems that could be evaluated on an unopened stretch of an expressway. Each consortium provided a DSRC gantry–based solution for evaluation and included the challenging requirement to incorporate motorcycles and scooters in the charging scheme. This placed additional requirements on the scheme for the detection, classification, and enforcement for two-wheeled vehicles, and the design of a suitable OBU. Figure 8.1 shows a design from the 1994 trials. These were the first comprehensive trials of multilane charging systems, and are described in more detail in [5]. A contract for the installation of the electronic road pricing system was awarded in 1995. A test program using a fleet of 250 vehicles and 12 gantries was undertaken to prove the performance of the selected system. The mass production of the IVUs (as OBUs are called in Singapore) and gantry equipment began after the extensive testing. The control center and central systems functions were awarded as a separate project [6].

Figure 8.1 OBU design for motorcycle from 1994 trials, Singapore. (Courtesy of Ian Catling.)

The ERP scheme utilized a DSRC solution that included an IVU with a smart card. Permanent IVUs are installed on all domestic vehicles, while foreign vehicles may use temporary IVUs. Two overhead gantries, located close to each other, are placed over each entry point. They carry antennas, vehicle presence detectors, optical detectors or line sensors, and enforcement cameras. Every time a vehicle passes through an entry point, the antenna interrogates the IVU in the vehicle, verifies its validity, and instructs it to deduct an appropriate entry charge from the stored value of a smart card inserted in the IVU. The IVU includes a liquid-crystal display (LCD) to show the current balance on the smart card. Figure 8.2 shows an OBU from the ERP scheme. Vehicles without an IVU, or with an insufficient balance on their smart cards, will be identified by means of cameras. Multilane operation is a requirement, so that vehicles at any point under the gantries are charged (at speeds up to 120 km/hr).

The Singapore solution utilizes a two-gantry solution, in which the transaction is initiated at the first gantry, and verified and completed at a second gantry some 30m further along the road, at which point enforcement may also be activated. The two-gantry solution is a pragmatic way of dealing with the time-critical processes required for a transaction (see Chapter 3), and of achieving accurate classification and enforcement (see Chapters 4 and 5). The solution selected for Singapore had to deal with mixed traffic that included scooters and motorbikes, which made these functions little more challenging. The aesthetics of the gantries are one of the very few criticisms of the scheme. Figure 8.3 shows a typical gantry arrangement on entry to the CBD.

The ERP scheme started in September 1997, with 680,000 vehicle owners being sent to one of the 200 authorized IVU installation centers. The scheme went live

Figure 8.2 ERP on-board unit, Singapore. (Courtesy of Ian Catling.)

Figure 8.3 ERP gantry, Singapore. (Courtesy of Ian Catling.)

in May 1998, with 98% of registered vehicle owners having an IVU installed. Figure 8.2 shows a typical IVU. The scheme was launched using the same price levels as the ALS. An administrative charge of $10 was levied on drivers who failed to maintain a balance on their smart card, and the fee for nonpayment was $70. The levels of the charge were set to try and maintain target speeds on key roads leading into the CBD and on some expressways. Target speeds from 20 to 30 km/hr were set for the roads leading into the CBD, and speeds from 45 to 65 km/hr were set for the expressways. If the speeds were too low, then it was assumed that traffic volumes were too high, and the charge could be increased; if the speeds were too high, then it was assumed that the charges were too high and too few vehicles were choosing to pay the fee, so it was decreased [7].

The success, vision, and leadership that Singapore has provided in introducing road user charging and other innovative demand management measures such as the ALS cannot be overstated. Singapore showed it was possible to introduce a large-scale ERP scheme in a congested business district and make it work with technology that was just emerging in the mass market. This successful scheme has demonstrated that charging is a powerful demand management tool that is able to balance traffic on different routes to achieve optimum flow.

8.2.2 London

London has considered the use of some form of road user charging to manage demand in the city on several occasions. In 1993, the Government Office for London (GoL), jointly with the U.K. Department for Transport (DfT), commissioned a major study on congestion charging options for London [8]. This report looked at the economic, traffic modeling, and technology options for a major charging scheme for London, and it was the first significant review and analysis of new technical options for road user charging in the 1990s. The report concluded that the technology was not yet sufficiently mature to support a fully automated road user charging scheme in London.

A new study into the feasibility and options of road user charging for London was initiated in 1998. The GoL examined road pricing in more detail, and considered two key factors: (1) legislation had been promised by the new government to enable local authorities to collect road user charges and private nonresidential (PNR) parking charges, enabling any new mayor of London to apply for such powers; and (2) one of the mayoral candidates was explicitly in favor of road pricing. The study, known as the Review of Charging Options for London (ROCOL), reported in 1999, with several options for schemes being considered, and a key recommendation that a fully electronic scheme using in-vehicle equipment (as in Singapore) or OBUs (as in Trondheim) was not feasible in the given time frame [9]. The proposed solution was to start with a small zone in the center of London, and to initially use a camera-based registration and enforcement scheme [10, 11]. This solution did not require vehicles to be equipped with in-vehicle boxes, thus dealing with regular and occasional users of the scheme in the same manner. Chapters 2 and 3 discussed this issue at length.

The scheme was specified and procured in a very short time frame of just over 3 years, and went live on February 17, 2003, 5 months after the United Kingdom's first scheme in Durham. The London scheme is the first in the United Kingdom to use an area charge as a demand management measure, and is presently the largest scheme of its kind in the world [12]. The Central London scheme is an important initiative that fulfills a goal in the government's 10-Year Transport Plan to introduce schemes to reduce congestion and to fund complementary public transport services. The scheme is a significant milestone in the development of effective, sustainable measures for reducing traffic congestion. The success of the scheme and the experience gathered from its operation adds to the debate about congestion charging solutions for other parts of the United Kingdom and around the world.

The London scheme charges drivers £8 per day if they enter a defined area around central London, known as the congestion charging zone (CCZ). The charge initially was £5, but was increased in July 2005 to maintain its restraining effect on traffic. Drivers using the zone are asked to register their intent to travel before they enter the zone, and to prepay the charge. The zone operates from 7:30 a.m. to 6:30 p.m. on weekdays. The zone entry points and the retail outlets that accept payments are identified with the U.K.'s national charging symbol—a white "C" on a red background (see Figure 8.4). Payment and registration can be done by visiting a shop or kiosk, on the Internet, by telephone, or by an SMS message service using mobile phones. Residents have a 90% discount, and registered disabled drivers and some other categories (e.g., taxis, emergency vehicles, some key workers, and some alternative fuel vehicles) are exempt. Discounted and exempt users total 30% of traffic (39,000 vehicles a day).

The congestion charging zone operates with monitoring by 688 camera units at 203 sites [13]. These sites act as a boundary on roads entering the charge zone, with cameras positioned on bridges across the River Thames, and with a number of mobile camera units. Each checking point takes both an image of the license plate and a wider context image that shows the location of the photograph, and

Figure 8.4 Congestion Charging sign, London.

vehicle make and color. ANPR automatically reads the license plate, and if the license plate matches a record of payment for that day, the image is discarded. The images are retained if no record of payment is found. Chapters 3 and 4 show examples of the camera locations and images generated. For drivers unfamiliar with the scheme or unable to register and pay in advance, there is an opportunity to register and pay up to midnight on the day of travel (since revised to permit "pay next day"). If no record of payment is made by that time, then a £100 PCN is sent to the registered vehicle owner. By July 2005, representations (a notification of opposition or disagreement with a PCN) had been made on approximately 20% of PCNs and appeals on approximately 2%. Figure 8.5 shows a typical entry point onto the congestion charging zone.

Figure 8.5 London Congestion Charging entry point.

The scheme has had a significant effect on the quality of life and travel in Central London. Three years into the scheme, overall congestion is reduced in the zone by 30%. Traffic entering the charging zone has been reduced by 18%, and car trips by 35%. Bus patronage has significantly increased, partly because 200 additional buses were introduced when the congestion charging zone went live. Bus travel has become more reliable, and journey times have improved. The scheme has annually generated net revenues in excess of £100 million, which, by law, must be reinvested into transport improvements in London. Using ANPR as the primary means of checking payments and enforcement was the initial uncertainty in the system. The technology seems to have worked well, but the need to manually read the images that the ANPR system cannot successfully read with the necessary level of confidence does put a financial burden on the system. The cost of processing a transaction is estimated at approximately 30% of the transaction itself, as compared to a multilane free-flow system, in which this figure is generally 10% to 15%. This concern about the cost-effectiveness and scalability of the technology, coupled with a commitment to extend the congestion zone to the Royal Borough of Kensington and Chelsea (the Western Extension), has led the TfL to explore the possibility of using other technologies.

The decision to develop a CCZ extension to Kensington and Chelsea was made in September 2005, with a planned opening date of February 19, 2007, four years after the start of the original CCZ. As mentioned in Section 2.3.3.5, the new cordon will have a free uncharged road running north-south through the extended cordon, and another in the northwest section of the zone, as illustrated in Figure 8.6.

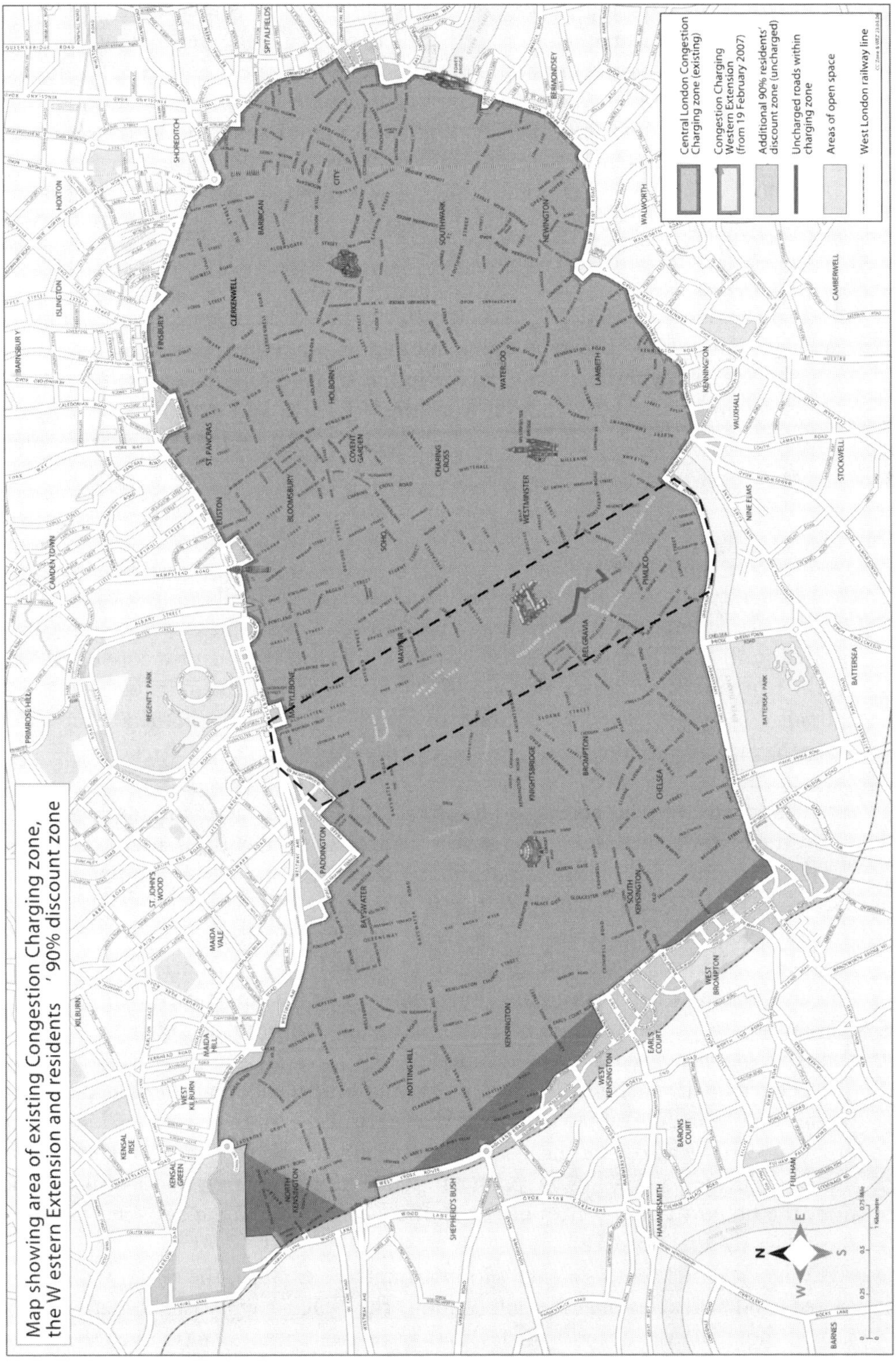

Figure 8.6 The existing and planned extension to the London Congestion Charging scheme. (© 2006 Transport for London. Reprinted with permission.)

The extension is expected to result in a 10% to 14% reduction in traffic, and a 15% to 20% reduction in congestion. Traffic on the boundaries is expected to stay the same, and the annual net revenues raised from the new extension to the congestion charging zone are expected to be approximately £30 to £50 million.

From 2004 to 2006, TfL has experimented with a number of road charging technologies, including GPS, DSRC, improved digital cameras, RFID, and 2.5G and 3G mobile phones [14]. TfL established GPS and DSRC trials in London in early 2006, with a major DSRC trial using two different suppliers in the London borough of Southwark. Figure 3.5 shows a photograph of the prototype urban gantry used for the DSRC trails in London. A decision to change technology will be made before 2007.

The London Congestion Charging scheme has been a huge success, and has reawakened interest in urban charging in other major cities. The success of London is due to the following factors: a simple initial scheme to reduce risk; an effective publicity team; a significant buy-in from the TfL staff; and probably most significantly, strong support of the scheme by the mayor of London, Ken Livingstone.

8.2.3 Durham

In October 2002, the City of Durham in Northeastern England was the first city in the United Kingdom to introduce an urban road charging scheme. The scheme tackled problems from unnecessary vehicle activity in the historic core of the city on the Durham Peninsula, which is a World Heritage Site due to its well-preserved cathedral and castle. The large number of vehicles detracted from the attractiveness of the area, caused congestion, and created conflicts with pedestrians and road safety issues.

The scheme limited access to one street (fed by three access roads), which led up to the historic peninsula. The access control scheme charges users £2 on exit from the cordon. The charge is payable between 10 a.m. and 4 p.m., Monday to Saturday. Entrance and exit from the area is free at all other times. Exit during the restricted period is controlled with an automatic bollard, which is linked to payment and permit detection apparatus (e.g., AVI tag and smart cards). The pay machine is a modified parking machine, with an added smart card reader and tag reader. Figure 8.7 shows a photograph of a zone exit point in Durham. Note the standard signage is the same as in London.

The payment machine also accepts exemption permits that are issued to users by the establishments on the peninsula that have access to their own parking space. Drivers who wish to exit the cordon must stop at the stop line and red traffic indicator located alongside the payment machine. Following a successful transaction, the bollard will lower, and the traffic signal will change to green, allowing the driver to safely proceed.

When a vehicle enters a safety loop around the bollard, the signal will change to red, and when leaving the safety loop, the bollard will rise. Experience has shown more than a dozen vehicles that have attempted to pass over the access points without paying the correct charge, which has led to the vehicles being damaged by the raising bollard.

Figure 8.7 Access charging scheme exit point, Durham. (Courtesy of Durham County Council.)

Drivers who fail to pay the charge will be permitted to proceed through the bollard system. However, a £30 ($50) charge notice is issued to the vehicle owner. Vehicles are recorded on the closed circuit television (CCTV) system, and owners traced through the Driver and Vehicle Licensing Agency (DVLA).

The significance of the Durham scheme is that it has shown that urban demand management could work in small historic cities, without using a particularly sophisticated form of charging technology (although the access points now enable transponder and smart card payment, as well as manual payments). The scheme has been successful in the sense that public acceptance is high, while the corresponding level of traffic using the roads has been reduced by almost 90%, which seems incredible, considering that the access charge is only £2 ($3.50) [15]. In this case a low charge regime has achieved a significant reduction in demand.

8.2.4 Stockholm

Stockholm first explored the possibility of introducing a toll ring around the city in the early 1990s, when the planned charges were to finance a series of ring roads. Extensive planning and technical research were undertaken. However, this scheme was cancelled in 1997 due to a change in the political makeup of the government. Studies through CIVITAS and at the national level began in 2001 to explore the possibility of introducing a road user charging ring to deal with increased congestion problems and traffic pollution. The successful launch of the London scheme encouraged planners in Stockholm, although political tensions between the city and the national government delayed the process. An announcement was finally made in 2004 of a trial for road user charging. The system would operate for 9 months, and then be followed by a local referendum to establish whether there was public support to continue and extend the scheme.

The trial congestion charging scheme went live in Stockholm in January 2006. The scheme is part of a wider transport policy with environmental benefits, such as free parking for "clean" vehicles (e.g., those running on electricity or ethanol). The aim was to reduce traffic by 10% to 15%, cut pollution, and improve the public's perception of the urban environment.

The toll system uses infrared cameras to identify the number plates of vehicles passing in and out of the city center. Vehicles traveling in certain areas (i.e., inside so-called "cordons") automatically pay tolls electronically using DSRC tags provided by Q-Free. An estimated 60% of payments use an in-car tag [16].

Early results from the ongoing project are impressive. After the first full week of operation, the peak hour traffic was reduced by more than 25%, with some links experiencing reductions as much as 35% when the toll charges were the highest. Moreover, queueing has been reduced by 30–50% in morning peak and emissions measured are reduced by 14%. This settled down to an approximate 20% reduction in traffic.

In September 2006, the city held a referendum to test acceptance for the permanent application of the new scheme.

The significance of the Stockholm scheme is to illustrate that, in Europe at least, the success of the London Congestion Charging scheme has encouraged other road and city authorities to follow suit. Sweden is of particular interest, since the road administration suffered a setback in the mid-1990s with the original Stockholm toll ring project. This makes the fact they have established road pricing so soon after the canceled scheme even more remarkable. The fact that Sweden has several DSRC toll sites, and is participating in the Nordic countries' EFC interoperability project, means that we could soon see charging cordons in Norway and Sweden, as well as toll sites in Norway, Sweden, and Denmark, using interoperable technical solutions. Section 8.4.1 discusses the possibility that this could be a significant benchmark for European EFC.

8.3 Small-Scale Toll Systems

8.3.1 Ålesund/Giske Bruselskap Tunnel

On the northwest coast of Norway, between Bergen and Trondheim, lie the communities of Ålesund (population 40,000) and Giske (population 6,600). The communities include a large number of islands connected by ferries. The local airport is also located on one of the islands. To improve economic ties and provide a more reliable link between the islands and Ålesund on the mainland, one of the world's longest undersea tunnel connections was commissioned, comprised of three tunnels with a total length of 11.5 km, and a bridge. The project, opened on October 20, 1987, by King Olav V, includes a six-lane toll plaza equipped with an automatic vehicle identification system integrated by Philips. This represented the world's first commercial, nonstop, ETC scheme [17], although other trials had been conducted previously in the United States and Europe.

The toll plaza was equipped with PREMID RFID roadside systems, developed and manufactured by Philips Kistaindustrier in Stockholm. Three 2.45-GHz RFID transceivers were initially installed on the side of each of the two dedicated ETC lanes that were located at the extremes of the six-lane toll plaza, as shown in Figure 8.8. The number of antennas was subsequently reduced to one per lane, mounted on a pole on the roadside, reading OBUs mounted either in the side window or on the windshield of subscribing vehicles.

Road users had the option to pay by magnetic card, by cash at a manual booth, or by prepaid ETC account. The AVI tag was programmable at short range, although in operation, the tag was used as a read-only device, transferring account information from the vehicle to the roadside system. The quoted time to read the binary coded decimal (BCD) encoded tag was less than 100 ms. As the driver approached the ETC lane at a speed up to 60 km/hr, an in-ground loop triggered the AVI system to search for a compatible tag. If a tag had not been found before the vehicle triggered a second loop, then a CCTV camera connected to a VHS recorder captured an image of the rear of the vehicle and its license plate. A member of the Ålesund/Giske Bruselskap staff viewed the videotape to check these license plates images against valid subscriptions. The system was designed for automatic identification of the vehicle owner by requests to the central vehicle registry and automatic issuance of violation fees.

Figure 8.8 Toll plaza configuration with ETC on the outermost lanes. (Courtesy of Ålesund/Giske Bruselskap Tunnel.)

After only a few months of operation, the traffic throughput had grown to 3,000 vehicles per day, with nearly 60% of all traffic using the dedicated ETC lane, instead of paying either by magnetic card or at the manual booth. As many as 90% of the vehicle passages were charged by ETC lanes during rush hours.

The scheme is still in operation at the time of this writing with the original 2.4-GHz OBUs. The rest of Norway has changed to a CEN-based system (5.8 GHz), but Ålesund will keep their system until the installation closes in 2009. Approximately 4,000 OBUs are currently in use. This example has been included as a case study, due to its historical significance as the first (and still effective) commercial installation of modern AVI-based toll collection.

8.3.2 Dartford

The Dartford Thurrock Crossing (DTC) is located east of the city of London, and provides a fixed road link across the Thames River. This is also the busiest estuarial crossing in the United Kingdom, with more than 130,000 daily passages carried via two two-lane tunnels carrying traffic to the north, and one six-lane bridge carrying traffic to the south. This link represents one of the most important sections of the busy 130-mile (208-km) ring motorway, otherwise known as the M25, which encircles Greater London. Figure 1.2 shows a photograph of the toll plaza at the DTC.

Dartford River Crossing (DRC) Limited was awarded the concession to operate the toll crossing by the U.K. Department of Transport; in return, DRC was permitted to collect tolls, according to prescribed rates, from every vehicle. As part of the concession, DRC was required to maintain the quality of the service and ensure adequate provision to accommodate the expected growth in traffic over the concession period. An international tendering process selected the PREMID TS3000 nonstop electronic toll collection system in 1991, as part of DRC's strategy to renew its toll facility and ensure that its commitments to its users could be met well into this century.

Each of the 27 toll lanes is equipped with the AVI system. Some of the lanes are defined as fully automatic, and include the electronic charging systems as well as an automatic coin machine, providing two means of payment. A transceiver located near the exit of each fully automatic lane reads the unique account identifier from the PREMID tag installed on participating vehicles. If the tag is authorized for use at the DRC system, and if the identified account is in credit, then the toll is automatically debited and a rapid-action barrier is raised, allowing the vehicle to leave the lane without stopping.

The ETC systems were upgraded in 2001 to be CEN-compliant. Le Crossing is the current operator, appointed by the U.K.'s Highways Agency. The significance of the tolled Dartford crossing, along with other estuarial crossings, is that it is getting the U.K. motorist accustomed to paying charges for using road infrastructure. Dartford, being the first U.K. toll facility to implement AVI and DSRC tolling, has defined itself as a reference for other operators that are planning new electronic technology for charging and its operation.

8.4 Regional and Interoperable Tolling

8.4.1 Norway

Norway introduced several urban toll rings in the late 1980s and throughout the 1990s. The installation of the toll rings had several purposes, including revenue raising, demand management (in the case of Trondheim), and environmental improvements. Many of the concessions to collect tolls have recently expired, or are nearing expiration, due to the fact the capital infrastructure projects for which they were designed to raise revenue have been paid off. This is raising interesting questions and future possibilities for the Norwegian road authorities.

The case studies have been included under regional and interoperable tolling, because of the significant work in Norway since the turn of the century in the development of a national interoperable specification for all the electronic road charging schemes. The three main cordon schemes will be briefly described, followed by a description of how these and the other toll facilities in the country have been linked.

8.4.1.1 Bergen

Bergen is a city of 220,000 inhabitants, with a further 110,000 residing in the Greater Bergen area. Access to the historical center of the city is by four roads, with a daily traffic volume of 65,000 vehicles. In order to improve access and traffic flow around Bergen, a new tunnel bypass, the Fløyfjells Tunnel, was planned some time ago. An introduction of tolls on all routes leading into the city was proposed to finance the tunnel and the other main roads. A restrictive parking control system was also introduced, and charges were levied for parking in the majority of the central area.

The toll system was initiated in January 1986 with parliamentary approval. The Floyfjells Tunnel was completed at the time of installation of the toll system. Additional funds were available to be applied to the completion of this project using toll revenue, which showed tangible benefits from the new system of tolls in the public's perception. This helped to gain public acceptance of the toll system. It is unlikely that introducing the toll system initially as a traffic restraint measure would have gained the necessary degree of public support.

The toll system operates from Monday to Friday, between 6 a.m. and 10 p.m. for all traffic entering the city. There is no charge on leaving the city. There is no charge on Saturdays, Sundays, and public holidays, and for public service vehicles, there is no charge at any time. The toll is a uniform fixed rate for each vehicle type (e.g., light, heavy, and motorcycle), regardless of the time of day. Tolls are paid either in cash or by ticket at the tollbooth, and there is a discount for prepurchased books of tickets. The toll collection system is very simple. There are six toll collection points located on the periphery of the city. Toll collection is entirely manual, via cash payment, or prepaid ticket or monthly pass. The toll collection points comprise either four lanes (two for passholders, two for those paying cash); or two lanes (one for passholders, one for those paying cash), with the cash payment lanes on the outside, and the inside lane reserved for buses. Unmanned lanes are equipped with video cameras in order to compare license plates with registered passholders. Nonpaying motorists are liable to fines, in a system similar to that for parking fines. Income lost through the system is considered to be approximately 1.2% of the total [18].

The Bergen toll ring, although entirely manual (until 2004), demonstrates that user acceptance increases if there are tangible benefits, such as the tunnel, and drivers can see that the revenue collected is being used for their benefit. It was the first toll ring in Europe to charge for access into a city, and this has led to a significant amount of groundbreaking economic and user behavioral research to be undertaken and published on the effects of the toll ring, in Bergen and others subsequently introduced in Norway. A number of the key references have been provided both here in this chapter, as well as in Chapter 2.

8.4.1.2 Oslo

The Oslo Toll Ring was opened in 1990, and was one of the first major toll collection facilities in the world with an electronic debiting system [18]. Nineteen toll plazas make up the framework of the system, with more than 120,000 subscribers to the automatic Køfri (later renamed Q-Free) system. All plazas have at least one nonstop lane for subscribers, and one separate lane with an attendant for manual payment.

The system comprises an AVI system in each of the nonstop lanes, one computer in each of the toll plazas, and an accounting system managed by a central computer. Each subscriber is issued with a tag with a unique tag number, relating to the subscriber's account. Money paid for a subscription is registered as a credit on the subscriber's account by the central computer. Valid tag numbers (i.e., accounts in credit) are then distributed online to all the plaza computers. The identification number of vehicles passing the plazas is swiftly checked against the list, giving automatic control and debiting.

Subscribers prepay for an unlimited number of passes within a certain period of time (1 month, 6 months, or 1 year), or a certain number of passes (25, 175, or 350) in an unlimited period of time. HGVs pay double the standard car fee, while motorbikes, disabled persons, and public transport users do not pay a toll [19].

The initial AVI system in all the Norwegian sites used a surface acoustic wave (SAW) technique with a spread spectrum 856-MHz radio frequency signal. The signal was emitted from an antenna mounted above the lane, and was reflected back to the antenna from an AVI tag mounted inside the car, giving the unique identification number of the tag. The tags have now been replaced by 5.8-GHz DSRC. Section 3.5.2.1 explains this in more detail.

Digital video pictures are taken of the front license plate numbers of every vehicle in the lanes. The picture is deleted if a valid AVI tag is registered. If the tag is invalid or if the vehicle has no tag, then the picture is stored and the owner of the car is billed later or fined by mail. The operating company Fjellingen has online direct access to the National Car Owners Register.

The significance of the Oslo system is that it was the first major scheme to use nonstop electronic tolling as one option for collecting the toll charges. It also illustrated that mixed manual and electronic means of toll collection could exist, and that if the package were put together correctly, then a high proportion of the toll facility users would see the benefits of using the electronic tags and choose to subscribe.

8.4.1.3 Trondheim

The Trondheim Toll Ring was opened in 1991. It is similar to the system in Oslo, with some different technical and operational concepts [20]. Ten out of the total of twelve plazas are unattended. These stations have an additional lane with self-service manual payment equipment for drivers without a valid tag subscription (usually situated adjacent to the toll lane). Manual payment by coins or a magnetic card is allowed. Drivers without cash or a card can receive a bill by mail, and drivers who need advice can call for help through an intercom.

Subscribers pay for a certain number of passes, with either prepayment or postpayment through debiting the subscriber's bank account. Drivers never pay more than once per hour in the morning peak period, and they do not pay more than 75 times per month. No charge is applied on weekends or between 5 p.m. and 6 a.m. on weekdays.

The Public Roads Administration County Roads Office is responsible for the total project, and has contracted the separate equipment elements to different companies. As in Oslo, Micro Design AS (now Q-Free) manages the AVI system, but in Trondheim, the company also manages video enforcement and integration with other equipment at the plazas. The Norwegian Telecom Authority is responsible for the network communications.

The significance of the Trondheim scheme is that it was the first operational scheme that used electronic tolling as its primary form of charging. Unlike Oslo, where toll plazas gave the option of payment via manual or automatic lanes, the Trondheim system was automatic, with a much less attractive option of manual payment via coin machines near the entry points but not in a toll booth layout, more like a coin machine in a lay-by. Moreover, the Trondheim scheme was introduced with demand management as a primary objective. Extensive measurements of traffic before and after the introduction of the scheme were made, as were studies of the behavior of drivers and travelers. This continues to provide an important reference material on the effects of a demand management scheme. Until its closure in December 2006 when the capital projects that it was designed to finance were paid off, the Trondheim scheme had proven to have provided important reference material on the effects of demand management.

8.4.1.4 Automated Free-Flow, Interoperable Tolling in Norway

In 2000, progress with the CEN DSRC standards (see Chapter 3) and the need for a technology review, spurred the Norwegian Public Roads Administration to study the EFC activities in the country. All the existing toll rings were upgraded with CEN-compliant DSRC systems, and Bergen and Tønsberg introduced two new electronic toll rings in January 2004. These toll rings, along with the interurban toll roads, constitute over 40 different toll projects in the country, run by 37 toll companies. The Norwegian EFC systems have been joined under the AutoPASS banner. Since 2004, all vehicles equipped with AutoPASS-compatible OBUs have been able to pass through all similarly equipped toll facilities, at full highway speeds, and receive only one consolidated invoice from their OBU issues. This has generated a significant ETC market in Norway and an effectively interoperable National Charging Scheme [21]. This has also encouraged authorities to consider demand management through road user charging as a means to continue operating their toll rings, as their existing concessions to raise revenue for local transport projects come to an end. For example, the Trondheim concession terminated in 2005.

The Nordic countries of Sweden, Denmark, and Norway launched an ambitious plan called NorITS in June 2005, in an attempt to specify fully interoperable EFC between the three countries [22]. This may make a significant contribution to EFC interoperability in Northern Europe, as ASECAP has in Southern and Eastern Europe.

The Norwegian schemes demonstrate that it is easier to launch national and international schemes based on local schemes. These existing schemes provide systems and technology experience, acclimatize drivers to the use of these systems, and highlight the subsequent benefits in terms of reduced congestion and improved infrastructure. As they grow organically, they may evolve into a national scheme.

8.4.2 Highway 407, Toronto

Highway 407, also known as Express Toll Route 407 (407 ETR), was the world's first all-electronic highway. Canada first suggested a new highway to bypass Toronto in the 1950s, but specific construction work on the highway (the largest single-contract infrastructure project in Canadian history) did not start until the mid-1980s. The route was finally completed at a cost of $1.6 billion, and opened to traffic in 1999, extending for 108 km (67.1 miles) from the Highway 403 and QEW interchange in Burlington, to Highway 7 in Pickering. There are 40 interchanges on Highway 407, connecting the road with the main transportation arteries in Greater Toronto. Usage of the road is seasonal, but in March and August 2006, there were 164 million and 200 million km traveled, equaling 320,000 and 380,000 trips per day, averaged over a month.

Shortly after construction, the provincial government signed a 99-year lease with a private concession contractor (ETR International Inc.) for $3.1 billion. The contract permits the concessionaire to revise toll rates, which provides maximum flexibility to meet its commercial objectives (although this flexibility has been publicly controversial).

A traditional toll road that requires space for toll plazas and administration buildings was not feasible. Instead, the Ontario government opted for a dual four- and six-lane MLFF electronic toll highway that would serve both regular and occasional users. A combined DSRC-based MLFF charging and enforcement scheme registers vehicles that enter any of the 29 interchanges [23] located along the 69 km (43 miles) tolled portion of the highway. The charges are based on closed toll road principles for all users, with or without OBUs. Users are charged for the distance traveled according to an advertised tariff table for each segment, with some adjustment according to the time of day. There are four types of accounts: OBU-based ($1 lease fee and $1 monthly account fee); preregistered occasional users; unregistered occasional users; and a prepaid, cash-based anonymous account [24]. OBU users when exiting are notified by a green light and four short beeps to indicate that the toll transaction has been successfully completed [25]. Yellow and red lights on the OBU signify low balance or insufficient funds/transaction failure, respectively. Occasional users are identified by the vehicle's license plate. Driving on the road triggers the registration process, and the highway operators then request owner details from the Ministry of Transportation (MTO) for Ontario plates, or from the Canadian Council of Motor Transportation Administrators for other plates. Not registering is not an offense, but, at this time of this writing, an additional charge of $3.50 was added to the invoice for all video toll charges incurred within the billing cycle, which is then sent to the registered owner. An additional $2.50 was charged for every month in which the vehicle was detected without an OBU, to further encourage the use of an OBU-based account.

The vehicle's license plate number (LPN) is linked to the OBU as part of the preregistration process. If the vehicle's LPN does not match the LPN originally allocated to the OBU, then two charges are applied—the standard OBU-based charge, and a video toll charge, both of which must be paid.

Each charge point includes a vehicle detection and classification (VDAC) system that uses laser scanners to create a time-sliced lateral profile of the vehicle. The height, width, and depth of each vehicle are used to build a profile, which is then compared with the vehicle classification declared by the OBU. A classification mismatch triggers the enforcement process to retain images of the rear of the vehicle. The same classification check is used when the vehicle exits the highway. Locator antennas are used to localize the OBU to the specific vehicle as part of the enforcement subsystem located at the charge point. The DSRC subsystem operates in the 902- to 928-MHz range.

Visible patrols by the Ontario Provincial Police (OPP) and Ministry of Transportation (MTO) Enforcement Officers enforce the scheme. OBUs are mandatory for heavy vehicles with a registered gross vehicle weight over 5 tons. Heavy goods vehicles not equipped with an OBU are liable to a fine under the Highway 407 Act if stopped by OPP or MTO Enforcement Officers.

Multiple images of the rear license plate of all possible violators are retained for enforcement action. Accounts overdue by more than 90 days may be sent to a collection agency, and are subject to a collection fee of $13.50. The Ontario Divisional Court in November 2005 required the Ontario Registrar of Motor Vehicles to deny the validation or issue of vehicle permits to road users who have failed to pay fees for the use of 407 ETR for more than 125 days. A Canadian vehicle owner is required to obtain and clearly display a new validation sticker every 1 or 2 years, to indicate that the vehicle registration is still valid. This plate denial scheme has contributed to high levels of compliance.

Law enforcement vehicles, firefighting equipment, ambulances, vehicles registered to the Department of National Defence, and vehicles bearing Ontario diplomatic license plates are exempt from fees.

The significance of 407 ETR is that the road would not have been built within such a short time frame using conventional sources of finance, since the authorities did not have a sufficient budget. 407 ETR has also served as an important model for a number of subsequent schemes, including the Cross-Israel Highway. 407 ETR is widely cited as an early reference for multilane toll collection, and the effective use of ANPR as a backup system to support occasional users.

8.4.3 TIS, France

Most motorways in France are equipped with conventional cash-based toll plazas at both entrances and exits. Each toll motorway company in France is independent, and is represented by two separate associations, called the Association des Societes Francaise d'Autoroutes (ASFA), and the Union des Societes d'Autoroutes a Péage (USAP).

More than 40 experiments using AVI have been installed in France in recent years. The first generation systems included read-only electronic tags, and were mainly intended for use by commuters on open-toll sections and specific closed-toll sections; early experimentation and trials began in 1989 [4].

In 1992, a project entitled Poids Lourds (Heavy Vehicles) was initiated with the objective of developing a standardized automatic tolling system for trucks using the French highway network. Two demonstrations resulted: one by SAAB-Combitech, SC, and CGE; and the other by CEGELEC/CGA and GEMPLUS. The lessons learned from the development of an interoperable scheme for HGV users were modified and extended to include cars and other vehicles using the toll road network.

Six years of trials and demonstrations were followed in 1998 by the first order for TIS in-car tags. The TIS initiative has now achieved interoperable electronic tolling on the French highway networks operated by ASFA's eight members, and there are plans to participate in the pan-European initiatives to extend this to European interoperability with the ASECAP members [26].

The significance of the TIS scheme is that the introduction of electronic tolling is easier when there is a tradition of plaza-based tolling. The benefit of electronic tolling the elimination of lines for paying tolls, and the subsequent reduction in pollution. The TIS scheme also demonstrates that, while waiting for the complexity of a European-wide interoperability standard to be developed, a coordinated national scheme that adopts the European standard is possible and able to deliver its own set of benefits at the national level.

8.4.4 New York, United States

Three New Jersey road agencies joined in 1995 with the Port Authority of New York and New Jersey to form a regional consortium for ETC. The Delaware Department of Transportation joined in 1996, after which a call for tenders was issued. The aim was to implement a regional EZ-Pass program (as the scheme was called), through the creation of a common customer service center and communications network to serve five agencies. This request also included proposals for equipment upgrades for the New Jersey Turnpike and the New Jersey Highway Authority's Garden State Parkway. The system allows a motorist to use one central account to travel on any of the EZ-Pass–equipped toll facilities: in New Jersey; on the Port Authority bridges and tunnels; in Delaware; on the New York State

Thruway; and on the bridges and tunnels operated by New York's Metropolitan Transportation Authority (MTA).

New Jersey and Delaware were the first states to capitalize on the provisions in the Telecommunications Act of 1996, which allows states to trade, sell, or barter their transportation rights-of-way to telecommunication providers. Consortiums were not required to place a down payment, and the agencies pay nothing for the system for eight years. If, at the end of that period, revenues are below estimates, then the remaining costs will be divided among the five agencies. This is an interesting new approach to the joint financing of road projects, and is being actively pursued by other agencies as a possibly significant model for the future financing of toll roads. The interoperable solutions being deployed by EZ-Pass and other toll system vendors in the United States are beginning to deliver large-scale, regionally interoperable toll collection schemes, in which a number of toll road operators across several states join the interoperable toll collection solution. Schemes in Illinois and California illustrate this trend.

8.4.5 Melbourne and Sydney, Australia

The Melbourne City Link Project (MCLP) is an example of the build, own, operate, and transfer (BOOT) approach to privately finance infrastructure projects. Transurban was awarded the concession in 1994 to build and operate a 22-km length of highway through the center of Melbourne. Transurban was contracted by the Victoria government to operate the road for 33 years, in return for constructing the highway, the interchanges, the Burnley Tunnel, and a new bridge over the Yarra River. The project cost of $1.4 billion was to be recovered through electronically collected tolls, without the necessity of stopping vehicles. Annual revenue was expected to be approximately $160 million.

The City Link was opened in April 1999, with 13 charging points and more than 40 lanes serving 600,000 vehicles per day. Kapsch TrafficCom AB (formerly Combitech Traffic Systems AB), based in Sweden, supplied the roadside equipment and the tags for the ETC scheme. This system was one of the first to be entirely based on the multilane free-flow principle. One requirement of the contract was that the toll collection system would not interfere with traffic flow. The use of gantries spanning the highway rather than tollbooths also ensures that the land occupied by the highway is more effectively used. The system uses CEN DSRC–compliant 5.8-GHz microwave communication between the toll gantry and an OBU installed by the user.

The maximum speed limit is 100 km/hr, but the system is required to have an operational performance up to 160 km/hr without degradation of the charging process. No lane separations are needed, and lane switching within the charging zones is allowed. The system is designed for an hourly average of 2,300 vehicles, and can handle a peak load of more than 900 vehicles in a 15-minute period.

Two methods of toll payment are allowed. The primary method is based on a centralized account linked to a subscription number stored on each tag (OBU). The number is transmitted to the roadside system when the vehicle passes through the charging area. An alternative account for occasional users was launched, known as the DayPass. The occasional user must contact the highway operator prior to using the road, by telephone (call center or IVR), to request authorization to use the highway for an agreed period of time. The user must provide their car registration number and the means of payment, such as a credit card number. This information is stored at each roadside unit, so the vehicles can use the system for the prearranged period. The DayPass system relies upon capturing the images of vehicle license plates from all vehicles that have not fulfilled a valid payment transaction. This account type has now been renamed as CityLink Pass, and is available for 24 hours from the first trip anywhere on the City Link (a lower price version for the Bulla to Flemington road segment), or as a weekend pass. The number of account types has expanded for intermediate frequency users (3 to 12 trips per year), and additional business user accounts differentiated by usage (see Section 3.5.1).

An array of ANPR gantry-mounted cameras captures images of vehicle number plates of vehicles that have not fulfilled their payments. Each toll point relies on close interaction between a VDC system based on stereoscopic cameras, a 5.8-GHz DSRC interface for passage-based ETC, and a vehicle registration (VR) system to capture images of vehicle number plates and the front of the vehicle. By spatially and temporally matching the location of OBUs with the locations of the vehicles tracked by the VDC system, the toll point can determine which vehicles have OBUs and which do not. This information then triggers the relevant VR camera to retain and forward the image of the number plate of a suspected violating vehicle [27] at the requested time of the event. A significant, although nontechnical difference between the Melbourne and Toronto systems is that unregistered users on Highway 407 are simply charged an additional fee, while unregistered users on the Melbourne CityLink are treated as violators (technically an infringement), and incur a fine.

Melbourne CityLink is also significant because it was one of the first to be entirely based on MLFF. Other regions in Australia have adopted the same interoperability specification (see Section 3.6.3), including the M1, M2, M5, M7, Sydney Harbour Bridge, Sydney Harbour Tunnel, and the recently opened Cross City Tunnel. The same account, known as e-Tag Roam, also applies on the Gateway Bridge and Logan Motorway in Brisbane. Interoperability for image-based accounts is less scaleable than for tag-based accounts, so the occasional-user product known as Roam e-pass has limited coverage in the Sydney area (see Section 3.5.5 on scalability).

8.4.6 Taiwan National ETC Scheme

The national toll collection project initiated by the Taiwan Area National Freeway Bureau (TANFB) is one of the largest ETC schemes in Southeast Asia. The islandwide project includes new and upgraded road infrastructure, partly funded by tolls, and will serve an estimated 6 million vehicles daily. The first plaza equipped with new infrared DSRC charging technologies opened on February 14, 2006, and it is envisaged that 2 million users will be equipped with OBUs in the first year, with subsidies from issuing banks. The OBUs are equipped with a smart card reader, which acts as an electronic account that is debited at each charge point.

According to [28], if the DSRC subsystems "... at the tollbooths do not function and fees are not deducted, the users will receive a text message notifying them to repay the fees at designated freeway service areas in 17 days. Those who fail to do so would also be fined NT$3,000 [$92]."

TANFB, through its contractor, aimed to ensure that OBUs could be acquired at a national network of outlets, including an initial 200 ETC service centers along the Zhongshan National (No. 1) Freeway and the Formosa National (No. 3) Freeway. At time of opening the first ETC plaza, the charging technologies were single sourced, although TANFB will be seeking a second manufacturer to introduce OBU supply competition.

The progressive upgrade of ETC facilities by 2010 aims to charge road users based on distance traveled. TANFB will lead more than 20 related ETC projects over this period, and will create jobs for more than 1,000 people. The initiative will introduce RUC and stimulate economic growth through better road links (i.e., better access for communities). TANFB has stated that their role is also to improve the territory's international competitiveness through the provision of a better road network.

Taiwan is included here as a small national scheme that is expected to grow quickly to an estimated 2 million users in the first year. It is also significant in the fact that it has adopted a technical solution that utilizes a DSRC solution using an infrared communications link. This technology has recently reemerged as a candidate communications technology for DSRC after a decade, as discussed in Chapter 9.

8.4.7 Japan ETC

Using tolls and public borrowing to finance infrastructure development is prominent in Japan, which has more than 1 million km of highways. A well-developed network of toll roads connects the major metropolitan centers in Hokkaido/Touhoko, Honshu, Shikoku, and Kyushu prefectures. The toll roads are currently in public ownership but the aim is to transfer operations to private ownership. Overall, one-quarter of all tax revenues is spent on roadway investments. That is four times the proportion spent by Germany, whose roads cover about the same land area as in Japan, and two-thirds of that spent by the United States, which is 25 times larger. Paying a toll for travel within and between metropolitan centers is largely accepted. All highways and major urban and interurban highways are tolled in Japan, compared with less than 10% in the United States. This can be partially explained by the population distribution, with 65% of the population condensed in only 3% of the land area. Greater Tokyo is home to 33 million people and the nation's most dense toll road network.

A small pilot at Odawara tollgate in Kanagawa Prefecture began in March 1997 [29], and by late March 2001, commercial operations commenced at 63 toll plazas in the Tokyo metropolitan area, Okinawa, and other locations in Japan. The planned deployment had a target of approximately 1,100 plazas nationwide, initially opening at a rate of 100 per month. Subsidies and prominent marketing at each toll plaza encouraged the rate of adoption of OBUs. By March 2005, approximately 1,300 toll plazas had been equipped for reading OBUs, equating to 90% of all toll plazas and exceeding the initial planned deployment. The penetration (as measured by transaction rate) was almost 60% of vehicles by April 2006.

The OBU population was scheduled to reach 10 million in December 2005, and had grown to 11 million by mid-April 2006.

ETC is now regarded as one of many ITS services promoted through mutual cooperation among ITS-related government agencies, including the National Police Agency, MITI, Ministry of Transport, Ministry of Posts and Telecommunications (MPT), and the Ministry of Construction. In 1999, the Organization for Road System Enhancement was given the responsibility of establishing a common approach to vehicle-roadside communications, OBU types, payment channels, and tariff structures as part of the national ETC program.

The primary motivation for the national introduction of ETC was to reduce congestion at toll plazas. Increasing the penetration of ETC-based accounts among road users was paramount to meeting this objective. Subsidies, proactive marketing, and the wide availability of OBU outlets nationwide has significantly reduced localized congestion. Statistics from the Tokyo Metropolitan Expressway showed that traffic congestion had been reduced by 70% in 2 years.

- *November 2002*: ETC utilization rate: 4.8%; congestion: 123.4 km-hr/day;
- *November 2004*: ETC utilization rate: 28.1%; congestion: 33.4 km-hr/day.

The approach to toll collection in Japan includes variable pricing and innovative discount schemes, to increase the adoption of ETC, to lower average operating costs per vehicle, and to minimize diversion onto alternative lower capacity routes. Discounts are offered to public users to encourage increased usage, to establish and maintain a prepaid account, to encourage local commuting during off-peak times, and to promote late-night use of expressways.

Three types of OBUs are available: the traditional single unit, with integral card reader; a unit with a separate antenna, allowing the main body to be located anywhere on the dashboard; and the preinstalled unit offered by audio equipment and automotive manufacturers. The payment card (called the ETC card) is also available in two types: dedicated toll account card, or toll account card combined with credit card [30]. The card contains account-specific data, and can be moved between OBUs that store only the vehicle-specific information. If the credit card option is chosen, the road user effectively switches billing and debt recovery over to the credit card company. All transactions at any of the ETC plazas are reported on the credit card bill. The OBU and ETC card are usually available separately, reinforcing the underlying philosophy that payment channels and OBU fulfillments are strategically separate operations, managed by separate organizations [31].

The situations in Japan and in Taiwan demonstrate that a national road pricing system is feasible, at least for major highways, but that the extension to all roads is more difficult. The emerging Asian markets for tolls will remain one of the main growth areas for the industry for the near future.

8.5 Charging for HGVs

8.5.1 Introduction to the Main European Schemes

There has been a growing acceptance in Europe that HGVs should be charged a more realistic cost for their use of the roads, in order to recover the costs of the congestion, environmental damage, and significant wear and tear to the road service made by HGVs. This is viewed as a significant issue in the transit countries of Central and Western Europe, where the vehicle owners are seen as not contributing to the cost of the road network, since they pay VEDs in the country in which the vehicle is registered. There is also a problem in countries like the Netherlands and the United Kingdom, which have a high vehicle duty, since freight operators buy their fuel more cheaply in other countries. The European Commission discussed the idea of the territoriality principle for HGVs in the early 1990s [32]. This essentially enabled individual countries to raise road use fees for HGVs using their road networks, in a similar way to the Alpine Tax system imposed by Switzerland on vehicles entering their country. Several European countries (Germany, Netherlands, Denmark, and Austria) also considered the implementation of a Eurovignette as a means of a national supplementary license for HGVs using their roads. This was deemed unworkable, unenforceable, and susceptible to legal challenges, due to Europe's internal open market philosophy.

The Swiss government also implemented research and subsequently the procurement for a system that could replace the fixed charge for using Swiss roads with a charge for the distance an HGV traveled on the road network. This system went live in 2000, and was the catalyst for a number of other European countries to use road user charging systems [40].

New systems are being discussed, which more clearly follow the principles set by the European Commission's green and white papers on fair and efficient pricing, as well as the EU transport policy "Time to Decide." The overall objective is that taxes and charges for road traffic should reflect the socioeconomic marginal costs, and should contribute to achieving the transport policy objectives [33, 34].

Most systems for charging heavy goods vehicles have been based, until recently, on a yearly flat fee, which gives the right to use the roads for transport purposes. The current developments are toward systems that charge the users for the distance traveled, which is regarded as more effective to managing demand.

The European Commission has been an active stakeholder in this development, advocating road user charges as a more fair and efficient pricing for the use of infrastructure. A more recent change has been for RUC to finance the building of infrastructure as well as controlling the use of infrastructure, in terms of type of roads, time of day, and type of vehicle. This has been the case in London for all vehicles, not just HGVs. A directive was issued in 2003, which outlined a framework for the introduction of a common European service for road user charges, and allowed for interoperability between different types of RUC systems. Several expert and decision committees were scheduled to complete the specific definitions in mid-2006, which will be ready for introduction for heavy goods vehicles in mid-2009, and in 2011 for all vehicles.

The transport infrastructure in Europe is currently underfinanced, and the European transport network has not achieved the planned (or desired) level of quality. A key prerequisite to the successful enlargement of the European Union is a high-quality transport infrastructure, connecting the new and old member states. This must be put in place within a fairly short timetable, but the financing of such infrastructure is (and will continue to be) a major challenge, particularly for the less well-off accession countries, the 10 states that joined the EU in 2004 and those that will join in the next couple of years. The European Commission is therefore viewing electronic fee collection as an innovative funding solution [35]. The directive "On Interoperability of Electronic Road Toll Systems in the Community" establishes the conditions necessary to ensure a European electronic toll service that is interoperable at the technical, contractual, and procedural level. The aim is to have a single contract between the users and the operators, and a set of technical standards that allows the industry to provide the required equipment in an open market with a significant number of potential systems and OBU suppliers, each able to deliver equipment to an open European specification. The directive describes the essential principles of the system, and a committee will work with the definition of the so-called European Electronic Toll Service (EETS), described further in Section 9.3.5.

According to the directive, all new electronic toll systems in Europe shall use one or more of the following technologies:

- Satellite positioning;
- Mobile communication using the GSM-GPRS standard;
- 5.8-GHz microwave technology, using dedicated short-range communication.

From an HGV-charging perspective, the advantages and disadvantages of these three techniques are as follows.

- Satellite positioning is an advanced technology able to distinguish which roads are being used as well as the distance driven. This technology uses a digital map to which satellite positions are matched, as well as a price list, as described in Chapter 3. The data is transmitted to the road operator through the use of mobile communication. Germany is the only country that has introduced this system, although only on its motorways (the Swiss scheme uses GPS for auditing and as a redundant backup to the odometer as the primary means of measuring road coverage). Problems with this technology include the fact that the accuracy of GNSS is not yet fully proven in the built-up urban environment. HGV charging may be installed in large vehicles with special map-matching and inertial navigation systems (INS) to augment the lack of accuracy with current generation GPS, but HGV charging is less acceptable for private cars. GPS works with submeter accuracy, and the size and cost for an entire vehicle fleet (i.e., a national charging scheme) are not yet realistically available. The Galileo system is anticipated to enhance the accuracy of the GNSS segment of the system, as well as to make more satellites available, thus improving the visibility of satellites in urban canyons.
- Mobile communication using the GSM-GPRS standard utilizes the positioning function of the mobile technology for both distance measurement and communication of the fees. GSM-GPRS is included in the German system, but only for communication purposes. Tests in several countries, including a major evaluation in London, have suggested that it is not yet realistic to consider the mobile phone as a viable road user charging on-board unit. Proposals have been made for tolling systems based on charging for entering

a radio cell, with the first trials being held on the A555 Köln–Bonn autobahn in 1996 [36]. Until recently, this option could be discounted, since phones could not offer sufficient accuracy in pinpointing its location. This may change with the potential locating function that is inherent in 3G mobile phones. The 3G companies claim an location service for business phone users with an accuracy of 10m (although this depends on location, coverage, and cell geometry), which is ample for road use charging purposes, but not for enforcement and prosecution. Extensive trials of current 3G phones, undertaken in London in from January 2004 to June 2004 to evaluate potential future technologies for the London Congestion Charging scheme, suggest a location accuracy of several hundreds of meters, which is not nearly enough to run a credible scheme and deliver credible evidence for the prosecution of nonpayers. Nevertheless, since mobile phones already have secure access and a central payment facility (as well as European interoperability), the technology needs only to provide a credible security and enforcement scheme to be considered in the future [37, 38].

- 5.8-GHz microwave technology using DSRC, as described in Chapter 3, demands roadside infrastructure with transceivers, enforcement systems, and other equipment. This infrastructure processes information at all charging points along the road or on each link of a road network (if distance-based charging is to be implemented) for communication with the on-board unit. Despite the infrastructure-heavy burden of DSRC solutions, many operators regard the technology as proven, safe, and feasible for both LRUC and general-vehicle road user charging. Austria has adopted a DSRC-based truck charging system, and most of the toll roads in Europe (and elsewhere) which have adopted an electronic system have pushed for some form of DSRC [39], although the most recent, Slovenia, has opted for GPS for LRUC.

Even with the advantages and disadvantages of each of the technologies, the European Directive hopefully will ensure that the equipment shall at least be interoperable and capable of communicating with all the systems operating in the member states, using one or more of these technologies. The universal on-board unit (UOBU) initiative, which is discussed further in Chapter 9, addresses this issue. The operation of the EETS system and its tariff principles are likely to be left up to the individual national authority in the immediate future.

8.5.1.1 Current and Planned Schemes in Europe

A number of implemented and planned lorry road user charging systems currently exist in Europe. Table 8.1 summarizes the key systems. This is not a complete list. At the time of this writing, the Czech Republic, Hungary, Slovenia, Belgium, and the Netherlands are also quite well advanced in their plans for some form of LRUC system. Others countries, such as France and Greece, have recently announced feasibility studies in this area, while the United Kingdom canceled their procurement in July 2005 and LRUC now forms part of their National Road Pricing.

Table 8.1 Overview of Implemented and Planned Road User Charges for Heavy Goods Vehicles in Europe

	Switzerland [40]	*Austria [41]*	*Germany [42]*	*Sweden [43]*
Status of scheme	Implemented system	Implemented system	Implemented system	Planning stage
Object of fee	Vehicle above 3.5 tons total weight	Vehicle above 3.5 tons total weight	Vehicle above 12 tons total weight	Vehicle above 3.5 tons total weight
Type of road to have fee	Public land fee (on all roads)	Public road fee (on highways and expressways only)	Public road fee (on highways only)	Fee on all public roads (no fee on private roads)
Tariff	Per kilometer	Per kilometer	Per kilometer	Per kilometer
Foundation of costs	Weight; number of axles; pollution class	Number of axles	Number of axles; pollution class	Vehicle type; number of axles; pollution class; time of day
Special aspects		Higher tariffs on sensitive roads		Different costs on different roads
Cost of OBU	Until 2004, free of charge; estimated unit cost of €1,300	€5	€300, to be used as fee credit	Not decided yet; (possible solutions: Internet, OBU, mobile phone)
Payer of OBU installation	Vehicle owner	Vehicle owner	Vehicle owner (for up to 4 hours)	Vehicle owner
Technology	DSRC	DSRC; GPS for control	GSM/GPRS; GPS	Not decided yet; (possible solutions: DSRC; GNSS/CN)
Traffic table	OBU	Server	OBU	Server, OBU
Prepayment or postpayment	Both	Both	Both	Both
Methods of payment	Cash; debit; credit	Cash; debit; credit	Cash; debit; credit	Debit; credit
Means of payment	CHF; Euro; major fuel and credit cards	Euro; Quick (electronic purse); major fuel and credit cards	Euro, and at terminal location official foreign currency; major fuel and credit cards	Not decided
Period for payment	60 days to send billing information; 1 calendar month for paying bill	Daily transmission of data from credit institute; bill sent out every fortnight	Daily transmission of data from credit institute; monthly check of credibility	Postpay up to several days; transmission for each performed HGV journey to tax authorities
Use of revenue raised	Financing of infrastructure; reducing external costs, improving railway network; shifting transport from road to rail	Financing of road infrastructure	Financing of infrastructure; user related costs/ payment system; increasing competition between modes; leadership road pricing systems	Compensation for roads exposed to wear caused by HGV; maintenance of road network; reducing emissions; reducing accidents

Source: [39].

Countries clearly have different problems and financing issues to address. There are different objectives and business models, ranging from pure revenue generation to systems that should support socioeconomic marginal costs.

The basis for introducing more advanced road user charge systems also differs between countries. As indicated earlier, several European countries have extensive transit traffic by trucks that use the roads without paying for them. Other countries have environmental concerns, and use the revenues for enabling the shift of goods to (supposedly) more sustainable modes. Other countries wish to finance the infrastructure investments via road user charges. In some countries, the legal basis for the road user charging system is taxation rather than road use.

DSRC and satellite-based technology systems differs in basic technology, and in their philosophy, where the preference for some operators is for simple OBU solutions with so-called thin clients, and for others, more complex on-board units. New, more innovative approaches to HGV charging are also being explored, and should not be precluded from future policy development. One of the most interesting approaches is the extension of HGV charging beyond only charging for the distance traveled by a truck, to also charging for the wear and tear costs on the infrastructure. Newcastle University has developed a prototype system for the U.K. DfT that dynamically measures the loading of the vehicles axle, and calculates a charge based on the load. Section 9.3.3 describes this further [44].

The European Commission has a political goal to achieve interoperability between the systems used in the member states. The directive "On Interoperability of Electronic Road Toll Systems in the Community" [35] deals with the implementation of differentiated road charges for heavy vehicles in Europe. It defines the conditions necessary to ensure a European electronic toll service (EETS) interoperable at the technical, contractual, and procedural level. The aim is to have a single contract between the users and all operators, and a set of technical standards that allow the industry to provide the required equipment in a competitive market.

In summary, the charging for road use by HGVs is likely to be a policy repeated in many countries of the European Union. This policy internalizes the costs exhibited on the road network by HGVs, many of which are registered in other countries, transiting a country and not contributing in any way (up to now) toward the costs of the road infrastructure. The challenges of delivering such systems at a national or European level are significant. Several technical solutions currently exist in the implementations in Switzerland, Austria, and Germany. The European Commission recognizes the need to make these systems interoperable, so that a single device is installed in a truck that can be used in many different countries. The problem is that there is no common European technical specification that would have the agreement of all nations involved. Moreover, countries are likely to implement their truck road user charging schemes in different ways (e.g., distance, time, flat-fee). Some of the schemes may be taxes and some may be use-based fees, which have very different legal bases in Europe. Furthermore, at the moment, there is no common specification for evidential records.

Thus, the challenge for Europe is to establish an interoperable technical specification along with the associated contractual, legal, and procedural agreements, to achieve cross-border interoperability, and to establish a system that can incorporate different forms of HGV charging and diverse fiscal regimes. If we consider that it has taken 15 years to move towards interoperability for EFC for toll roads in Europe, then we cannot be fully confident that an LRUC system would take any less time at the European level. However, if the opportunity is missed (and the time window for success is not that wide), then we will be leaving the legacy of disparate and inefficient piecemeal implementations of LRUC across many parts of Europe.

8.5.2 HGV Charging Schemes in the United States

The main HGV charging schemes in the United States are known as preclearance schemes. Here, precertified HGVs can bypass the manual inspection stations on highways, particularly at interstate borders, by using an electronic tag to provide information to roadside interrogation stations about the vehicle, its load, and weight.

The system evolved out of one of the first real developments of AVI equipment in the world, the Heavy Vehicle Electronic Licence Plate (HELP) Project in the 1980s [45, 46]. The oldest and geographically largest of these schemes is called PrePass. HGVs are fitted with a DSRC tag that enables the vehicles to communicate with the roadside as the HGV passes through the toll infrastructure, which includes both automatic vehicle classification and weigh-in-motion platforms (WIMPs). At the approach to a tolled section of a motorway, or to an interstate border crossing, the WIMP determines whether the vehicle is complying with axle and gross vehicle weight limits and safety requirements for HGVs. Toll charges based upon the number of passages the HGV makes of the toll road are billed monthly. The PrePass system has more than 259 currently operational sites in 25 states, and almost 300,000 equipped HGVs. A map of the existing toll sites and state partners in PrePass can be found at http://www.prepass.com/map.htm. Similar systems are operated in other states, such as the NORPASS system, whose coverage can be found in the map at http://www.norpass.net/coverage%20map.htm. This scheme currently has 59,000 equipped HGVs, and operates largely in states not covered by PrePass (eight U.S. States and three Canadian provinces). NYSTA and HELP have recently signed an agreement to make their electronic clearance systems interoperable (using EZ-Pass and PrePass technologies, respectively), significantly expanding the system to some 400 toll facilities [47].

8.5.3 New Zealand

Although not strictly an electronic system, the charging system in New Zealand is included here, as an example of the first effective national HGV charging systems that charges by the distance traveled per weight of load (measured in ton-kilometer). New Zealand introduced a road user charging system for HGVs in 1977 that charges road users in proportion to the damage they cause to the road network. The charging mechanism did not employ the use of fixed annual charges and fuel taxes. Instead, HGV charges were calculated based on the actual distance traveled on the New Zealand road network.

The system involves equipping each HGV with a hub odometer, a distance-measuring device fixed to the hub of one of the driven wheels. Charges are paid per ton-kilometer using a distance license fee, which varies by gross vehicle weight, axle configuration, and distance. Paper-based distance licenses (in multiples of 1,000 km) are prepurchased at New Zealand post offices. The system also includes a rebate on fuel taxes paid by HGVs, as well as supplementary licenses for excess vehicles loads. Discussions are underway in New Zealand to examine how the introduction of a national electronic toll collection system could incorporate in some way the National HGV charging scheme, and what upgrades and modifications to the system would be possible.

8.6 HOT and HOV Lanes, United States

HOV and HOT lanes are a widely used and interesting variant of road tolling and traffic demand management measure in the United States. By its very nature, this is a complex policy, and there is much debate about it, primarily in Europe and the United States. The HOV approach offers the individual a choice (i.e., to use the HOV lane if one has a passenger on board, but not otherwise).

Many examples of HOV and HOT lanes exist in the United States, with few elsewhere in the world. The difference between the two approaches is:

- *HOV:* The lane can only be used by vehicles with more than a prescribed number of people in the car.
- *HOT:* The lane can be used free of charge by high-occupancy vehicles, but vehicles with a lower occupancy would need to pay a toll.

Current U.S. policy has encouraged the design and implementation of HOV lanes for use by carpoolers. The lanes are usually separated from the general traffic, and they allow vehicles that meet the required occupancy level to avoid congestion and to travel at a speed faster than vehicles in the normal lanes. More than 700 miles of HOV lanes are presently in operation in the United States, and several more are in various stages of planning and design. Some HOV lanes are underutilized, while the adjacent regular freeway lanes are congested for many hours of the peak period. Some HOV-2 systems (i.e., two individuals in the car) have traffic levels, such that speeds are below free-flow, reducing their attraction [48].

HOT lanes have proven to be a useful answer for nonefficient use of HOV lanes. The idea is to convert underutilized HOV lanes, allowing congestion pricing to be introduced. HOT lane projects include the 91 Express Lanes in California, the I-15 reversible lanes in San Diego, and the Katy Freeway in Houston. Some of the public have been unhappy with the conversion from HOV lanes to HOT lanes. Their argument is that investment in HOV lanes was to increased capacity, and as a means of reducing cars on the network while maintaining similar levels of people movement. When an HOV lane is converted to an HOT lane, individuals who can afford the single-occupancy rate can use the new infrastructure, whereas other lower income groups may not have the choice. However, studies have shown that in reality HOT lanes are used equally by all social classes and income groups [48]. There is some concern over whether HOT lanes will discourage carpools if people can pay for the same time savings that carpoolers receive. On the other hand, there may be an incentive to form a three-person carpool to avoid paying the toll. An example of a highly successful HOV facility in California is described below.

8.6.1 SR91 Express Lanes in California

The SR 91 Express Lanes (with four lanes, two in each direction), a privately funded and operated 10-mile toll facility built in the middle of State Route 91 in Orange County, California, opened to traffic in late December 1995. It is in one of California's most congested highway corridors, and offers a faster alternative to motorists who are willing to pay a toll that is varied by time of day. The toll was \$2.75 in 1997 for traveling in the peak periods, which was raised to \$3.20 in 1998. Under the existing schedule, late night and early morning tolls remain at 75¢, while toll prices for most other time periods range from \$1.25 to \$4.25, depending on time and direction of travel. Tolls for the four busiest hours on the 91 Express Lanes, which are eastbound on Thursdays between 5 p.m. and 6 p.m., and on Fridays between 4 p.m. and 7 p.m., are \$4.25 per trip, as established by the California Private Transportation Company (CPTC) in 2000. As an incentive to carpools with three or more occupants, who currently pay between \$1.45 and \$1.65 per trip for their morning commute, the 91 Express Lanes will lower the toll price to \$1 per trip during the rush hours between 5 a.m. and 8 a.m., Monday through Friday. Originally cars with three or more passengers traveled free.

The purpose of SR 91 Express Lanes is to reduce corridor congestion, while generating revenue to finance the deployment of electronic toll collection systems and operation of the toll lanes. It has been noticed that when tolls are increased, a temporary drop in demand follows, although demand is generally increasing. The introduction of the Eastern Toll Road has resulted in a noticeable drop in demand on the SR 91. One-half of the vehicles use the facility only once per week. HOVs accounted for 40% of these before imposition of tolls, but this has decreased as a result of the tolls.

It has been noted that 30% of low-income users frequently use the service, compared with 50% of high-income users (there is a low proportion of low-income households in the area). Surveys indicate that two-thirds of users are male and younger, and that older people are less likely to use it. A major benefit in terms of acceptance is that users pass those drivers waiting in congested parallel lanes, and the operators suggest that this helps them more effectively market the service. Most people using the service save from 12 to 20 minutes, although they tend to overestimate their savings. It is worth noting that initial public approval was not forthcoming, although approval has increased with time.

One-third of regular users in a sample taken by Parkany [48] were found to carpool, but only one-fifth of infrequent users. Parkany also found that nearly 40% of HOV-2s never use the Express Lanes, which is less than the 50% of solo drivers who never use the Express Lanes. Approximately 10% of the solo drivers and 25% of the drivers who always drove in a two-person carpools increased the number of passengers in their vehicle between November 1995 and mid-1997. However, the regional surveys determined that other carpools reduced the number of passengers in the 18-month period, by 40% of the cars that had been previously traveling with three or more per vehicle.

The study concluded that the SR91 HOT lanes have encouraged drivers to pool their car use and offered an effective means of managing demand through encouraging car sharing. Such an approach is yet to be tried in Europe on interurban roads, although currently the use of HOV lanes is extremely limited and confined to urban and peri-urban roads.

8.6.2 The Eastern Toll Road in California

The Eastern Toll Road is 17 miles of tolled highway in the Riverside area of California. The opening of the road meant that 41.5 miles of toll roads exist in California, and the new facility offers two additional lanes on the Riverside Freeway. The cost of the Eastern Toll Road was $765 million. The price of a single trip is $3.25, and the travel time savings is 13 minutes each way during normal conditions. The system uses FasTrak transponders, which charge the driver's prepaid account. Lane entry sensors determine the toll by weight and type of vehicle. If a toll is paid, then the signal turns green. If a vehicle leaves while the signal is red, then a camera photographs the license plate. Violators are ticketed by mail.

It is likely that the success of HOT and HOV lanes in the United States will encourage others to implement this approach to demand management. The automatic enforcement of such systems is currently difficult, but EFC tags have been used to declare the entitlement to use such lanes [49]. With many no-car lanes appearing in European cities, the possibility of using these lanes to offer HOV incentives for carpooling is an opportunity not to be wasted.

8.7 Significant Trials and Pilots

8.7.1 Hong Kong

The Hong Kong government introduced in 1983 a 2-year trial of electronic road use charging, in an attempt to reflect road usage more as a demand management strategy rather than the fixed costs of motoring. The latter was regarded as unfair, since it allegedly penalized the poorer Chinese vehicle owners who lived in the New Territories, and who rarely brought their vehicles into the congested CBD of downtown Hong Kong and Kowloon [50, 51].

The Hong Kong ERP experiment from 1983 to 1985 involved fitting a sample of 2,500 vehicles (mainly government and some business) with electronic number plates (ENP), which were welded to the underside of the vehicle. The ENP was about the size of a VHS videocassette, and operated as a passive radio wave transponder operating in the 30-MHz band, communicating with specially configured inductive loops buried below the surface of the roads. The ENP transponder was energized by the buried loops; that is, it derived its electrical power from the received radio waves. The ENP transponder then communicated a unique identification number to the inductive loop and antenna of the roadside system. Roadside microcomputers installed at selected charging points then relayed the vehicle's identification code to a control center. Car owners were sent monthly billing statements that detailed the amount of actual road use subject to ERP. This is similar in operation to many AVI systems discussed in Section 2.2.2. Figure 8.9 shows a photograph of the inductive loop arrangement at an ERP entry point in Hong Kong.

Closed-circuit television cameras photographed vehicles that either lacked electronic number plates or had defective ones [52]. The ERP pilot experiment proved to be an overwhelming technical success, with 99.7% reliability (see Section 2.3.3.6). The experimental scheme operated with five charging bands, covering the morning and afternoon peaks, the interpeak, and the shoulder peaks (exceeding the 99% target specified by the government), with the prices in each band reflecting the level of travel demand. The system operated with 18 entry points defining a number of different zones. The benefits include the savings in travel time to those who stay and pay under the ERP scheme, and the vehicle operating cost savings from less congestion, as well as the penalties to those who are priced off the route to avert the ERP charge. Overall traffic in the CBD was reduced by 20%, and peak traffic by 13%.

Although the experimental scheme was deemed a technical success, there was no decision to move forward with a full-scale ERP scheme at the end of the trial in 1985. This was due to several reasons, including a suspicion by the public about the use of the revenue (there was an 8:1 revenue-to-cost ratio), and widespread

Figure 8.9 Hong Kong ERP entry point. (Courtesy of Ian Catling.)

concerns with having ERP tags fixed to vehicles that could potentially enable authorities to track citizens' movements. This fear was highlighted with the signing in 1984 of the Sino-British Joint Declaration on the future of Hong Kong after July 1997. The arguments in favor of and against a decision to proceed to a full scheme were finely balanced, and the newly established District Boards narrowly voted against its introduction. Had the vote gone the other way, road pricing may well have been introduced elsewhere in the world much sooner than actually transpired [53].

The Hong Kong government again studied the issue of road pricing in the late 1990s. The Electronic Road Pricing Feasibility Study in Hong Kong, from 1997 to 2000, analyzed the potential impacts of road pricing in the territory, developed a recommended scheme, and demonstrated the technical viability using two different technical approaches [54]. The project was notable because one of the demonstrated technologies was based on GNSS, or vehicle positioning systems (VPS) as it was called in these trials. This was the first time that urban road pricing using this technology had been shown to be viable other than in small-scale trials, such as those undertaken in Newcastle in 1995–1996. Both technologies, VPS and DSRC, were successfully demonstrated, and the cost analysis showed that there was little to choose between the two: the higher costs of VPS in-vehicle equipment were largely offset by savings in roadside infrastructure. Although the technology demonstrations were successful, and the analysis again showed that road pricing would produce substantial benefits, no decision to proceed to a full system was taken.

The trials provided valuable technical performance data, and although the systems were deemed to work well, the conclusion was that there were still technical barriers to overcome. Moreover, since China had only just resumed sovereignty over Hong Kong, it was felt that politically it was too soon to implement such a system. Details of the trials can be found in [54]. Figure 7.3 shows a photograph of the off-road trials, and Figure 8.10 shows a photograph of a bridge equipped with the DSRC and enforcement equipment.

The Hong Kong government tendered for a new study in 2006, to again investigate the feasibility of road pricing.

8.7.2 Cambridge, United Kingdom

The Cambridge Congestion Metering trial formed part of the highly successful ADEPT project, which was funded under the European Commission's Dedicated Road Infrastructure for Vehicle Safety in Europe (DRIVE) II research program. ADEPT is recognized as the research project that proved and demonstrated as far back as 1991 the first multilane DSRC system, OBUs that were integrated with smart cards, and generic equipment used for a range of urban and interurban applications. The project was also the basis for the original CEN DSRC standards [55–57].

The field trial of the ADEPT system at the Cambridge test site from October 1993 to November 1995 was the first demonstration of congestion metering and road use pricing technology in the United Kingdom. This system differs somewhat in operation from the other variants of the ADEPT system that were implemented in the other field trials, since this system required the additional input of a distance-

Figure 8.10 Gantry arrangement for communications and enforcement, Hong Kong. (Courtesy of Ian Catling.)

measuring sensor in the vehicle, so that the congestion encountered or the distance traveled within the predefined charging cordon around the city of Cambridge could be measured and charged for [58]. Other variants of the ADEPT system generally used the transponder alone for multilane tolling, or parking reservations and payments. Figure 8.11 shows a photograph of the in-vehicle transponder and meter, and Figure 8.12 shows the roadside equipment arrangement.

The congestion meter operates by using the ADEPT microwave beacons to activate the in-vehicle equipment as it enters a predefined cordon around the city [59]. Once switched on, the distance sensor provides information to the ADEPT transponder, enabling calculation of the time taken to travel a unit distance. If this time is greater than a predetermined value, then it is presumed that the vehicle is traveling on a congested stretch of road, and a charge is deducted accordingly. The underlying principle is that a vehicle that encounters congestion is also contributing to it, and thereby imposing costs on other vehicles in the vicinity. Electronic credits to reflect these costs are deducted from the user's smart card. If the vehicle is traveling on an uncongested road, then no charge is deducted. This process continues until the vehicle leaves the cordoned area and the in-vehicle equipment is switched off by an exit beacon at the roadside. No charging can therefore occur outside the cordon. Figure 8.13 provides an example of the congestion charging algorithm for a particular vehicle speed profile.

Figure 8.11 Cambridge congestion metering: in-vehicle equipment.

Each time a charge is levied and credits are deducted from the smart card, the amount of the charge is displayed. When the meter is switched off using the exit roadside beacon, the total number of credits consumed within the area is displayed. Other elements of the Cambridge demonstration included the following:

- The provision for traffic information to be conveyed over the microwave link to the driver, and to be displayed on a prototype in-vehicle LCD [60];
- An online enforcement video system at the entry beacon points (Figure 8.11);
- A PC-based smart card "service station," which provides the facility to write personalized data to the smart card, add credit to it, or review its balance and transaction history;
- The demonstration of other charging algorithms: actual distance-based charging, open toll-collection (single-point), and closed toll-collection (entry-to-exit point charge), all of which reside in the transponder and can be automatically activated and controlled by the roadside beacon.

Figure 8.12 Cambridge congestion metering: roadside entry point equipment.

Researchers from the Transport Operations Research Group at Newcastle University undertook a detailed technical evaluation of the system. This involved the testing of the system under different environmental and traffic conditions on a test site within the grounds of the University of Newcastle upon Tyne. Many thousands of test runs were logged and the results analyzed. A separate contract awarded by the U.K. Department of Transport assessed the performance and relative merits of the different charging algorithms. The U.K. Department of Transport and Cambridgeshire County Council jointly funded a study of behavioral responses [61]. Political support for any further development of the congestion monitoring and pricing concept was uncertain. The U.K. government embarked on a trial in late 1994 of multilane tolling systems for possible implementation on the U.K.'s strategic road network, to copy somewhat the German A555 trials of 12 systems from 1994 to 1996. Interest in the United Kingdom shifted from charging in the urban road environment to experimentation on interurban roads, so it would be almost a decade before political interest in urban congestion schemes reemerged.

The significance of the Cambridge trial is that it showed that with relatively simple DSRC equipment and some additional sensors, quite sophisticated forms

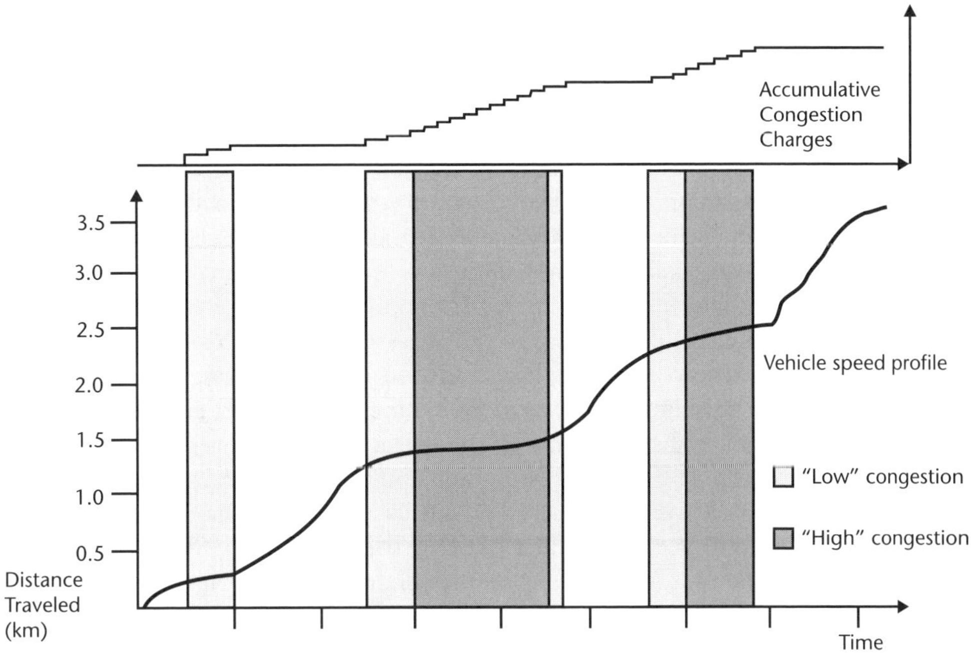

Figure 8.13 The charging algorithm used in the Cambridge congestion metering trials.

of road user charging could be implemented. The Cambridge Trials demonstrated all of the following: simple cordon-based charging (levying a charge for crossing a cordon threshold); congestion-based charging (using the measurement of congestion by the vehicle itself as the means of calculating the charge); distance-based charging (charging the vehicle for the distance it travels within the cordon); and time-based charging (charging vehicles for the time they are traveling within the cordon) [62–64], a level of charging flexibility that is now reemerging on the political agenda of a number of government and road authorities.

8.7.3 Timezone

Another trial in the United Kingdom, around the time of the Cambridge trial, in the Kingston and Reading areas used the time parameter, in a system developed by GEC-ESAMS called Timezone. The time-based model and initial trials in Reading suggested that a time-based charge was not a particularly good basis for a road user charge, influencing driver behavior in a negative way, in the sense that drivers attempt to drive within the cordon as quickly as possible, to minimize traveling time and reduce charges. This leads to road safety implications. Nevertheless, the Cambridge congestion metering trial did illustrate that the measurement of distance could be a significant parameter in future charging schemes, as it is for truck road user charging in Germany, Switzerland, and New Zealand.

8.7.4 The Netherlands

The Netherlands has made a number of attempts since the mid-1980s to introduce road user charging and other demand management measures to deal with the chronic congestion on many of the country's arterial routes [65]. This has led to a wealth of experience with the options for road pricing and the technical challenges to implement such a system. Successive Dutch Ministers of Transport over the last decade have seriously considered the implementation of road pricing schemes, perhaps more so than in other EU countries, with the exception of Norway, Sweden, and the United Kingdom. The motivation to explore road use pricing in Holland is due to the geographic and spatial makeup of the central Netherlands. This Randstad area, including the major cities of Amsterdam, Rotterdam, The Hague, and Utrecht, and significant parts of the province of Noord-Brabant, is among the most densely populated areas in the Western World. This is reflected in the severe levels of traffic congestion, especially during commuting peak hours. A second reason for this interest is economic, since the Dutch economy is significantly dependant on transport, logistics, and related trade services, all of which could be impaired if the level of congestion and length of travel times make the country less attractive in the use of such services. A third issue is quality of life and environmental considerations as a by-product of congestion and the associated traffic pollution.

Since the first interest in road user charging in the Netherlands, three types of road pricing schemes have been researched and experimented with [65]. These are:

- Peak-hour permits ("spitsvignet");
- Toll plazas ("tolpleinen");
- Peak-hour cordon charging, using electronic charging systems ("rekeningrijden").

The idea of cordon charges (rekeningrijden) was introduced in 1987. The government saw the scheme as a useful alternative to increasing taxes and toll charges in the "Randstad Accessibility Plan." In 1988, Rijkswaterstaat (The Dutch Ministry of Transport, Public Works and Water Management) appointed a "rekeningrijden" project team. Major research projects were undertaken to explore the challenge of electronic pricing, with national research being augmented by collaboration in EU-funded projects, such as the DRIVE I projects (VITA, PAMELA, and HADES); the DRIVE II projects (CASH, ADEPT, and the original CARDME project); extending into such projects as MOVE-IT, A1, and ADVICE in later research programs. None of the projects and initiatives was ever implemented, but it did help to build up a significant level of expertise in Europe in the complex issues of road user charging.

The most recent proposal, rekeningrijden, probably came closest to actual implementation, with significant trials of on-road technology [66] supported by an off-road research program with some unique pieces of research, such as the toll transaction simulator developed by the University of Aachen.

Following a request for participation from industry, a number of consortia were retained to develop trial systems for the rekeningrijden. The five awarded contracts for the first phase of the evaluation were [67, 68]:

- CGA, MFS Transportation Systems, Inc., and GSZ;
- Bosch/Philips Rekening Rijden consortium;

- SSSL/Combitech consortium (Siemens Nederland NV, Combitech Traffic Systems AB, Siemens Nixdorf, SICE, and Lockheed Martin IMS);
- MAC SpA (Alenia + Marconi Communications SpA);
- Micro Design ASA.

Following an initial simulation phase, the first phase of the on-road tests took place from February to March 1999, on the A12 at De Meern in the direction of Utrecht. The system test covered all the technical aspects of fee collection systems under Dutch weather and traffic conditions. The experience accumulated during the system test resulted in improvements in system design, in the areas of electronic payment and payment based on the license numbers. ANPR technology developed by Philips in collaboration with R and H Technology, Ltd., had the most equipment-rich and complex gantry arrangements ever seen anywhere in the world.

The results of the system test were satisfactory. It appears that both the EFC systems and ANPR system were able to handle the large amounts of data. During the 6-week test period, the systems detected and registered approximately two million vehicles. For the purpose of detailed analysis, 80,000 vehicle detections and 40,000 license number registrations were used. The license numbers were not linked back to any national database that could identify the owners of the vehicles. The test vehicles of 400 volunteers were equipped with an in-vehicle payment unit. They made a total of more than 5,000 electronic payments during passage through the fee collection points, all of which were analyzed.

The test showed that automatic payment by means of an in-vehicle unit worked in accordance with the requirements set for the test, with a reliability of 99.99%.

The Dutch ministry put forward a new proposal in 2000 to explore a charging system based upon a kilometer charge "kilometerheffing" [67]. The concept is moved toward a variable road use charge, rather than reliance on the current taxation regime of fixed costs through VED and vehicle duty taxes, and variable costs related to the duty paid on fuel (which has some relation to usage). Electronically installed odometers could record the number of kilometers driven. The initial proposal was to charge a flat rate for the distance traveled, but it is expected that the technology would have allowed differentiation of taxes according to the location and time of road usage. For such a complex differentiated distance-based charge, three possible technologies have been explored in the Netherlands:

- Electronic odometer: developed for HGVs, but possible for the whole vehicle fleet;
- Differential GPS (DGPS) systems: to give data about the geographical position of the car, which seems accurate enough for kilometer recording, and possibly as backup for odometer-based distance measurements, as is the case with the Swiss HGV charging scheme;
- DSRC beacons: to switch on a distance-measuring device in the vehicle within a charging cordon, as demonstrated in the Cambridge congestion metering scheme in 1993–1994, and/or to provide part of the enforcement function for either of the above two options.

The scheme developments were suspended and replaced by a new initiative in May 2006.

8.7.5 DIRECTS Trial, United Kingdom

The U.K. DfT has been researching aspects of RUC since 1994.

The United Kingdom embarked on a 2-year series of trials of equipment to investigate the feasibility of electronic tolling for the motorway network, beginning in 1996. After detailed investigations and trials, the recommendation of the tolling project was not to proceed to a live implementation without first undertaking a pilot trial of a complete solution [69]. There was a change in government during this period, and a shift in emphasis from interurban tolling to urban road pricing schemes [70]. This led to the conception of the DIRECTS project. This was the largest of the DfT research programs in this area, and completed its trial phase in April 2006.

The DIRECTS project was configured to provide insights into the operation and performance of a free-flow system with on-board, roadside, and central equipment being provided by a range of manufacturers whose equipment has been proven to work together in an interoperable and robust way.

DIRECTS has simulated the requirements of a national charging scheme, and was designed to facilitate operations in both urban and interurban road contexts, ranging from OBU issue to the production of invoices and penalty charge notices, while meeting challenging end-to-end performance targets [71].

A key output from this program is the creation of an OMISS to provide the basis for the possible procurement of charging systems within the United Kingdom (and potentially beyond), whether for congestion reduction, tolled crossings (e.g., tunnels and bridges), or to support other future schemes [72].

The DIRECTS contract was awarded to the Fareway consortium (KBR, Thales, and Atkins) in May 2001. The solution included trials of mobile positioning system (MPS)–equipped vehicles in both Leeds and Bristol, and the establishment of a number of DSRC gantries in and around Leeds. MPS is the DfT term for GNSS-based charging units.

To facilitate national road user charging, or a profusion of local and regional schemes as enabled by the U.K.'s 2000 Transport Act, the U.K. DfT created a business model with an explicit separation of account management functions from the provision of local roadside systems. It was perceived as cost-inefficient for each scheme to have its own back office interface with users, when the capability to manage user accounts already exists in the market. The model assumed four groups of entities having a role in the road user charging model:

- On-road services providers: Suppliers of on-road systems constituting the local scheme (four provided in total);
- Payment services providers: Suppliers of prepay and postpay off-board user support and accounting services (two provided, representing two towns and cities);
- Data clearing operators: Operators between the on-road service providers (ORSPs) and the payment service providers (PSPs), essentially providing

privacy of users' movements on invoices, by separating user details from location details, as well as minimizing the number of contractual arrangements;

- Users and vehicles: Approximately 550 volunteers from the business community.

Figure 8.14 illustrates this arrangement.

Leeds was selected to be the primary location for the project, following offers to host the trials from a number of local authorities. A number of different road layouts were selected to demonstrate operations in a range of traffic environments, including the M621 urban motorway.

The Fareway consortium hired four subcontractors to deliver microwave solutions for charging. Each subcontractor supplied one or more microwave OBU designs, providing a total of six device types, two of which supported on-board charging either via an integral account or via a smart card (see Figure 8.15). Two of these companies also supplied roadside equipment that was installed on six gantry locations (see Figure 8.16). For the DSRC trials, 550 vehicles were equipped with OBUs, and during the trial period, more than 800,000 transactions were successfully recorded. In addition to the DSRC OBUs, a trial of 50 MPS units was also undertaken, initially in Bristol as part of the PROGESS project, and then in Leeds to complement the DSRC trials [73].

The data from the trials is currently being analyzed, and it is expected that the results of the project will be published before the end of 2006. The key output, however, of the trials has been the publishing of Open Minimum Interoperability Specification Suite (OMISS). The United Kingdom is committed to a research program to investigate the feasibility of a national road user charging scheme to complement the local congestion charging schemes already in place in London and Durham [74]. There seems to be a growing consensus that some form of national

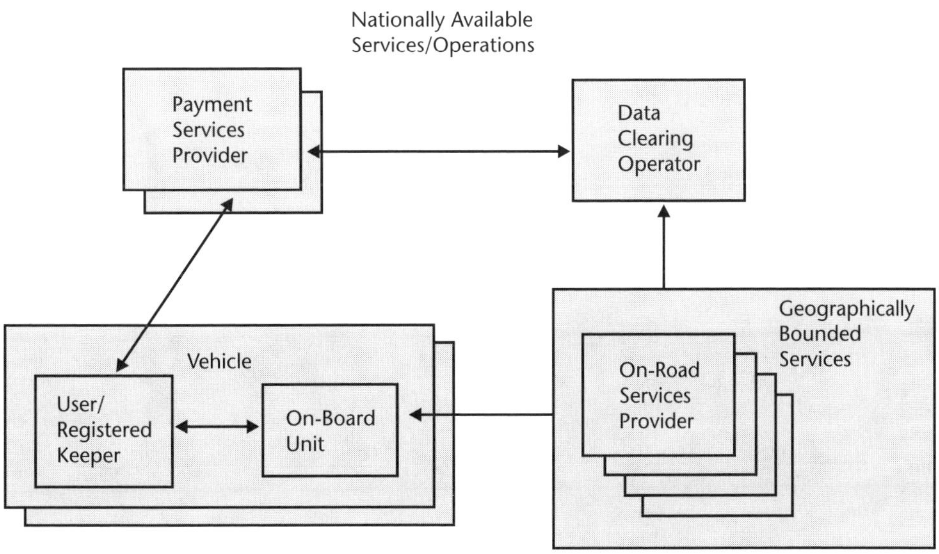

Figure 8.14 The DIRECTS business model. (Courtesy of the U.K. DfT.)

Figure 8.15 MPS OBU from the DIRECTS trials. (Courtesy of the Fareway Alliance.)

Figure 8.16 Roadside gantry (DIRECTS). (Courtesy of the Fareway Alliance.)

charging system will be introduced in the United Kingdom within a decade. Extensive research into solutions for distance-based charging and more targeted congestion and environmental charging is currently being undertaken. The success of the large-scale road pricing scheme in London, coupled with the results of the DIRECTS project, are building a foundation of experience and evidence-based knowledge in the area of road pricing to enable the United Kingdom to further its plans for charging. A number of local authorities, including Durham and Cambridge, also have been awarded funding from the U.K. government's Transport Innovations Fund to explore innovations in demand management and road user charging. It is expected that this fund will eventually provide annual funding up to £200 million for the next decade, allowing local authorities to explore innovative demand management measures [75].

8.7.6 AGE A555 Technology Trials, Germany

A 7-km stretch of road on the A555 between Bonn and Köln was equipped in 1994 with toll gantries for the demonstration and evaluation of 10 automatic toll systems. Figure 8.17 shows a photograph of all the trial systems of in-vehicle units installed in a single vehicle. Various systems were included in the trial, ranging from simple AVI tags to transponder/smart card–based systems, to wide area satellite/GSM–based systems.

The significance of the trials was that a variety of system types were competing to fulfill the requirements of a multilane interurban tolling system for Germany. Several DSRC systems were tried, using either 5.8 GHz, infrared, and 2.45 GHz. Figure 8.18 shows a mobile enforcement vehicle using infrared DSRC. Some radio

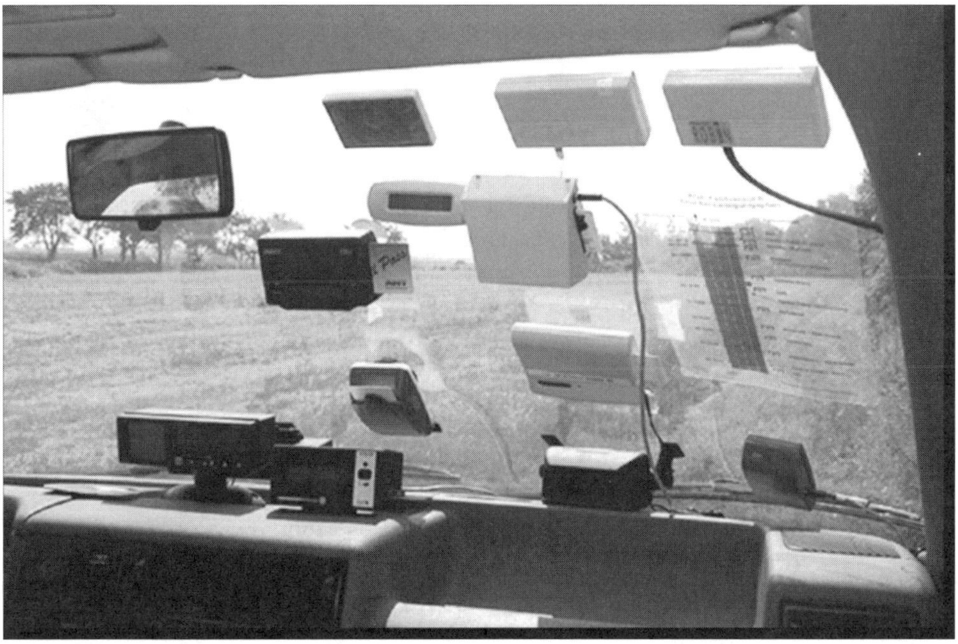

Figure 8.17 OBUs under test from the AGE A555 trials.

Figure 8.18 Mobile enforcement vehicle with roof-mounted infrared transceiver.

frequency wide area systems were also tried, using cellular radio from DT Mobile, and a GPS-based solution, using the ROBIN system from Mannesmann (now Vodafone).

The trial evaluation was extensive and led to the conclusion that technology to introduce tolling on the autobahn network of Germany was at least a decade away [36]. However, the final report did suggest that it would be feasible to introduce a system for HGV charging in a shorter time frame. This would be due to the lower number of vehicles to be equipped, and the possibility of using larger, costlier, and more sophisticated in-vehicle equipment than would be acceptable for the general vehicle fleet. This led to the introduction of HGV charging in Germany, initially planned for launch in 2003, but delayed until 2005 due to technical difficulties [42, 76].

References

[1] Wilson, P. W., "Welfare Effects of Congestion Pricing in Singapore," *Transportation*, Vol. 15, No. 3, 1988, pp. 191–210.

[2] Toh, R., "Road Congestion Pricing: The Singapore Experience," *Malaysian Economic Review*, Vol. 22, 1977, pp. 52–61.

[3] Menon, A. P., et al., "Singapore's Road Pricing System: Its Past, Present and Future," *ITE J.*, Vol. 63, No. 12. 1993, pp. 44–48.

[4] Blythe, P. T., *State of the Art Review of Automatics Debiting and AVI Systems*, DRIVE Project V1030 PAMELA, Commission of the European Communities, Brussels, 1991.

[5] Blythe, P. T., "Electronic Road Pricing in Singapore—Report on Demonstration Phase," *RTI Focus Report*, September 1994.

[6] Menon, A. P. G., and K.-K. Chin, "The Making of Singapore's Electronic Road Pricing System," *Proc. Intl. Conference on Transportation into the Next Millennium*, Singapore, 1998.

[7] Lew, Y-D., and C. Tee, C., "Singapore's Experience with Road Pricing: From Manual to Automatic," *5th ASEAN-Japan Workshop on Urban Transportation*, 2000.

[8] Government Office for London, "London Congestion Charging Research Programme: Principal Findings, Report," *HMSO*, London, U.K., 1995.

[9] ROCOL, "Road Charging Options for London: A Technical Assessment, Report," *HMSO*, London, U.K., 2000.

[10] Blythe, P. T., J. Walker, and P. K. Knight, "Is It Feasible That Automatic Licence Plate Recognition Systems Can Be Used as the Primary Means of Road-User Charging?" *Proc. 8th Int. Congress on Intelligent Transportation Systems*, Sydney, Australia, October 2001.

[11] Hills, P. J., and P. T. Blythe, "The Future for Road User Charging in the UK," Ingenia, *Journal of the Royal Academy of Engineering*, No. 14, 2002.

[12] Livingstone, K., et al., "Driving Through Change—Introducing Congestion Charging in London," *Planning Theory and Practice*, Vol. 5, No. 4, December 2004.

[13] Evans, J., "Congestion Charging in London," *Proc. 10th World Congress on Intelligent Transport Systems*, Madrid, Spain, November 2003.

[14] Dix, M., "London Results and Reactions. Three Years After the First Implementation of the Scheme," *Proc. Engineer Conference, EU Road User Charging 2006*, London, U.K., January 2005.

[15] Leromonachou, P., M. Enoch, and S. Potter, "All Charged Up—Early Lessons from the Durham Congestion-Charging Scheme," *Town and Country Planning*, Vol. 72, No. 2, February 2003, pp. 44–48.

[16] Schelin, E., "Road Charging in Sweden: An Update," *Proc. IEE Road Transport Symposium*, London, U.K., December 2005.

[17] Waersted, K., and K. Bogen, "No Stop Electronic Toll Payment Systems," *Proc. 2nd Intl. Conference on Road Traffic Monitoring*, London, U.K., February 1989, pp. 128–132.

[18] Hoven, T., "Experiences with the Norway Toll Ring Project," *Proc. IBC Conference Electronic Payment Systems in Transport*, Amsterdam, 1996.

[19] Ramjerdi, F., "The Norwegian Experience with Electronic Toll Rings," *Proc. Intl. Conference on Advanced Technologies in Transportation and Traffic Management*, Singapore, May 1994.

[20] Trondsen, J., "The Autopass Contract—An Implementation of the CESAR II—EFC Interoperability in Norway," *Proc. 10th World Congress on Intelligent Transport Systems*, Madrid, Spain, November 2003.

[21] Gjerde, A., "Implementing Fully Automated Free-Flow Toll Stations in Norway," *Proc. ITS World Congress*, San Francisco, CA, November 2005.

[22] Waersted, K., "How to Achieve Public and Political Acceptance for Urban Tolling: Experiences from Seven Norwegian Cities," *Proc. IBTTA Organization Management Workshop and Leadership Summit*, Orlando, FL, April 2005.

[23] Canadian Highways Infrastructure Corporation, *Project Sheet—407 ETR*, 2006, http://www.chichwys.com/407_EN.pdf.

[24] Information and Privacy Commissioner (Ontario), *407 Express Toll Route: How You Can Travel the 407*, 2006, http://www.ipcon.ca.

[25] Government of Ontario, *Highway 407 Act*, 1998, http://www.e-laws.gov.on.ca/DBLaws/Statutes/English/98h28_e.htm.

[26] Gambard, J.-M., and F. Malbrunot, "Raising Awareness of the MEDIA Project and Its Links to HGV Tolling and France and Europe," *Proc. Engineer Conference, EU Road User Charging 2006*, London, U.K., January 2006.

[27] Davis, G., "Electronic Tolling, Payment and Enforcements," *Proc. 10th World Congress on Intelligent Transport Systems*, Madrid, Spain, November 2003.

[28] MyEgov Taiwan Headline, "Electronic Toll System to Begin Operation," April 13, 2006.

[29] Organization of Economic Cooperation and Development, *Country Update: Japan*, 2004, http://www.oecd.org/dataoecd/18/9/2004782.pdf.

[30] Organization for Road System Enhancement, *How to Participate in ETC*, 2004, http://www.orse.or.jp/english/howto_participate.html.

[31] Organization for Road System Enhancement, *ETC in Japan*, 2004, at http://www.orse.or.jp/english/info.html.

[32] Blythe, P. T., "Is It Feasible to Introduce a Distance-Based Road User Tax for the Haulage Industry in the U.K.?" *Proc. 9th World Congress on Intelligent Transportation Systems*, Chicago, IL, September 2002.

[33] CEC, "Commission of the European Union, European Transport Policy for 2010: Time to Decide," white paper, 20010912, Brussels, 2001.

[34] EMCT, European Conference of Ministers of Transport Council of Ministers, *Reforming Transport Taxes and Charges*, CEMT/CM, 2003.

[35] Commission of the European Union, "Proposal for a Directive of the European Parliament and of the Council Amending Directive 1999/62/EC on the Charging of Heavy Goods Vehicles for the Use of Certain Infrastructure," 2003.

[36] Stappart, K.-H., "Field Trials on Electronic Fee Collection on Motorways in Germany," *Proc. IBC Conference on Electronic Payment Systems in Transport*, Amsterdam, March 1996.

[37] Birle, C., "Use of GSM and 3G Cellular Radio for Electronic Fee Collection," *IEE Seminar on Road User Charging*, London, U.K., June 9, 2004, http://www.iee.org/oncomms/pn/auto.

[38] Schelin, E., I. Gustafsson, and P. T. Blythe, "European Road User Charging for Heavy Goods Vehicles—An Overview," *Proc. 12th World Congress on Intelligent Transport Systems and Services*, San Francisco, CA, November 2005.

[39] Blomberg, I., and R. Poersch, "Tango Collect Report 3: Analysis of the Current Situation on Road Charging for Heavy Goods Vehicles in Germany, Austria, Switzerland, Sweden, Great Britain and Netherlands," Blekinge, 2004.

[40] Balmer, U., "Demand Management by User Charging—The Swiss Experience," *Proc. 10th World Congress on Intelligent Transport Systems and Services*, Madrid, Spain, November 2003.

[41] Egeler, C., and M. Bibaritsch, "Enforcement of the Austrian Heavy Goods Vehicle Toll," *Proc. 10th World Congress on Intelligent Transport Systems and Services*, Madrid, Spain, November 2003.

[42] Charpentier, G., and G. Fremont, "The ETC System for HGV on Highways in Germany: First Lessons After System Opening," *Proc. 10th World Congress on Intelligent Transport Systems and Services*, Madrid, Spain, November 2003.

[43] Gustafsson, I., and E. Schelin, *Tango Collect—Differentiated Kilometer Charges as a Driving Force for Implementing Telematics for Heavy Goods Vehicles,* Final Report, Blekinge Institute of Technology and SWECO, 2004.

[44] DoDoo, N. A., and N. Thorpe, "A Pavement Damage-Based System for Charging HGVs for their Use of the Road Infrastructure," *Highways: Cost and Regulation in Europe*, Bergamo, Italy, 2004.

[45] Davies, P., and F. K. Sommerville, "The Development of Heavy-Vehicle Electronic License Plate Concept," *Transportation Research Record*, No. 1060, 1986, pp. 121–127.

[46] Henion, L., and B. Koos, "The Heavy Vehicle Electronic License Plate System Development Program: A Progress Report," *Proc. Transportation Research Forum*, Vol. 27, No. 1, 1986, pp. 188–192.

[47] http://www.roadtraffic-technology.com/projects/ny_thruway/.

[48] Parkany, E., "Can HOT Lanes Encourage Car Pooling," *Transportation Research Board 78th Annual Meeting*, Washington, D.C., 1988.

[49] O'Mahony, M., and P. T. Blythe, "Charging for Road Use: A Comparison of U.S. HOV and HOT Lanes with the European Approach to Road User Charging," *Proc. World Conference on Transport Research*, Seoul, Korea, July 2001.

[50] Dawson, J. A. L., and F. N. Brown, "Electronic Road Pricing in Hong Kong 1 A Fair Way to Go?" *Traffic Engineering and Control*, Vol. 26, No. 11, 1985, pp. 522–525, 527–529.

[51] Catling, I., and B. J. Harbord, "Electronic Road Pricing in Hong Kong: 2 The Technology," *Traffic Engineering and Control*, Vol. 26, No. 12, 1985, pp. 608–615.

[52] Dawson, J. A. L., "Electronic Road Pricing in Hong Kong: 4 Conclusions," *Traffic Engineering and Control*, Vol. 27, No. 2, 1986, pp. 79, 81–83.

[53] Dawson, J. A. L., and I. Catling, "Electronic Road Pricing in Hong Kong," *Transportation Research*, Vol. 20A, No. 2, 1986, pp. 129–134.

[54] Catling, I., and J. Opiola, "The Hong Kong Electronic Road Pricing Feasibility Study," *Research Seminar on Experiments in Road-User Charging*, University of Newcastle upon Tyne, November 1997.

[55] Blythe, P. T., and P. J. Hills, "Electronic Road-Use Pricing and Toll Collection: The Results of the ADEPT Project," *Proc. Intl. Conference on Advanced Technologies in Transport and Traffic Management*, Singapore, May 1994.

[56] Blythe, P. T., and P. J. Hills, "The ADEPT Project: 1. Overview," *Traffic Engineering and Control*, Vol. 35, No. 2, February 1994.

[57] Blythe, P.T., and M. J. J. Burden, "The Technical and Institutional Issues Associated with the Enforcement of Multi-Lane Debiting Systems," *Proc. IEE Colloquium on the Enforcement of Traffic Regulations*, London, U.K., November 1996.

[58] Oldridge, B., "Congestion Metering in Cambridge City, United Kingdom," in *Road Pricing: Theory, Empirical Assessment and Policy*, B. Johansson and L.-G. Mattsson, (eds.), Boston, MA: Kluwer Academic Publishers, 1994, pp. 131–140.

[59] Clark, D. J., et al., "The ADEPT Project: 3. Congestion Metering—The Cambridge Trial," *Traffic Engineering and Control*, Vol. 35, No. 4, April 1994.

[60] Blythe, P. T., A. Rourke, and D. J. Clark, "Results of the ADEPT Trial in Cambridge," *Proc. IEE Intl. Conference on Road Traffic Monitoring and Control*, London, U.K., April 1994.

[61] Clark, D. J., et al., "Implementation and Evaluation of the ADEPT Congestion Metering System in Cambridge," *ADEPT Contract Deliverable V2026:ME-XII*, CEC DG-XIII-C6, Brussels, June 1994.

[62] Ison, S., "A Concept in the Right Place at the Wrong Time: Congestion Metering in the City of Cambridge," *Transport Policy*, Vol. 5, No. 3, July 1998.

[63] Blythe, P. T., "Demand Management: An Overview of European Research Activities," *City Trans: 4th Intl. Conference on City Planning, Transportation and Traffic Engineering*, Singapore, September 1997.

[64] Santos, G., "Road Pricing on the Basis of Congestion Costs: Consistent Results from Two Historic U.K. Towns," *Transportation Research Record*, 2000, Vol. 1732, pp. 25–31.

[65] Van Wijk, D. P., "Pan-European Automatic Toll Collection by Means of a Standardized Short Range Communications Link," *Proc. Intl. Conference on Advanced Technologies in Transport and Traffic Management*, Singapore, May 1994.

[66] Van Nifterick, W., "RekeningRijden: Free Flow Multilane Fee Collection as a Demand Management Tool," *Proc. 5th World Congress on Intelligent Transport Systems*, Seoul, Korea, October 1998.

[67] Ubbels, B., P. Rietveld, and P. Peeters, "Environmental Effects of a Kilometer Charge in Road Transport: An Investigation for the Netherlands," *(Elsevier Science) Transportation Research Part D: Transport and Environment*, Vol. 7, No. 4, June 2002, pp. 255–264.

[68] Dutch Minister of Transport, "Contours of Implementation of Congestion Charging (Rekening Rijden): Abstract of a Letter to Parliament from the Minister of Transport," June 23, 1995.

[69] Department of Environment, Transport and the Regions (DETR), "Paying for Better Motorways. Issues for Discussion," 1993.

[70] Department of Environment, Transport and the Regions (DETR), "A New Deal for Transport, Better for Everyone," 1999.

[71] http://www.dft.gov.uk/stellent/groups/dft_localtrans/documents/page/dft_localtrans_503865.pdf.

[72] Matheson, D., et al., "Open Minimum Interoperability Specification Suite (OMISS) for U.K. RUC," *Proc. IEE Road Transport Symposium*, London, U.K., December 2005.

[73] Jones, K., K. McGhee, and D. Makinnon, "DIRECTS Road User Charging Demonstration Project: Results," *Proc. IEE Road Transport Symposium*, London, U.K., December 2005.

[74] Matheson, D., "Towards a National Specification for Interoperable Road User Charging," *Proc. 12th World Congress on Intelligent Transport Systems*, Nagoya, Japan, October 2004.

[75] Blythe, P. T, "Congestion Charging: Technical Options for the Delivery of Future U.K. Policy," *Transportation Research Part A*, Vol. 39, No. 7–9, August 2005, pp. 571–587.

[76] Kossak, A., "Tolling Heavy Goods Vehicles on Germany's Autobahns," *IEE Seminar on Road User Charging*, London, U.K., June 9, 2004.

CHAPTER 9

Future Developments

9.1 Introduction

As seen from the previous chapters of this book, the ideas, concepts, and technology for road tolling and road user charging have evolved dramatically over the past 20 years, from humble beginnings to a profusion of systems now available commercially or being demonstrated and trialed. The worldwide market for the equipment exceeds $10 billion, with estimates placing the U.S. market at $3 billion [1]. The successful implementation of demand management schemes in Singapore, London, Trondheim, Stockholm, and Durham have also seen technologies for charging and enforcement evolve into the urban environment to manage traffic demand. This evolution created the need for interoperability, interworking systems, the introduction of truck tolling charging, and probably national pay-as-you-drive schemes. It is likely that we will see the introduction of some new technologies for charging, new means of introducing the charge, and new ITS systems that complement road user charging. This chapter introduces some of these concepts and discusses road charging in the context of current concerns about congestion, energy availability, environmental pollution, greenhouse gas emissions, and climate change. The perspective of this chapter on future developments extends from the short term to 2055.

The future developments described here are not comprehensive. The selection gives examples of what could impact the policies and technologies that will shape the future of electronic toll collection and road user charging.

9.2 New Communications and Location-Based Technologies

9.2.1 Vehicle Infrastructure Initiative

Looking at future developments in communications and technology that may have implications for the road tolling and road user charging markets, one of the most interesting is the vehicle infrastructure integration (VII) initiative, led by the U.S. DOT as part of the national ITS program.

The VII initiative is not primarily aimed at the tolling and road user charging market, but rather for safety and anticollision applications. Nevertheless, the goal of harmonizing infrastructure to vehicle communications for ITS applications could have a significant impact on the future of charging policy and technology developments in the United States [2].

VII is essentially a partnership between the ITS industries, the automotive industries, the government, and state and local transportation agencies. If plans develop as envisaged, VII could result in the installation of a 5.9-GHz communications module as original equipment in every new car manufactured for the U.S. market. This would be significant, because this has been a goal of European governments and road agencies for the past decade, but with little movement from the European car manufacturing sector. Thus, if VII is successful in the United States, then it could generate the necessary momentum to achieve a similar result in Europe and elsewhere.

A key feature of the VII program is the concept that there could be significant improvement in crash prevention if vehicles could communicate with each other and with the highway infrastructure, if it were technically feasible to do so [3]. This is the top priority of the Federal Highway Administration (FHWA) and state highway authorities, since there is a clear objective to prevent crashes at intersections, which account for 9,000 deaths and 1.4 million injuries, of which more than one-third are serious. This equates to the killed and serious injuries (KSI) benchmark as used in Europe. The total accident rate in the United States is also high, with more than 2.5 million accidents annually. If ITS communications technology could be pervasively deployed at potential accident hotspots, then vehicles could cooperate with each other and exchange information and warnings with the local infrastructure. Similarly, to avoid a variety of crashes in which the vehicles are forced off the road, which account for more than 13,000 deaths annually in the United States, drivers and highway operators could benefit greatly from communication between roadways and vehicles. These two classes of accidents account for more than 50% of U.S. highway fatalities.

The U.S. DOT is also focusing on the management and operation of the existing transportation system and the related development of new tools and approaches, including interactive traveler information. Over the past 15 years, the amount of new roads constructed in the United States has increased by 2%, while the vehicle miles traveled have increased more than 80% [2]. Thus, congestion remains a significant social, economic, and environmental issue. For example, the total amount of vehicle delay reached 3.7 billion hours in 2003, and the American Association of State Highway and Transportation Officials (AASHTO) estimated that the unnecessary consumption of fuel while in traffic jams was approximately 2.3 billion U.S. gallons. Other harmful effects of congestion also include increased noise and air pollution.

Using VII to provide real-time information on network performance, and enabling traffic signals and controls to be operated "smarter," incidents could be detected and responded to more quickly, work zones could be better planned, and motorists and commercial vehicle operators could be better informed of adverse travel conditions, such as congestion, construction, and poor weather. If vehicles could communicate with the roadway using VII, then they could even act as anonymous traffic probes, informing transportation operators of the road status, as has been used in Singapore (ATMS system), and in Sweden, Germany, and the United Kingdom.

The third ITS activity that has driven the VII initiative is the use of 5.9-GHz DSRC for electronic tolling. The U.S. DOT, working in cooperation with state and local governments and the toll industry, has obtained 75 MHz of spectrum in the 5.9-GHz band, a frequency that distinguishes VII from existing toll tags, the majority of which operate in the 902- to 928-MHz band in the United States, whereas in Europe DSRC at 5.8 GHz is the norm. The Federal Communications Commission (FCC) allocated this new spectrum for both public safety and private applications in 1999, and issued the licensing rules for its use in 2004, which is the catalyst for the new VII thinking. The transportation community, led by the toll industry, has simultaneously completed a draft set of standards for using the allocated spectrum. In its licensing order, the FCC requires the use of these standards (IEEE 802.11p, IEEE 1609, and IEEE1556) must cooperate and coexist with the CALM standards (see Section 9.2.5), as illustrated in Table 9.1.

The premise of the VII initiative is that the new standards established and licensed by the FCC will reduce the risk of installing 5.9-GHz equipment in vehicles by vehicle manufacturers. The VII initiative would also expect a GPS receiver system to be provided in the vehicle and integrated with the DSRC. This is beginning to be installed as original equipment manufacturer (OEM) in higher range vehicles, while many tens of millions of retrofitted navigation systems already exist in the U.S. automobile fleet. State authorities also would be required to introduce 5.9-GHz data communications nodes at the roadside and in other locations.

The DSRC roadside transceiver, which will provide medium-range broadcast capability, is currently under development, with a nominal range of 300m, creating hotspots around vehicles and roadside installations. It is initially anticipated that roadside units will be located at a significant number of signalized intersections in urban areas, and along the primary highway networks in rural areas throughout the United States. This would create a nationwide series of hotspots. The intent is not to establish a contiguous communications system, akin to the cellular phone network. Rather, a nationwide network of communication portals or hotspots will be established in areas of likely accidents or congestion, and sensors and traffic information beacons will be strategically placed at certain intersections and toll facilities.

In addition to the medium-range systems required for traffic information and tolling, wider area systems are being developed in the same frequency band to support safety information systems and emergency services. Figure 9.1 summarizes these systems.

The development of the vehicle-infrastructure communications element of VII, sometimes referred to as the wireless access in vehicle environment (WAVE) [4], will significantly enhance the introduction of 5.9-GHz road-to-vehicle communications technology. Using a standard, licensed communication channel should increase the prospects for interoperable DSRC tolling in the United States with many other value-added services, supported at the state and federal levels, and should take vehicle roadside charging technology in the United States to a new level. However, it is not clear whether the charging technology will remain with short-range DSRC, over 10m to 30m, or will migrate to the 300-m medium range. This would add technical challenges for the system, especially if the 300-m range were used for more than just broadcast communications and actually formed part of the transaction process, due to classification, localization, and enforcement requirements.

In the United States, the OmniAir Consortium, an independent, nonprofit association of DSRC operators, manufacturers, integrators, and transaction service providers, is actively working toward the twin goals of interoperability and certification.

Table 9.1 The Current Allocation of the 5-GHz Band

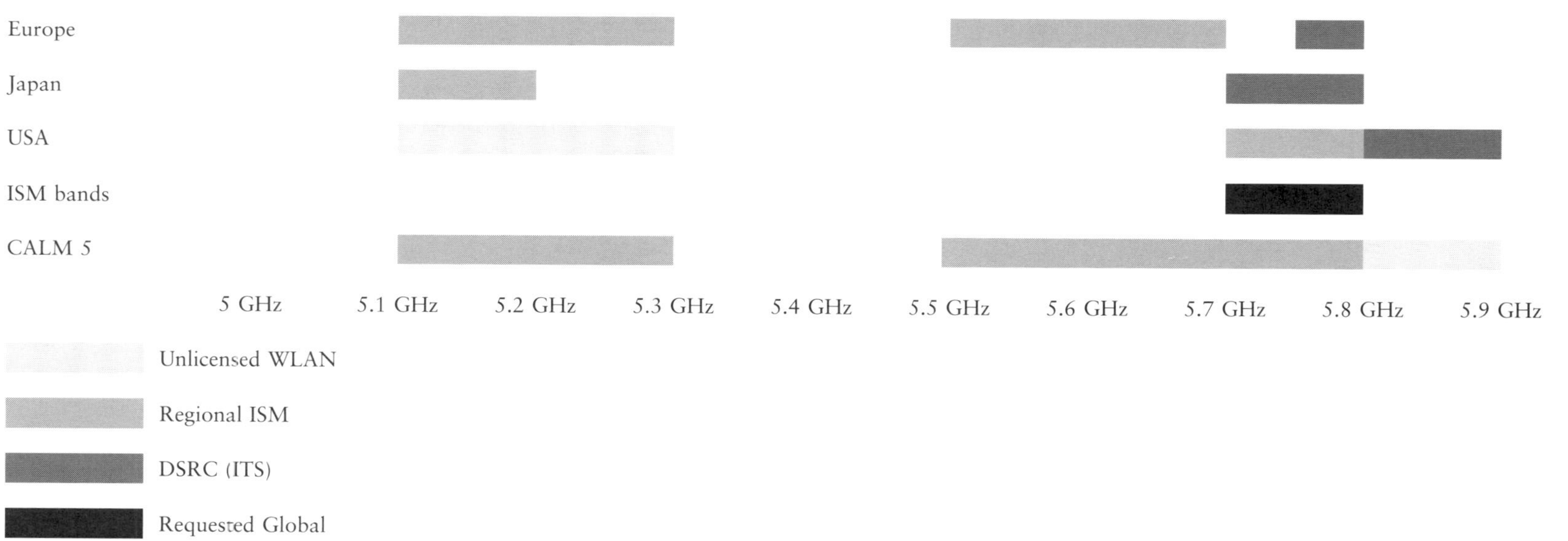

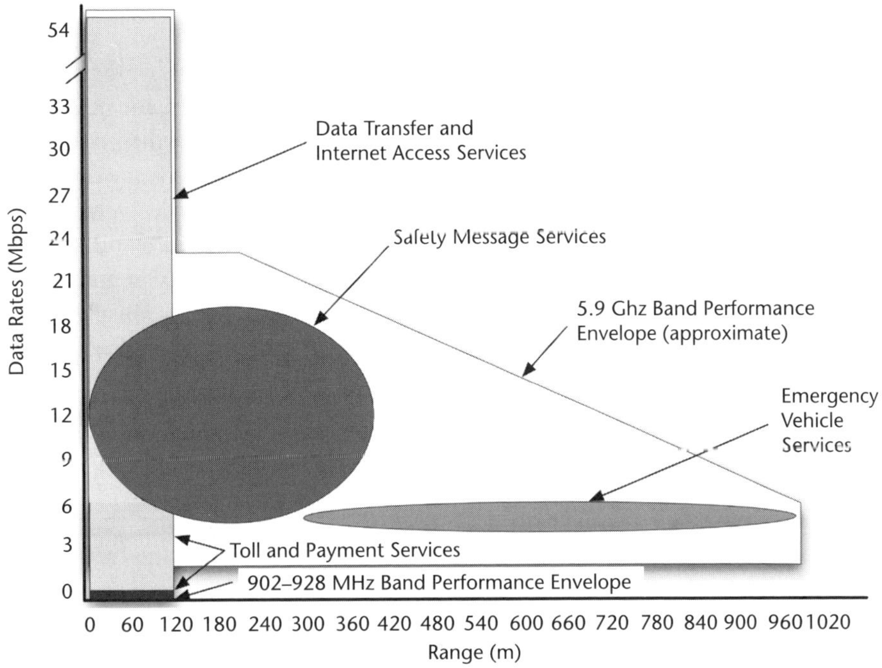

Figure 9.1 Service applications in the 5.9-GHz band.

9.2.2 Location-Based Services

9.2.2.1 Background

DSRC provides an accurate way to determine and record that a vehicle has passed a particular point on the road network. This technology supports a number of road pricing policies, particularly estuarial crossings, entry and exit to arterial interurban routes, and entry to urban areas where a cordon is defined. However, it is expected that road pricing will need to be increasingly more adaptable in the future. Systems based on Global Navigation Satellite Systems (GNSS) coupled with a wide area communication facility and short-range interface (DSRC) for enforcement are seen as the long-term solution, as described in Section 3.5.3.

There is much debate about whether GPS provides sufficient accuracy to enable an enforceable road pricing technique. GPS certainly works on interurban routes, but its use in an urban environment is less certain. Urban canyons restrict the number of visible satellites, and with much closer spacing of streets, the errors in the line-of-sight GPS signal mean that it may not always be possible to unambiguously identify the road. Section 3.5.3 described the technology behind augmentation methods at a systems level. In general, there are two ways to overcome this problem:

- Accept the limitations of GPS, and adapt the pricing principles to be compatible with its capabilities;
- Apply techniques to improve GPS performance.

9.2.2.2 Clarification of Location Requirements

To consider what GNSS systems can possibly offer in terms of location performance, it is worth considering why and what we require from a localization subsystem.

High accuracy is not generally required if GPS is used to calculate the overall distance traveled. High accuracy generally implies an accuracy of less than a meter. GPS is also sufficiently accurate to determine whether the vehicle is traveling on a highway or an adjacent primary road. Some complex road designs may have roads running parallel to each other, with only one road segment chargeable. In Japan, one of the arguments used against GPS systems in densely populated urban areas is not that two roads run side by side, but rather that one road runs underneath another road. Four or more GPS satellites can distinguish height as well as spatial location, but this system has not been tested much in the urban environment, where there are additional problems due to the reflection of satellite signals from buildings, changing the time and phase of the signal received from the satellite.

It may also be necessary to identify the specific lane that was used on an interurban route where high occupancy and toll (HOT) lanes are implemented (see Section 8.6). It would be possible to isolate HOT lanes from the other lanes if they were separated by a solid barrier (such as a Jersey barrier) and then monitor entry and exit from this lane at a limited number of places. However, use of this kind of barrier would reduce the overall capacity of the route and would have undesirable consequences. A completely open road would allow drivers to use the HOT when necessary, but would require vehicle location to be determined with precision greater than 0.5 to 1.0m. This level of precision in an urban environment would allow bus lanes to be adapted as bus and toll lanes, for example. Rather than to penalize a road user for using the bus lane, it might be better to charge a premium for using that lane. This is a possible future elaboration of a charging system, which is unlikely to be introduced until the public becomes familiar and comfortable with the current, relatively simple forms of charging.

When using GNSS systems, covered car-parking facilities raise additional challenges. If the vehicle is seen to approach a car park entrance and then "disappear," only to "reappear" outside the exit a few hours later, is it absolutely certain that the vehicle was in the car park?

9.2.2.3 Augmentation of GNSS Performance

There are several techniques that may be used to improve the GNSS system performance and increase the accuracy of the GNSS location function.

Two main methods are available for the current GNSS services, delivered through GPS:

1. *GPS satellite data correction.* GPS data contains many errors, such as satellite clock accuracy and ephemeris data, which can be detected at fixed, precisely known locations, so that corrections may be calculated from this data and applied to all receivers. Ionospheric and atmospheric effects introduce an error that varies according to the local weather conditions, so corrections must be applied from a local fixed tracking station.

2. *Carrier phase detection.* Carrier phase is a technique that can give accuracy of less than 1m, and is used autonomously to allow relative positions (e.g., for surveying) to be measured with high precision. The technique can be extended by comparing the phase data at the moving receiver with that at a fixed tracking station. This powerful technique is referred to as real-time kinematics (RTK) and requires a reliable communications channel between the two devices. Further work is needed to determine whether this could be implemented cost-effectively for ITS applications, including RUC.

9.2.2.4 Galileo

In December 2000, the European Commission and the European Space Agency agreed to fund the development of its own GNSS, Galileo, which is often touted as the European alternative to the U.S. GPS. The availability of an additional 30 Galileo satellites (27, plus 3 spares) to complement the GPS satellites will have far-reaching benefits for the user, both in the availability of satellites and in the services supplied.

Key augmentations that Galileo will deliver to the road user charging sector include the following.

- Galileo open service (OS) is the free-to-air service that will provide a number of improvements to the current version of GPS, including multiple frequencies, which will improve the multipath performance. The inclusion of satellite integrity information will allow receivers to choose the satellites that will give the best position fix, which may be critical in areas where current coverage may not be sufficient to offer the decimeter-range accuracy required for some aspects of road user charging.
- Galileo commercial service (CS) will offer further improvements in the location accuracy, plus a service level guarantee. These features should be attractive to the developers of road pricing applications, but may prove to be too expensive. Performance and cost details are not yet fully determined. However, the premise that a system will give a guaranteed service level is critical for road charging. If the GNSS service cannot be guaranteed, then there will always be an inherent risk in collecting fees through such means.

Galileo will have an important role to play in supporting future road user charging schemes. There will be approximately 60 GPS and Galileo satellites, with interoperability between the systems built into the receivers. A receiver will track approximately 10 to 20 satellites, and perform integrity monitoring of the signals to check on the signal quality. This will enable users to associate a higher reliability to the data than presently available, with a guarantee of service provided to the subscriber. The increased number of satellites will enhance positioning measurements in urban environments when buildings obstruct satellite signals' visibility. There will be less time when the receiver drops out, due to the lack of contact with the required four satellites. Studies [5] have shown that a 20-m positioning accuracy can be achieved 80% of the time in a high-rise urban canyon environment, using 28 GPS satellites plus 27 Galileo satellites. This compares to only 15% of the time with the current GPS constellation under the same conditions. The impact for traffic control and charging in urban areas is obvious.

In other applications, such as precise positioning in the offshore industry, the additional signal frequencies and satellites will reduce the time necessary to resolve the phase measurement ambiguities, which is required for the most precise sub-decimeter kinematic positioning. The user may not be interested in which satellites system is being employed at any time and may just utilize the highly accurate real-time positioning for his or her objective, whether it be navigation, surveying, leisure, or defense.

Several users of GPS have asked about the benefits of Galileo. Some of these benefits of Galileo have been outlined above, such as improved signal acquisition, signal integrity, and real-time positioning, along with overall improvements in urban canyon environments. The guaranteed safety of life (SoL) aspect in aviation, maritime, and road navigation applications, should be added. This is particularly important when we consider the vulnerabilities of any road user charging system that exclusively relies on the GPS system. Galileo will not substantially reduce the risk due to jamming. The low power of the received signals by a GNSS receiver do make them vulnerable to hackers or terrorists, who may wish to disrupt the charging process by covertly using a transmitter at the same frequency as the GNSS signals, which may interfere with the signals received by the GNSS receivers. This may have an economic repercussion on the scheme operator, notwithstanding the scheme credibility issue this may raise.

This problem is being actively investigated, and it is safe to say that the GPS and Galileo combined services will provide a more secure supply of GNSS, as well as an enhanced localization function that may overcome some of the concerns of using GNSS for road user charging in urban areas. After the successful launch of the first Galileo test satellite on New Years Eve 2005, it is expected that the program will begin delivering satellites for the Galileo constellation into orbit in 2008, with an expected date for a fully operational constellation sometime around 2012.

With the current emphasis on the development of GNSS solutions for HGV charging and possible solutions for national schemes for road user charging, it is likely that Galileo will have a major role in delivering such services. This premise is supported by several EU developments in the transport sector, including the conceptual universal on-board unit (UOBU) described later in this chapter.

9.2.3 Active Infrared

Infrared (IR) communications technology has had a long association with ITS. Autoguide was the brand name of an experimental dynamic route guidance system evaluated in Berlin and London in the late 1980s and early 1990s. This system used roadside-mounted IR beacons at intersections to collect route transit time information, with optimum routes calculated by a central traffic model, which were then downloaded to vehicles to provide navigation and guidance services with real-time updates. Technology has significantly advanced since then, with massive investment in IR technology for the entertainment and computer markets, particularly in visual displays, optical fiber communications systems, and reading and writing to optical media (CDs and DVDs). This has led to cost reductions and significant technical developments in infrared components and systems, some of which have been applied to the toll collection sector.

After the Autoguide [6, 7] trials in the United Kingdom and the similar ALI Scout trial [6] in Germany, IR was largely discounted for road-to-vehicle communications applications for a decade. In the early 1990s extensive technical evaluations in the Netherlands [8] and the United Kingdom suggested that microwave-based DSRC had significant performance advantages over the available infrared technologies, which at the time led to the concentration on microwave DSRC as the dominant technology for electronic toll collection, IR being largely discarded at the time. However, active IR has gradually reemerged in the late 1990s as a key ITS communications technology, and is extensively used in some countries, such as Japan, for infrastructure-to-vehicle information alerts. IR is emerging as a viable technology building block to support road user charging and tolling schemes. Two features of infrared communications: (1) the relatively large data bandwidth that the systems can support, and (2) the beamforming of the communications envelope (see Figure 9.2), which can be wide for broadcast applications or very narrow for vehicle (tag) localization and possibly anticollision and safety applications.

IR systems either complied with the Infrared Data Association (IrDA) standard, or were proprietary (i.e., did not support competitive supply or interoperability). The drafting and issuance of an international standard for the use of infrared for ITS applications have overcome these problems. ISO 21214 defines the physical and link layers for IR systems. The active participation of several companies led to this standard, which is part of the CALM [9] family of protocols. ISO 2121 can also be used to support dedicated applications (see Section 9.2.5).

IR is currently limited in its tolling and road user charging applications, but some key examples currently exist, and it is expected that, in Asia at least, the use

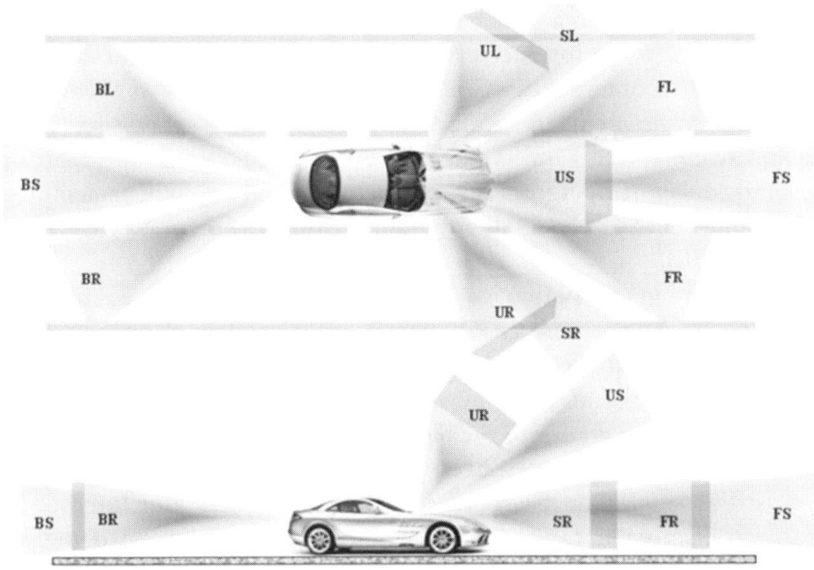

Figure 9.2 Directional communication for infrared within CALM. (*From:* [14]. © 2006 Bob Williams. Reprinted with permission.)

of infrared communications will increase as an alternative to microwave-based DSRC. No regional standards exist for such technologies at the moment in Asia, as opposed to in Europe and the United States, where great efforts continue to standardize DSRC systems in the 5.8- to 5.9-GHz range.

IR DSRC systems utilizing self-contained infrared OBUs are currently used for electronic toll collection in Japan, Malaysia, Taiwan (see Section 8.4.6), and South Korea. The German Toll Collect truck tolling enforcement system (see Section 8.5.1) uses IR communications to allow localized and directional interrogation of the on-board unit to determine that the system is operating correctly. In this project, enforcement is carried out in three ways:

- Fixed gantry based installations;
- Dedicated mobile enforcement vehicles, at normal traffic speeds, including interrogation of vehicles approaching in the opposite direction;
- Handheld units, from the roadside or on bridges, with ranges of up to 50m.

With several reference applications for IR communications for both toll collection itself and as an enforcement enabler for HGV charging, IR is likely to emerge as a viable technology option for road charging applications.

9.2.4 Wireless Ad Hoc Networks

The attention of the mobile communications research community is now focused on fourth generation (4G) systems. 4G systems will not in themselves be a new technology; rather, they will integrate a number of existing technologies, such as 3G cellular, digital audio broadcasts (DABs), and wireless local area networks (WLANs), into heterogeneous wireless networks to provide access to an increasing range of services. Data will be transported through 4G networks using packets that conform to the Internet Protocol version 6 (IPv6) standards. Mobile devices will be able to connect to a 4G network through the nearest WLAN hotspot access point (AP). The ability for mobile devices to access generic services via WLANs will make users totally independent of the mobile network operator. Local authorities and transport operators seem to favor this technology as a short- to medium-term solution for personal communications over local area network (LAN) distances [10].

The trend in computing is towards embedded data processing devices in everyday objects. As the density of such computing devices increases, so does their need for communications. Despite advances in 3G cellular networks, they are not infinitely scaleable, and will never provide sufficient bandwidth to support truly pervasive computing, due to the high cost of infrastructure and the limited capabilities of the embedded devices. Even 4G networks will have their limitations. WLANs are used as single-hop bridges between mobile users and Internet access points, so the transmission range of the technology limits access to the Internet. New ways of communication that do not rely so heavily on fixed infrastructure are needed [11].

A mobile ad hoc network (MANET) is a collection of mobile computing devices that cooperate to form a dynamic multihop network without using a fixed or completely fixed infrastructure (i.e., some devices in fixed locations, other devices are mobile, in cars or carried by individuals). WLANs in 4G networks provide a single-hop access to the Internet when a mobile device is within range of an AP. The devices in a MANET provide the routing services, so that a device can access the Internet even where no direct wireless connection exists between the device and an AP. One consequence of adopting a MANET architecture is that computing nodes become an integral part of the communications infrastructure, bypassing traditional network operators and allowing third-party access to mobile devices and their users. The evolution of current computing devices, such as motes [12], smart lumps, and SmartDust (using nanotechnology), is likely to revolutionize wide area communications in the next decade and provides a range of extremely low-cost wireless sensors that can measure a wide range of specific parameters. These sensors could measure pollution, noise, temperature, speed, direction, and vehicle presence, as well as provide pervasive vehicle-to-roadside communications, which will open up new possibilities and foundations for wireless road user charging systems. Studies at Newcastle University are evaluating this technology, with a trial of prototype devices mounted in vehicles and in the roadside infrastructure (e.g., lampposts and bus stops). See Figure 9.3.

Research into current SmartDust technology called motes (wireless devices), and their configuration into MANETs using on-road trials in the ASTRA project [13], is funded by the U.K.'s DfT as part of the Horizons Research Program. See Figure 9.4. The project aims to include future applications of the technology and protocols to transportation, particularly for development in the mobile environment (i.e., vehicle-mounted devices, interacting with other vehicle-mounted devices or infrastructure-mounted devices). Such devices that could form an "intelligent corridor" where vehicles and pedestrians are always connected with the infrastructure and may have a significant role to play in future intelligent infrastructure. Moreover, the devices may fill in gaps where GNSS or DSRC-based charging is not viable.

The project has already shown that MANETs represent a flexible new technology that can offer dynamic solutions to meet complex traffic scenarios and

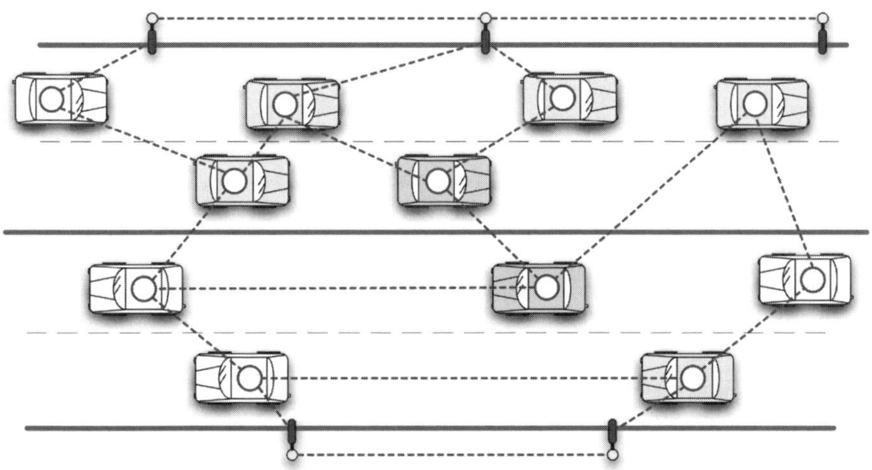

Figure 9.3 Mobile ad hoc communications.

Figure 9.4 Current SmartDust prototype using the Zigbee Communications Standard. (Courtesy of Newcastle University.)

innovative demand management strategies. Prototypes have been evaluated using the 800- to 900-MHz, 2.45-GHz, and 5.8- to 5.9-GHz frequency bands. Future in-house prototypes will migrate to a suitable frequency for widespread, rather than experimental, use.

The pervasive nature of the technology enables cars to be always connected to the infrastructure, in the same way that broadband users enjoy always-on Internet access at home, thus allowing an intelligent, configurable ITS infrastructure that will be available for a range of services to support travelers. Road users will perceive direct benefits from the introduction of the technology, increasing user acceptance. The costs of building and maintaining the infrastructure could be amortized over many such services delivered by third-party providers. Road user charging will be just another application, as far as the technology is concerned.

The research proves that motes, and their future nanoscale successors known as SmartDust, could easily supplement microwave DSRC, GNSS, and cellular charging techniques, and can quite possibly replace them, although this is probably an impractical use of the technology in the near future. The "always connected" nature of SmartDust motes means that they can provide the network with information on the location of every vehicle on the road. Air pollution sensor modules using SmartDust may measure the levels of air pollution due to the current level of traffic and congestion in real time. SmartDust could also form part of the future suite of communications technologies currently being specified in the CALM initiative.

9.2.5 CALM Communications

One of the major worldwide initiatives to bring together a number of communications technologies in the road and vehicle environment is CALM, which is currently being standardized in the OSI Standards body TC204, TC 37, and WG16 (see Sections 3.5.2 and 3.6.2).

The fundamental principle of the CALM concept is to bring together various communications media to meet the demands of continuous communications in the road environment. The architecture and standards are predicated on the principle of making the best use of the available resources. The resources are the various communications media available, and "best" is defined by the objectives to be achieved and their relative cost, while making the architecture flexible enough to enable future communications technologies to be added.

The CALM concept is therefore to provide a layered solution that enables continuous or quasicontinuous communications between vehicles and the infrastructure, or between vehicles, using the wireless telecommunications media that are available in any particular location. The CALM concept also requires the ability to migrate to different media as they become available. Media selection is at the discretion of user-determined parameters.

CALM is bringing together essentially four communications technologies into a suite of interoperable standards: 2G and 3G cellular phone technologies; short-range infrared communications, and DSRC-based communications that use the 5.8- to 5.9-GHz frequency band. All of the technologies will be familiar to operators of tolling and road use charging systems. CALM also includes the use of millimeter wave communications in the 62- to 63-GHz region for anticollision applications and mobile wireless broadband, which meets the HC-SDMA, IEEE 802.16e, and IEEE 802.20 standards [14]. See Figure 9.5.

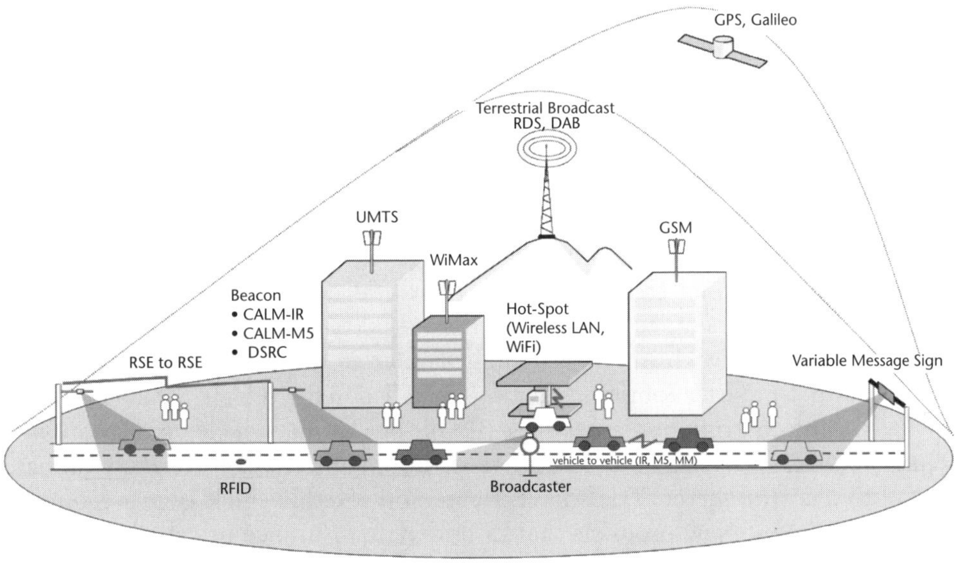

Figure 9.5 CALM Media. (*From:* [14]. © 2006 Bob Williams. Reprinted with permission.)

CALM will be used for infrastructure-to-infrastructure communications, but the primary use will be for vehicles communicating with other vehicles, or in the case of information and road user charging services, with the infrastructure. Therefore, all media must have a means of transparently networking from one cell to another, just as a cellular telephone does. This is a fundamental difference between CALM and the VII initiative previously described, in which the 5.9-GHz communications is in discrete, rather than continuous, locations. Transfers within a medium will be transparent to the user. Transfers between media may not be transparent, and may have brief interruptions, but the service will be automatically resumed on the new medium as long as it fits within the customer preferences. This is a key challenge to the industry over the next few years, but since the CALM consortium has selected the IPv6 protocol as its basis, it will undoubtedly deliver the desired performance.

The network protocols support the handover of a session between a point and a mobile station to another point using the same media, or the handover of a session between a point and a mobile station to another point using a different CALM medium.

Vehicles will be fitted with an ITS continuous digital communication device (whether or not this is CALM), which will be used as part of a road pricing and tolling application, and would logically replace a traditional (European) DSRC or U.S. UHF OBU. One must be realistic about the timeframe, so DSRC and UHF OBUs still have a few years as the dominant technology for tolling and probably road user charging applications. However, the CALM standard was adopted by ISO in June 2006.

9.3 Systems Innovations

9.3.1 Pay-As-You-Drive Insurance

Several countries have tested ITS systems that offer dynamic pay-as-you-drive (PAYD) insurance products, including Progressive in the United States, Lloyd Adriatic in Italy, Axa in Ireland, and Norwich Union in the United Kingdom [15]. What makes this development of potential interest for road user charging and tolling is that the equipment installed in vehicles to measure road use for insurance purposes could be modified to offer a wider road use charging capability.

These systems work by allowing drivers to pay insurance premiums based on the distance they travel, the time of travel, and the way that they drive, with a small additional charge to cover theft and damage insurance premium when a vehicle is parked.

The Norwich union scheme is similar to the trials undertaken in the United States by a sister company. A unit is installed in the vehicle, which measures mileage via the controller area network (CAN) bus of the vehicle, and location via a GPS receiver (see Figure 9.6). This data is stored and transmitted daily (in batch mode) to a central server. Trials have focused on several vehicle owner groups, including fleet users. New products that differentiate insurance pricing for young drivers in the United Kingdom discourage them from using their vehicles between 11 p.m. and 6 a.m., since it is found that young drivers are significantly more likely to

Figure 9.6 PAYD in-vehicle unit. (Courtesy of Norwich Union Insurance.)

have an accident during the evening and overnight periods [16]. Where GPS is employed, the insurer can advise drivers which roads to avoid, such as roads that are known to have many accidents. This data is generally shared by the insurance industry. A significant number of cars with PAYD boxes installed may gather data anonymously to sense the network and provide information on travel times and speeds for traffic management purposes.

Canada uses a slightly different approach. Aviva Insurance has launched their Autograph pilot in Canada. Motorists purchase a £35 microchip device that plugs into the diagnostic port of the car. The motorist receives up to a 25% discount from their insurance bill for downloading the data collected by the autograph device from the engine management system over the Web to the insurer. The data tracks mileage and speed of the car, and if the customer falls within certain risk management parameters, they will receive additional monthly discounts.

The technology required for PAYD may provide an alternative means for getting highly sophisticated telematics into a vehicle. These distance-measuring and location-recording functions could possibly enable other services, such as tolling and road pricing, to be integrated with the insurance service. Countries that are discussing the possible introduction of road user charging schemes, along with the costs and logistical issues associated with equipping tens of millions of vehicles, may be attracted to this integration with existing technology. This may equally apply to commercial navigation systems, as well as PAYD systems. Indeed, this piggy-backing principle is currently being seriously considered by several governments.

9.3.2 Universal On-Board Unit (UOBU)

The profusion of on-board units and tolling systems already in use in Europe, or those proposed for introduction over the next few years, has raised some worrying concerns for the European Commission, regarding the future interoperability and flexibility of such systems. The emergence of the European-led Galileo system in the early part of the next decade has encouraged the EU to launch several initiatives to examine the possibility of developing a generic, open specification for vehicle OBUs. The successful development of this specification would have a major impact on the road transport and haulage sectors and would cover a range of ITS services and systems. Tolling and road user charging would obviously be a major application domain for such a unit.

The European Commission has several initiatives that support transport policy, contained in the white paper "Time to Decide" [17]. The number of initiatives is encouraging, but the DG-TREN is concerned that the different requirements, and timetables for delivery may result in disparate developments and inefficient solutions, which ultimately could translate into increased cost to users. The UOBU project aimed to find an approach that could reduce the time to deploy services, which was considered to be key to the achievement of lower-cost European transport objectives [18].

DG-TREN awarded the study on the feasibility of a common OBU for Europe in January 2005 to SEA, a U.K. engineering and aerospace consultancy. The aim of the project, as stated by DG TREN, was "to investigate and define the functionality, constraints, and systems architecture, and to assess the benefits of a telematics platform integrated within the vehicle or as a single core vehicle unit" [19]. It was stated that the UOBU could potentially be used in all vehicles, to deliver a range of ITS services; and on the European transport systems, to supply a wide range of interoperable functions supporting road user charging, driver services, traffic management, enforcement, and emergency calls. The UOBU specification therefore must accommodate a wide range of users, including policymakers, statutory safety bodies, government bodies, automotive manufacturers, and road operators.

The project was split into three phases:

- *Requirements analysis:* Identification of user needs, system requirements, implementation scenarios, and business case considerations;
- *Core concept:* System design, and feasible component technologies;
- *Technical feasibility:* Overall UOBU feasibility, roadmap, and deployment recommendations.

The benefits of an OBU to stakeholders would only be visible if they supported the objectives of policymakers, sustained a business case for users and service providers, and achieved a critical mass for deployment within vehicles of all types. The feasibility assessment was therefore not only limited to technology capability and availability, but included deployment feasibility. Three potential implementation options were identified:

- Common on-board interface, an interface providing the minimum common functions that can be used by all applications;

- Upgrade of existing on-board equipment, a development of one of the key applications;
- An all-in-one UOBU, integrating of the applications in a single unit.

The study concluded that the common on-board interface (COBI) was the only option that met the stated objectives, was the least complex to install, and was expected to be available at the lowest cost and acceptable risk. Mandatory installation of this interface in all new HGVs initially, and eventually in all new light vehicles, was expected.

Figure 9.7 highlights the capability of the UOBU to deliver time, location, power, communications, and vehicle identification functions, through a common in-vehicle interface with the UOBU. This approach will support and accelerate delivery of strategic pan-European services, such as EFC, e-Call, and Digital Tachograph, and will facilitate private sector services. The UOBU would be made available for installation by vehicle manufacturers and for retrofitting by installers, using any combination of applications. The applications' respective capabilities depend on the UOBU, rather than the other way around. In fact, the UOBU would work quite well by itself with no application receiving its data.

The UOBU study and recommendations for its introduction was completed in mid-2006 [20].

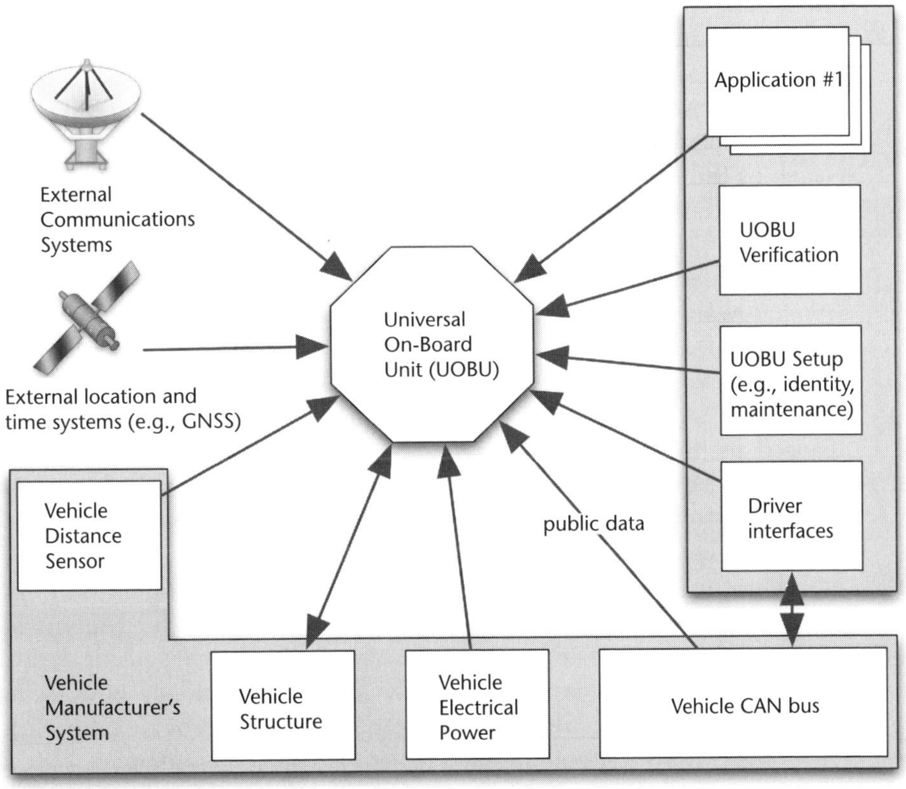

Figure 9.7 UOBU context diagram. (*After:* SEA, © 2006. UOBU contract awarded by the European Commission.)

9.3.3 Dynamic Heavy Goods Vehicle Charging

As discussed in Chapter 8, the use of road user charging as a means of recovering some of the external costs caused by the movement of heavy goods vehicles on roads is a trend that will continue to grow, as national governments recognize the benefits of such a demand-management and revenue-raising approach. The opportunity to modify HGV charging into a more targeted charging system will increase as technology develops. The HELP system in the United States dynamically measures the weight of vehicles using a weigh-in-motion platform, so the charge for crossing an interstate boundary relates to the load [21], and thus the likely wear and tear on the road.

The European Commission's 1995 green paper "Towards Fair and Efficient Pricing in Transport" identifies the principles of marginal cost pricing as a key requirement of a fair and efficient transport system. Marginal cost pricing involves the principle that a vehicle owner/driver should pay a fee for using the road that reflects all the costs of that journey, both the internal costs of fuel and vehicle operating costs, and the external costs of the journey. Such external costs include delays to the driver of the vehicle and to other drivers caused by capacity limitations, noise and air pollution, and the wear and tear of the road surface associated with the vehicle's use of the road. One cost that is clearly significantly higher for HGVs than for the general vehicle fleet is the physical damage caused by vehicles to the road infrastructure. HGVs are responsible for almost all of the structural damage to road pavements; private cars contribute little to this. Damage to the road surface is proportional to the fourth power of the weight. For example, a truck that is twice as heavy as a car causes 16 times as much damage, and a truck that is 10 times as heavy as a car causes 10,000 times as much damage. This is borne out by the fact that on the outside lanes of the highway, where in some countries (such as the United Kingdom) HGVs are banned, the wear of the road surface is significantly reduced. Various mechanisms for internalizing these costs, such as taxes and user fees, are currently used worldwide. The extent to which they are able to reflect the actual cost of damage has come under scrutiny as part of the move towards fair and efficient pricing in transport (see Section 8.5). The relative amount of damage caused by individual HGVs varies widely, depending on a number of vehicle, pavement, and climatic characteristics. One example of dynamic charging that considers a range of parameters to assess actual pavement damage caused by the HGV is a system being developed by researchers at Newcastle University, United Kingdom. They have developed and demonstrated a proof-of-concept system [22] for allocating pavement damage between individual HGVs for road user charging purposes (see Figure 9.8).

The components of the system included continuous data on position, speed, and distance traveled; a device to measure continuously dynamic wheel loads, on-board data processing and storage, two-way vehicle/roadside data communication, a digital road map database that combines with the positioning system to provide vehicle route information, and a database to provide information on pavement characteristics; and a model to estimate pavement damage [23]. Instruments for an on-road demonstration were installed on the front axle of an 18-ton, two-axle HGV (see Figure 9.9) in mid-2004, with data being recorded by the system

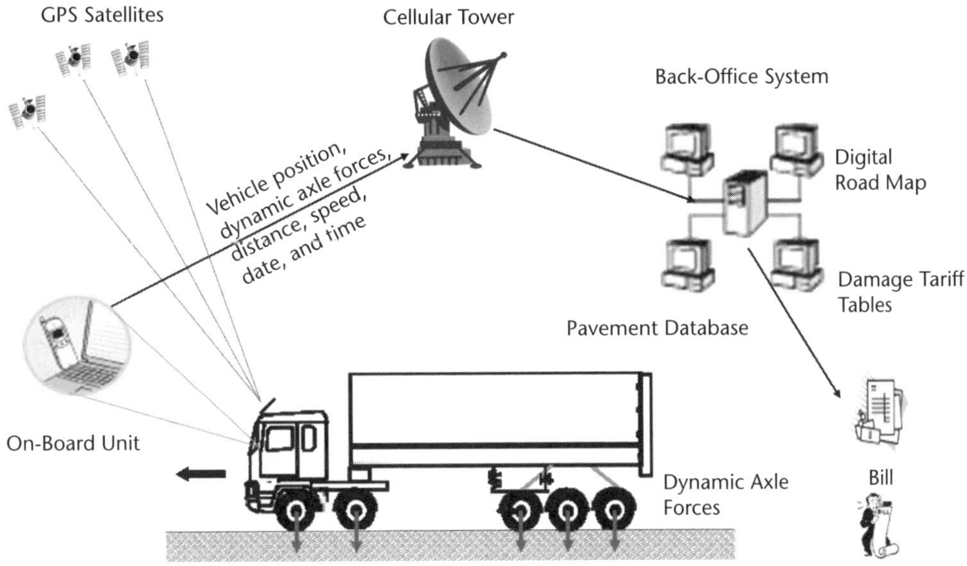

Figure 9.8 Schematic of HGV dynamic charging system. (Courtesy of DoDoo and Thorpe, Newcastle University.)

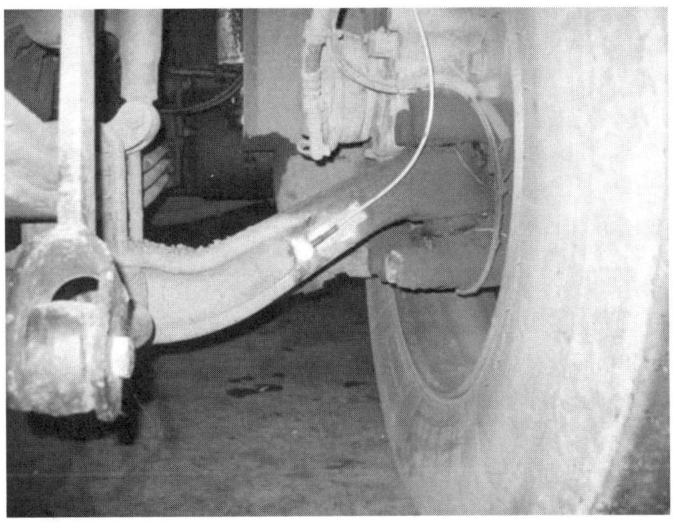

Figure 9.9 Dynamic axle weighing sensor fixed to demonstrate HGV. (Courtesy of DoDoo and Thorpe, Newcastle University.)

as the vehicle performed its normal daily activities throughout northern England over a 60-day period.

The potential impacts of this new system relate mainly to HGV operation, with charges being affected by the amount of load carried, and by other factors, such as the route taken, average speed, or season of operation. The system could encourage changes in fleet composition (e.g., to HGVs with higher numbers of axles), and in the selection of routes (e.g., to more durable pavements). Changes in route (e.g., from a thin to a thick pavement) will depend on factors such as congestion levels, travel time, and the location of freight distribution terminals. The system recorded route and used a national pavement database to know what sort of pavement on which the vehicle was traveling and the corresponding level of damage that the truck would cause. A further issue to be resolved is how to reconcile the opposing objectives of encouraging higher vehicle load factors to reduce traffic and emission levels by charging HGVs based on their maximum carrying capacities (e.g., in the Swiss and German systems), and of encouraging lower axle loads to reduce pavement damage. The appropriate vehicle configurations (e.g., size and number of axles) will provide the highest possible vehicle load factors and the lowest possible individual axle loads. See Figure 9.10.

The potential benefits of the widespread implementation of this system include an improved method for the recovery of pavement damage costs from individual HGVs; improved knowledge on where and how much pavement damage is occurring within the network to improve maintenance activities and targeting of resources; and improving loading practices such as reduced axle (and vehicle) overloading. Such a scheme is not presently on the agenda of any government, but the fact that technology can offer this improved charging policy enables policymakers to consider this approach as a future option for recovering costs from HGVs.

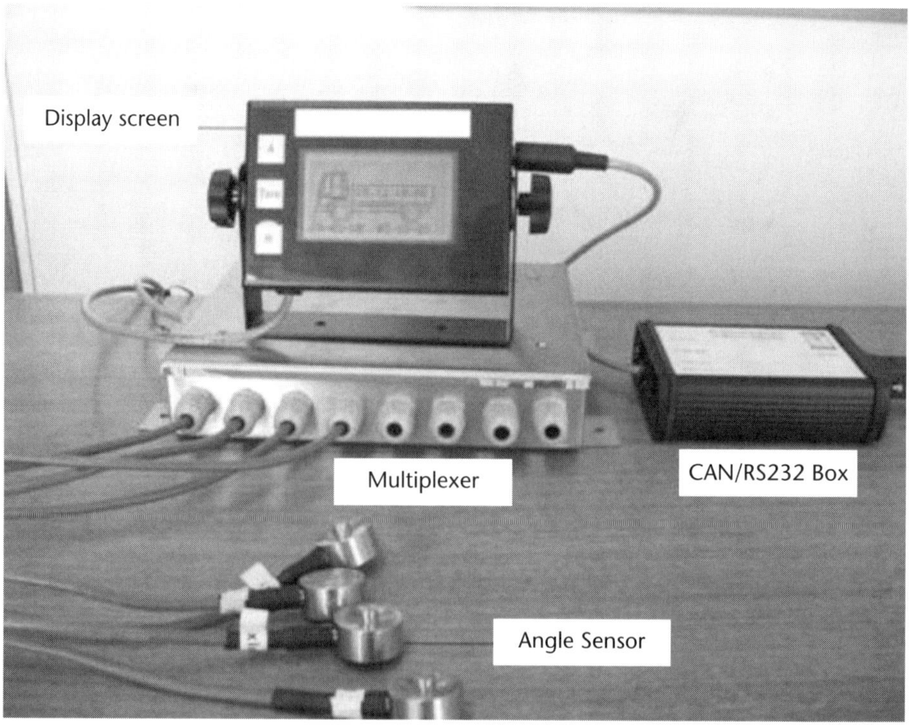

Figure 9.10 OBU with its dynamic weigh-in-motion axle sensors. (Courtesy of DoDoo and Thorpe.)

9.3.4 European Electronic Toll Service

9.3.4.1 Background

The Directive 2004/52/EC of the European Parliament and of the Council of April 29, 2004, on the interoperability of electronic road toll systems in the Community [24], outlines an European electronic toll service (EETS) that aims to provides technical, contractual, and procedural interoperability for equipment used to pay charges within the European Union. One of the primary objectives of EETS is to facilitate access to, and use of, chargeable roads and zones for road users with a single account, single contract, and single OBU. EETS will be phased in according to vehicle category, initially with trucks and long-distance coaches July 1, 2009, followed by passenger vehicles 2 to 3 years later.

9.3.4.2 Requirements

EETS overlays any existing toll or road user charging scheme, but requires scheme operators (referred to as "Toll Checkers" in the draft decision) to support EETS subscription on request, at no additional charge to road users. The draft decision [25] explains the approach:

- All scheme operators are required to accept an EETS-compliant OBUs on their networks.
- All scheme operators are required to offer road users, on request, a subscription to EETS and necessary EETS-compliant OBUs.
- To ensure interoperability with other scheme operators within the EU, it is proposed that the OBU support two microwave interfaces (CEN DSRC-compliant; and Italy's UNI 10607 parts 1, 2, and 3), CN/GNSS complying with ISO 17575, and minimum levels of reporting via GSM to ensure interoperable reporting.
- The CN/GNSS OBU shall be based on intelligent client architecture, and, in the absence of a standard, on-board maps shall be specified by operators whose charging policy is primarily based on CN/GNSS. The mechanism for agreement between operators is not defined.
- An OBU may offer additional applications to EETS. A European Network of Certification Centers (ENCC) will confirm EETS product compliance, ongoing supply assessment, correct installation onto vehicles, and correct operation on identified chargeable road segments
- The OBU will declare the vehicle's classification, according to a set of parameters (initially relating to heavy goods vehicles) that is common to all member states.
- The architecture of the OBU shall be able to support modular reconfiguration without modification of primary interfaces, to allow application upgrade from a competitive application and service supplier.

9.3.4.3 Expected Impact

The draft decision will be further revised during 2006 and 2007 and incorporated into national legislation within 2 years of final publication. The draft decision already sets expectations for these future logistical, operational, and contractual obligations on all charging scheme operators in the EU, except in those operators that have local charging schemes in which the cost of compliance is greater than the benefit afforded by EETS to road users.

The overall intention is to permit charging schemes to continue, and to impose requirements to ensure that road users have an OBU that will be technically and contractually accepted by all scheme operators (with the exceptions above):

- An EETS-compliant OBU is required to support all interfaces, but the scheme operator needs to provide either EETS-compliant DSRC or CN/GNSS infrastructure and related processes.
- A scheme that employs charging technology that is not EETS-compliant must accept an OBU that is EETS compliant in the future, potentially requiring an upgrade of the existing charging system.

The requirement to encode the vehicle's license plate number, and, at least for goods vehicles, mandatory installation, means that the OBU will offer additional support to the enforcement process, since the vehicle's license plate (as well as other vehicle specific parameters) will be declared as part of the charging transaction [26].

The path from draft decision to directive and then to local legislation clearly aims to establish multiple levels of interoperability that already meet the user's expectations similar to that of the roaming of mobile phones. The decision identifies the new roles of EETS Provider and Certification Centers that have not previously been identified. These roles will need to align with any existing national architectures of the EU's member states and with local plans, such as OMISS in the United Kingdom.

9.3.5 Convergence of DSRC and GNSS Charging

Local, regional, and national road user elements are required to meet the varied challenges of future road user charging schemes. Current systems have a way to develop before delivering a robust and workable solution to all charging domains.

There is a need to harness the best features of the primary charging technologies to achieve this. It is likely that a future solution for road charging will be delivered through a convergence of DSRC and GNSS systems, although this is at an early stage of development. To some extent, this is beginning to happen: the VII initiative assumes that vehicles will at some time in the future be manufactured with both a 5.9 GHz transceiver and a GPS unit in the vehicle as standard equipment [3]; CALM is developing a range of communications technologies that can coexist [14]; the UOBU project is moving toward an on-board unit with a generic specification to support a range of ITS applications [20]; and the European EETS directive explicitly names DSRC and GNSS as the generic solution types [25].

Such convergence is still far in the future, since the DSRC and GNSS manufacturers have traditionally operated with different business models, and have naturally been fierce competitors. However, this convergence will inevitably happen, whether by design, pragmatism, or by accident [25].

Short-range communications, delivered by DSRC, will be the solution for road tolling for the foreseeable future. It is unthinkable that toll plazas will not still exist 30 years from now. Innovations in charging, particularly distance-based charging and regional or national charging (whether for a specific subset of vehicles, such as HGVs, or for the entire vehicle fleet), will eventually harness GNSS. This is primarily due to the flexibility of GNSS, and the attraction of not having to install gantries (or other configurations of roadside equipment) at every on-road charging location. GNSS systems will always require an additional communications link for enforcement, although cellular telephone communications has been utilized in some cases, but in most existing systems, a DSRC link has been used for enforcement and checking (e.g., microwave DSRC in the case of the Swiss Lorry charging scheme, and infrared DSRC in the case of the German Toll Collect scheme). Integrating both technologies into a single package that can support the charging process, and with DSRC also supporting the enforcement process in some standardized way, will be particularly clear to national charging system operators and to vehicle owners. The calculation and distribution of charges should not overly concern the vehicle's owner, if the charges are understandable and the correct charges are levied. The driver should ideally have a relationship with only one OBU issuer. All charges incurred on any operators' networks will be billed through that one operator. The analogy is again to a mobile phone contract, in which the user has a relationship with one operator, but all charges from third-party operators are consolidated in a single bill. Such an integrated system could be widespread in many parts of the world within 10 years, if the trend toward road user charging schemes continues at the present pace.

9.4 Intelligent Infrastructure

9.4.1 Overview

New technologies that may emerge must be considered in the future of road user charging and congestion charging schemes. The incorporation of more intelligence into the transportation infrastructure is one possibility. This intelligence, and the blurring of the relationship between the infrastructure and the vehicle through major ITS communications initiatives, may deliver opportunities for new paradigms in road use charging. These may be based upon cost recovery, congestion management, and the environmental impacts of a car journey. Such a future exercise was undertaken in the United Kingdom from 2005 to 2006, under the Office of Science and Technology Foresight program.[1] Its intelligent infrastructure systems (IIS) explored how science and technology could bring intelligence into infrastructure over the next 50 years to meet demanding and sometimes conflicting objectives [27].

The project found that intelligent infrastructure could help meet these objectives and perhaps do more, such as stimulating growth, rather than simply supporting it. Intelligence could also support and promote a more inclusive society.

1. The Office of Science and Technology (OST) Foresight Directorate has given kind permission for original text from their reports to be used, where appropriate, in Section 9.4 [27, 28].

The process of looking 50 years into the future creates challenges for any project. It is very difficult to see how information technology might develop beyond a 5- to 10-year time horizon, let alone half a century. Businesses, being realistic in terms of profits and R&D investment, do not naturally look at such time frames; indeed, beyond 5 years is challenging to many. Government occasionally takes a long-term view in formulating policy, such as in energy availability and climate change. To deal with the uncertainties of future planning, the future of IIS was investigated in three complementary ways:

- Leading researchers wrote "state of research" reviews, which speculated on what all areas of science, including psychology, the physical sciences, and technology could deliver within the next few years. The research reviews covered areas as diverse as artificial intelligence, data mining, how information affects our choices, and the psychology of travel. These reviews are available in summary and as full papers at the Foresight Web site [28].
- The Development of a Technology Forward Look reviewed existing developments and applications of technology, and considered how IIS might shape business in the longer term [29, 30].
- The production of a set of scenarios may provide a range of credible and coherent pictures of the technology to invest in, and how society might react to those investments [31].

9.4.1.1 Technology Capabilities of the Future

In summarizing the key points of the study, advances in science and technology could provide:

- Information, so that individuals and those delivering transport-based services can make better-informed decisions, using all available, relevant information to them;
- The ability to manage the delivery of the services in real time, through the collection of information on the origin and destination of the user, real-time predictive modeling, and traffic demand management strategies;
- The ability to control the movement of goods and people, with vehicles connected to each other and the surrounding infrastructure, so they become an integral part of an intelligent system (the future intelligent corridor will connect people, vehicles, and information into a single entity);
- Infrastructure that is intelligent, so that it adapts itself to the needs of users;
- An integrated system that includes all modes of transport, both public and private;
- Integrated and intelligent supply and logistic chains that adapt continuously, to provide the most efficient path from supplier to user;
- Viable and sustainable alternatives to moving goods and people.

A number of new technologies would underpin these capabilities, described next.

9.4.1.2 Distributed Networks of Sensors

Networks of tiny and inexpensive sensors (see Section 9.2.4) could collect data on the position of just about anything and everything, and allow the monitoring of the flows of people and goods.

9.4.1.3 Data Mining

Software could analyze masses of data collected from monitoring the location of people and objects. It could detect patterns that allow an understanding of the behavior of complex systems in particular humans and how they interact with technology and transport to make informed travel decisions.

9.4.1.4 Agent-Based Software

Software agents could become the modern electronic equivalent of the butler, executive assistant, or broker, taking instructions and venturing out into the connected world to perform various tasks. The agent could find the best financial packages, negotiate deals, and help manage time.

9.4.1.5 Modeling and Simulation Technology

There is a growing use of computer models of complex systems to support decisions using principles similar to agent-based software. This approach makes it possible to test ideas for investment decisions prior to spending the money, in a way that reflects more closely the possible behavior of the main actors (e.g., commuters).

9.4.1.6 Advances in Communications Technology

The transfer speed of information will continue to increase. Wireless telecommunications technology that can download a movie in the time that it takes to drive past a video store is already under development in the United States. The suite of CALM technologies and VII deliver communications capability into every vehicle (see Section 9.2).

9.4.1.7 Speech Interface

The ability of computers to translate human speech is rapidly developing. The development of computers that can "understand" speech is still a major challenge, but there is broad agreement in the science community that this will be developed in the next 20 to 30 years. Speech recognition should become the primary mode of interaction with some information technology (IT) systems, a development that could be especially significant in transport, either to communicate with information systems or to provide instructions to our cars.

9.4.1.8 Self-Monitoring Complex Information Systems

Bringing together complex systems raises a number of issues in which software can play an important role. Software could anticipate unusual behaviors that can develop when complex systems interact. These behaviors can have affect stability, or may bring unexpected but welcome (or unwelcome) benefits. In the transport domain this may lead to micromanagement of networks. Software can already watch for signs of instability in complex systems, and could perhaps even develop the ability to repair or stabilize the system when emergent behavior could lead to failure, shut down systems that are causing problems, and break reinforcing loops that could cause damage.

9.4.1.9 How Can Technologies Meet Our Objectives?

The key issues surrounding these technical capabilities are the investments in them, and the reactions to them by various social groups around the world. For example, would we cede control of our cars to a central system if doing so would see the end of congestion? Would we be willing to let the system have information on the origin and destination of all of our travel, as we do now for air travel?

Historically, when transport systems have been improved and costs reduced, people have traveled more in distance (although generally not in time). Patterns of behavior have changed to reflect the increase in ease of travel, for example, living further away from places of work, developing cities and shopping facilities that are based around use of the car, and traveling for leisure on a national and international basis. This has supported economic growth, but it has also led to congestion, rising costs of maintaining the existing infrastructure, and increased environmental costs.

A key issue is how to use the technologies to improve efficiency, and deliver sustainable and robust solutions.

Some of the project's research reviews provide important insights, as described next.

Psychology of Travel

People appear to have a need to travel to find resources and to socialize. Individuals have spent between 55 and 65 minutes, on average, traveling a day since records were first kept. Travel decisions are based on cost (in time and money), on the travel-related activities, and perceptions about the mode of travel (is it reliable, safe, and pleasant?). Different people have different priorities when they travel. Reliability is the most important factor affecting the travel choices of most people, while some travelers like to explore and enjoy the uncertainty of a new route or mode of travel. Travel is embedded within long-established patterns of life, making changes difficult. Whichever category people belong to, most want to use the minimum amount of energy when thinking about traveling.

How Information Is Used to Make Travel Decisions

People make two types of decisions: daily decisions on how to travel and long-term decisions on where to live and work. The cost of property and the ease of access to places of work, leisure, and retail affect the long-term decisions. The decision on where to live establishes a travel pattern in the medium term. Decisions change from whether to travel to how to travel. Daily travel decisions tend to follow habit: the selection of a route that is good enough rather than the optimal route.

9.4.1.10 Economics

The use of technology often evolves in unexpected ways. For example, individuals may invest in in-car navigation technology to find their locations and the best route for travel. Over time, that very same technology could become part of a system to charge for road use. This could mean that the cost of introducing road charging might be less than we expect, since people already use and trust the required technology.

In order to capture and sustain these wider benefits, choices must be provided to the individual. Ways to support changes in the very pattern of social life must be found, with the realization that such changes take time.

9.4.1.11 Choice

There are four broad ways to introduce choice in trip making: the willingness to pay for a journey, and the decisions on where to live, work, and travel in the future.

Spatial Planning

An important way to support choice is to minimize the need to travel, so that people can live nearer to their place of work or education. Simple steps, such as redevelopment of city centers, construction of safe bicycle lanes into growth areas, and parking facilities near to public transport networks, might play a part in this. There are also many ways in which IIS can work with urban design to minimize travel and make it more efficient. Designing the urban environment with the best available technology will be important, building in resilience in case a shock that affects the freedom to move.

Virtual Communications

Information technology allows people to hold sophisticated virtual meetings instead of traveling, and evidence suggests that people use e-mail and telecommunications to maintain more geographically dispersed social networks. This can actually increase the required traveling distance. People still need some face-to-face contact, so they travel to see the people in their social networks from time to time, creating demand for longer trips. There has been a slow increase in working at home, but evidence so far suggests that this changes travel patterns rather than reducing travel. There might be a shift in travel patterns if there were a combination of increased telecommunication capability and increased travel costs through realistic road user and environmental charging.

Intermodal Choice

The ability to choose between different modes of transport can give people a more responsive and flexible service, which, in turn, can reduce the numbers of vehicles on the road. There are already signs of innovations in this area. A taxi company is using a text messaging system, with users sending their origin and destination in advance, so that the taxi company can manage its fleet more efficiently and provide better service. Another company operates buses to a timetable in busy periods, and in a more flexible demand-based way for the rest of the day, when people use their phone to request bus service, thus ensuring responsive services for rural areas. Car pooling and interactive traveler information also play a role here.

Local or Agile Production

Making things locally can reduce the need to move goods. The increase in the use of communications technology replaces some commodities, notably music and software, which people increasingly download rather than buy in a physical form. There is some shift from an ownership to an access economy. Developments in rapid prototyping have allowed complicated objects to be printed in three dimensions for commercial processes. As the cost of this technology falls, it opens the possibility of local manufacturing or even home manufacturing (i.e., the individual simply downloads the design and then prints the product at home). Laboratory-on-a-chip technology could offer a similar capability for the local production of medicines.

9.4.1.12 Supporting Behavioral Change

There are two broad ways to support changes in future social practices.

Information to Users

Providing easy-to-use information allows the traveler to select the optimal route and modes of travel, rather than a route that is just good enough. Providing information on that new route also reduces the stress of trying something new.

Full-Cost Recovery

Ensuring that people pay the full costs for each journey would make people aware of the real costs of travel. There are a number of options, from charging per kilometer traveled, to selling slots for journeys. A more radical option might be to give each person a carbon allowance, which would apply to all their activities, not just travel. In-vehicle displays provided with up-to-date information through VII or CALM technologies could give a dynamic breakdown of all elements of the travel costs, both approximately before one travels to facilitate planning the trip, and precisely while actually on the trip.

To deliver intelligent infrastructure that is sustainable, robust, and safe, we need to invest in intelligence on four levels.

- Intelligent design, minimizing the need to move, through urban design, efficient integration, and management of public transport and local production;
- A system that can provide intelligence, with sensors and data mining, providing information to support the decisions of individuals and service providers;
- Intelligent infrastructure, processing the mass of collected information and adapting in real time to provide the most effective services;
- Intelligent use of the system in which people modify their behavior to use the infrastructure in a sustainable way.

The scope for additional automated pricing of road travel is at the heart of the conclusions of the study. These are considered further in the scenarios used to support the Foresight project work. It is likely that road pricing will form the backbone of future intelligent infrastructure development, since it offers both the opportunity to deploy the technology in the infrastructure and in vehicles, and a mechanism to raise the revenue to pay for the deployment. The types of systems that may be deployed to deliver road pricing will also support a range of possible value added services that could utilize the technology.

9.4.2 Scenarios for 2055 and the Future Role of Road Pricing

Scenarios were developed to explore what future IIS may look like under a range of different conditions. These scenarios do not set out to predict what will happen or to suggest a preferred future. They are stories that offer various possible, even extreme, outcomes. The scenarios are designed to stimulate thought, to highlight some of the possible future opportunities and threats, and to inform today's decisions. The full details of the scenarios can be used to judge the risks and opportunities of policy relating to the future management of intelligent infrastructure.

The future is unlikely to look like any of these individual scenarios, and may well contain elements of all four, but the scenarios discuss how certain combinations of events, discoveries, and social changes could change the future. As such, the scenarios display the possibilities and opportunities that might arise [31].

For convenience, the scenarios have been labeled as:

- Good Intentions;
- Perpetual Motion;
- Tribal Trading;
- Urban Colonies.

It is worth pointing out that the names given to these scenarios are designed simply to help people to remember them. They are shorthand labels that capture the essential feature of each possible future. Shorthand names are also essential if the scenarios are to become part of a strategic conversation between an organization and its internal and external stakeholders.

The uncertainties about the future of intelligent infrastructure systems include developments in science and technology, the role of business and government, and social attitudes. The scenarios were underpinned by the development of systems maps. These allow a look beyond the detail of each scenario, and to think about why the trends and forecasted events might happen.

Each scenario envisaged some form of road user charging as a means of managing traffic demand, paying for infrastructure, and calculating the environmental costs of future transport and climate change.

9.4.2.1 Good Intentions

Good Intentions describes a world in which the need to reduce carbon emissions constrains personal mobility. A tough national surveillance system ensures that people travel only if they have sufficient carbon points. Intelligent cars, manufactured from recyclable, renewable, and energy-efficient materials, monitor and report on the environmental cost of journeys. In-car systems adjust speed to minimize emissions. Traffic volumes have fallen, and mass transportation is more widely used. Travel is priced in terms of road use and environmental pollution, and is pervasively collected from an IIS infrastructure that is not too different than the VII infrastructure delivered in 2009.

Businesses have adopted energy-efficient practices. They use sophisticated wireless identification and tracking systems to optimize logistics and distribution. Some rural areas pool community carbon credits for local transport provision, but many are struggling.

There are concerns that the world has done too little to repair the damage caused by decades of human activity. Airlines continue to exploit loopholes in the carbon enforcement framework. The market has failed to provide a realistic alternative energy source. See Figure 9.11.

Good Intentions is a world initially hamstrung by trying to satisfy all interests. Consensus on action to minimize environmental impact is lacking until 2025, when extreme weather has become so common that economic well-being is undermined by the impact on the environment. By 2050, drastic action becomes necessary, and it becomes a struggle to maintain the previous levels of economic activity. Over time, technology systems become essential to deliver efficiency and allow use of individual CO_2 allowances. The world becomes dominated by carbon budgets in the absence of cheap low-emission energy. The importance of designing the urban environment for less travel and more efficient use of resources slowly achieves sufficient importance.

9.4.2.2 Perpetual Motion

Perpetual Motion describes a society driven by constant information, consumption, and competition. Instant communication and continuing globalization have fueled growth in this world, and demand for travel remains strong.

Figure 9.11 Future scenario: Good Intentions. (*From:* [31]. © 2006 Foresight IIS. Reprinted with permission.)

New, cleaner fuel technologies are increasingly popular. Road use is causing less environmental damage, although the volume and speed of traffic remains high. Aviation still relies on carbon fuels that remain expensive, and is increasingly replaced by telecommuting for business, and rapid trains for travel.

Integrated, interoperable payment for all transport modes and associated services is collected through road use, value charging, and a range of service, energy, and infrastructure charges. See Figure 9.12.

A precondition of the always-on world of Perpetual Motion is energy supply, emission-free and preferably low-cost. The use of hydrogen is explored as an energy solution. The benefit of zero emissions at point of use that hydrogen gives is a major advantage in this future.

Technology in all its aspects is a large but not exclusive part of the picture, and the human capacity to cope with such a world resists full-scale adoption. In this scenario, technology achieves levels of interoperability, resilience, and ubiquity that renders it effective and trustworthy.

The strong economic position reflects a return on the investment made to deliver this technology and an energy-rich world. Problems still exist, arguably because technology is applied without regard to the design of the physical environment or its waste footprint. People are also too busy to think about efficient use. Crime adapts to a more connected world, as does law enforcement.

9.4.2.3 Tribal Trading

Tribal Trading describes a world that has been through a sharp and savage energy shock. The world has stabilized, but only after a global recession has left millions unemployed. The global economic system is severely damaged and infrastructure is falling into disrepair.

Long-distance travel is a luxury that few can afford, and for most people, the world has shrunk to their own communities. Cities have declined, and local food production and services have increased. Canals and sea-going vessels carry freight. The rail network is worthwhile only for high-value, long-distance cargoes and trips. There are still some cars, but local transport is typically by bicycle or horse.

Figure 9.12 Future scenario: Perpetual Motion. (*From:* [31]. © 2006 Foresight IIS. Reprinted with permission.)

There are local conflicts over resources; lawlessness and mistrust are high. The state does what it can, but its power has eroded. Toll charging reverts to the payment for protection of secure travel routes, as it did two millennia ago. See Figure 9.13.

The world of Tribal Trading is overwhelmed by shocks, initiated by a sharp energy shock and global competition for resources. Intelligent infrastructure is not on the agenda. It seems to be a world of opportunities not grasped and challenges ignored until too late. Some places fare better than others, but the universal focus is on making the most of the available resources, particularly local resources, and on being patient. Recycling is not just a good idea, but an economic necessity. Technology is limited to that which is robust and able to cope with fluctuations in energy supply. Legacy infrastructure is patched and patched again. Society starts to recover eventually, but it is a long, hard path.

9.4.2.4 Urban Colonies

In Urban Colonies, investment in technology primarily focuses on minimizing environmental impacts. In this world, good environmental practice is at the heart of economic and social policies. Sustainable buildings, distributed power generation, and new urban planning policies have created compact, dense cities. The use of environmentally led road user charges is accepted as the norm.

Transport is permitted only if green and clean. Car use is still energy-expensive and is restricted. Real-time information about transport is available in the cities. Public transport, electric and low-energy, is efficient and widely used. See Figure 9.14.

Competitive cities have the IT infrastructure needed to link high-value knowledge businesses, but poor integration of public systems means that private networks are most trusted. Rural areas have become more isolated, effectively acting as food and biofuel sources for cities.

Consumption has fallen. Resource use is now a fundamental part of the tax system, and disposable items are less popular.

Figure 9.13 Future scenario: Tribal Trading. (*From:* [31]. © 2006 Foresight IIS. Reprinted with permission.)

Figure 9.14 Future scenario: Urban Colonies. (*From:* [31]. © 2006 Foresight IIS. Reprinted with permission.)

Urban Colonies demonstrates that improved urban design, organizing to minimize the need for travel, has an important contribution. This is partly a response to environmental concerns and climate change, but it is also driven by suspicion about intelligent technologies, which creates a reason to find alternatives to travel. Cleaner technologies and low-emission energy create an environmental benefit, but the overall economic focus is more city-based than global, with medium economic growth. Societal benefits accrue from a society integrated more at the local level. Because of technology resistance, safety benefits are limited, and systems resilience is uneven, reducing global competitiveness. Clearly, people in this scenario are environmentally aware and more careful in their use of resources.

As seen from the scenarios, it is likely that road pricing will form the backbone of future intelligent infrastructure development, since it offers both the opportunity to deploy the technology in the infrastructure and in vehicles, and a mechanism to raise the revenue to pay for the deployment. The types of systems that may be deployed to deliver road pricing also support a range of possible value-added services that could utilize the technology.

9.4.3 Smart Market Protocols for Future Road Pricing

The Intelligent Infrastructure Systems project has largely focused on the transport domain, and includes consideration of how the operators of future transport infrastructure may harness the opportunities offered by enhanced ICT and intelligence to manage the competing claims on the transport infrastructure. A significant challenge is the management of road demand, particularly as ownership of private cars and the use of heavy goods vehicles for the distribution of goods continues to grow.

Despite the implementation of a range of innovative traffic management and demand mitigation strategies, and the growth of the use of ICT in transport, there is a growing consensus that some form of road pricing is needed for effective demand management.

Only a handful of countries have introduced road user charging and urban congestion charging schemes. In the United Kingdom, Foresight noted that the schemes now in place in the London and Durham were in the vanguard (along with Singapore and Norway) in promoting road user charging. These schemes grew from a legacy of almost no recent experience in charging for road use, except for a small number of tolled estuarial and river crossings, and some innovative flirtations with congestion charging trials, such as the Cambridge congestion charging scheme in the mid-1990s. The U.K. government is also now actively considering the feasibility of introducing a national road user charging system to fully or partially replace fixed car taxes and fuel duties. Other European, American, and Asian countries are actively considering similar schemes.

It is reasonable to predict that charging for road use could be implemented in some innovative way not yet considered by the transport community, particularly over the 50-year time horizon of the IIS project. That possibility motivated the project to bring together the expertise on smart markets, road pricing, transport modeling, and complex systems. The paradigm to be considered is how new technology, such as those investigated in the Foresight project, could enable new forms of road user pricing for demand management purposes [32].

Technological advances, such as in the speed of computer processing, in the real-time access to information, interactive communication in a parallel and distributed fashion, and in pervasive computing, have altered the scope and design of markets. Smart market solutions can now use electronic network technology to implement market protocols that can offer a new auction-based approached to bidding and paying for road use. As a consequence, Foresight funded a project to investigate future innovations in road user charging. The solution investigated by research teams from Essex, Cranfield, and Newcastle Universities was a road user charging regime that investigated and modeled a "Dutch auction" approach to road user charging, under the Smart Market Protocols for Road Transport (SMPRT) project.

The "cap and trade" solution was investigated for SMPRT, since it is increasingly used as means of controlling and pricing negative externalities from economic activity. The core of the Smart market in road slots is a capacity to obtain bids from potential road users that represent their maximum willingness to pay for a limited or capped supply of travel slots, in a given time slice through a cordon area of the congested road network. The parameters that determined the cap were derived from the VISSUM traffic microsimulator, which was used to probe traffic efficiency of the road system and define an optimal level of flow for the efficient operation of the road. In the future, more intelligent infrastructure and increased knowledge of traveler preferences, habits, and profile, could have a greater impact on the willingness to pay for a particular trip, thus opening up opportunities for new approaches to road user charging.

The "cap and trade" approach is increasingly being used as a means of controlling negative effects by assigning property rights to the negative economic activity. A landmark application of this concept arose with the Title IV of the 1990 Clean Air Act Amendments in the United States, aimed at reducing sulphur dioxide (SO_2) emissions from coal- and oil-fired electric generating plants [33–35].

In the case of the smart market protocol for road transport, the cap refers to the optimal level of congestion, which determines the fixed supply of travel slots.

This is determined by a road traffic microsimulator that can simulate traffic flows in real time, using measured historic data from the test network of Gateshead in Northeast England. In a real implementation of the system, this would be based on real-time data of the road network on a link-by-link basis. Figure 9.15 illustrates all the bids received by drivers to use the road network in a particular time period. X denotes the number of vehicles that bid and would be expected to use the road if there were no cap. X* represents the number of vehicles slots allowed in that time period. The price set for the cap is the lowest price bid by a driver that is above the capped level. All successful bidders will pay this P* price, irrespective of how much they may have bid above this level to ensure they have won a slot. An alternative approach (not tested here) would be to charge everyone the price they bid for a slot, which would curb high-price bids and make drivers think more precisely about the value of the journey slot for which they are bidding.

The cap approach in the SMPRT uses an auction-based protocol to manage demand, by setting a cap on the quantity of vehicles entitled to use the roads, by enabling potential users (drivers) to bid for a limited number of slots. The advantage of this approach is that the operator can control exactly the quantity of users on the network at any one time, thus delivering a quality service for the road network to all drivers who have won the auction. Such an approach is clearly unrealistic at the moment, but in the future, when road user charging is mainstream, innovations to basic road pricing may be desirable to meet a particular policy objective. As London has demonstrated, it may be necessary to increase the price of the

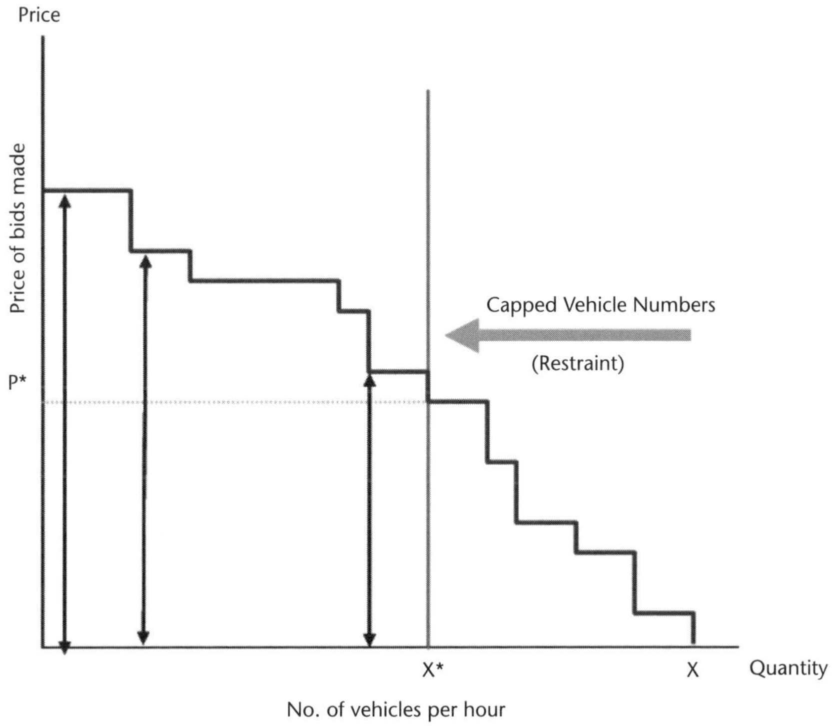

Figure 9.15 Smart markets cap and trade level, price versus quantity (of vehicles). (*Source:* P. Blythe, A. Markosi, and B. Allen, 2006.)

congestion charge (£5 to £8 within 2 years) to ensure that the demand restraint effect of the scheme is maintained.

Potential drivers submit an electronic (sealed) bid of what they are willing to pay to travel in a particular slot the following day. For example, if X drivers wish to travel, the cap is set to a lower value X*, then the price the bidders would pay is the lowest successful bid at the X* threshold. All the bidders who submitted a price above this threshold would pay this fixed lowest bid above the cap level. By understanding the demographics of the area, it is possible with some degree of accuracy to predict the proportion of drivers of different socioeconomic groups and forms of employment that would be affected by such a system. Different forms of bidding could be experimented with, including those that consider other external factors, such as environmental costs, in the setting of the charge [35].

This potentially gives the road operator an opportunity to cap the traffic on the road network at an economic optimum, or some other manageable or acceptable value either below or above the optimum, and to allow the market to set the monetary value of the cap.

The results of the project suggest that in the future, as technology develops, new, innovative, and more targeted charging regimes could be introduced in a practical manner, utilizing innovative algorithms and future developments in intelligent infrastructure.

9.5 Summary

Chapter 3 presented the three main competing technology approaches for future charging systems. Each has different attributes, advantages, disadvantages, requirements for in-vehicle equipment, and roadside infrastructure. For many years, short-range OBU/tag-based systems have been preferred, due to their simplicity of operation, potential for supporting additional services for vehicle users, and, most importantly, ease of comprehension and use. Technological advances have opened up new opportunities for innovative charging schemes using new wireless networks and new communications standards that may become mainstream in future generations of charging equipment.

Wide-area charging schemes, which rely on the in-vehicle equipment determining the location of the vehicle and charging the vehicle accordingly, are attractive, and offer new possibilities for charging without the main disadvantage of short-range charging systems (i.e., the associated roadside infrastructure at every charging and enforcement point). Some infrastructure is still required for enforcement purposes, but this can be situated in locations where aesthetics are not a prime consideration. Effective operation and enforcement using GPS-based systems was demonstrated in the Hong Kong charging trials (1998–1999), and in the operational German Lorry Charging scheme provided by the TollCollect consortium, as mentioned in Chapter 8. Looking to the future, there is likely to be a trend in many countries to gradually adopt some variant of the distance-based taxation of heavy goods vehicles, which could probably only be efficiently implemented using some form of wide-area charging, and probably linked to vehicles' digital odometers, as in Switzerland or by using GNSS, as in Germany.

Video-based charging is a very recent innovation, with London being the first large urban area to adopt such an approach. In Norway, video was used as the primary charging means in the cities of Kristiansand and Bergen; however, this was on a very small scale in comparison with London. For central London, the scheme has required a very complex back office clearing and management system, to register on a daily basis all those who wish to pay to use the charged area within the cordon, and to record and process the images of all vehicles recorded as entering the charged area, but who have not registered and paid. The main issue of the ANPR approach seems to be scalability of the solution.

Looking ahead, despite the general competition between suppliers, the short-term future evolution of charging is likely to be a fusion of DSRC and wide-area charging systems, which will be able to support several different charging configurations with one set of in-vehicle equipment. Most drivers seem to prefer a system that they can actually see working through some sort of display in the vehicle.

In the longer term, there is a school of thought evolving that is suggesting that advances in communications and wireless mobile networking technologies may actually cause a radical rethinking of how vehicle-to-vehicle and vehicle-to-infrastructure communications may evolve.

It has been more than 40 years since the U.K. government published its work "Traffic in Towns," known to professionals as "The Buchanan Report." This predicted massive growth of road transport, when the average level of car ownership in Europe was just over 100 (now 500) cars per 1,000 inhabitants. Nevertheless, the report indicated that urban design and new technologies may have a role to play. The Smeed Report on the Technical and Economic Possibilities of Road Pricing was published by the same government ministry a year later, in 1964. The intelligent infrastructure study and many of the new innovations in road user charging, described in this book, show that we still have a challenge to meet to efficiently manage traffic, collect charges for road use, and deal with the disadvantages of travel, congestion, and pollution. There is also the pressing need to mitigate the effects of greenhouse gases and their contribution to climate change. Along with possible future energy and fuel shortages, these are two of the most pressing problems (both with a transport component) the world must face over the coming decades. Pricing has a role to play, but how we take it a step further to try and wean us off our love affair with the car in the West, and to help the emerging economies to deal with traffic demand (which seems to be inextricably linked to economic growth) is a challenge in which those dealing with road pricing and innovative demand management schemes have a clear and leading role to play.

The challenge will be met, and this book will hopefully help some aspiring champion to come up with new innovations in road user charging and demand management. However, as the foreword to the Buchanan Report by Sir David Crowther so eloquently put it, 43 years ago [36]:

> We are nourishing a monster of great potential destructiveness. And yet we love him dearly . . . the motor car is clearly a menace that can spoil our civilisation. But translated into terms of the particular vehicle that stands outside the door, we regard it as one of our most treasured possessions or dearest ambitions, an immense

convenience, an expander of the dimensions of life, an instrument of emancipation, a symbol of the modern age. To refuse to accept the challenge it presents would be an act of defeatism . . . we must meet it without confusion of purpose, without timidity over means, and above all without delay.

References

[1] Frost and Sullivan, "North American Toll Collection Market," *Frost and Sullivan Market Analysis Report*, Pub ID: MC499392, May 2001. Authors' note: There are a number of unpublished sources which also provide input on the market size of ETC; however, for purposes of a publicly accessibility this particular reference is provided.

[2] Panianti, J., "Vehicle Infrastructure Integration: A New Initiative That Could Change the Face of Electronic Tolling in the United States," *Mapping the Future*, Ch. 4, Tollways, Winter 2005.

[3] Opiola, J., "The Vehicle Infrastructure Initiative," *Proc. IEE International Conference on Automotive Electronics*, London, U.K., March 2005.

[4] Kavener, D., T. McGuckin, and J. Crawford, *The Great Enabler? The Promise of WAVE/ DSRC*, Tolltrans, 2005, pp. 38–42.

[5] http://europa.eu.int/comm/dgs/energy_transport/galileo/intro/future_en.htm.

[6] Walker, J., (ed.), *Mobile Information Systems*, Norwood, MA: Artech House, 1990.

[7] Walker, J., (ed.), *Advances in Mobile Information Systems*, Norwood, MA: Artech House, 1998.

[8] Bruinsma, A. J. A., J. C. Henkus, and F. C. van der Mark, *Experiments with Prototype Infrared and Microwave Data Transmission Equipment for Rekening Rijden*, TNO Delft, the Netherlands, Tech. Rep. TPD-HAI-RPT-90-31, 1990.

[9] Schalk, A., "CALM Standard and Its Applications to Roadside to Vehicle Communications Systems," *Proc. IEE Automotive Electronics Conference*, London, U.K., March 2005.

[10] Prasad, R., and L. Munoz, *WLANs and WPANs Towards 4G Wireless*, Norwood, MA: Artech House, 2003.

[11] Prasad, R., and M. Ruggieri, *Technology Trends in Wireless Communications*, Norwood, MA: Artech House, 2003.

[12] Motes, 2004, http://www.willow.co.uk/html/wireless.html.

[13] Blythe, P. T., A. Tully, and G. Martin, "Investigating Next-Generation Wireless Technology to Deliver Pervasive Road User Charging," *Proc. 12th World Congress on Intelligent Transport Systems and Services*, San Francisco, CA, November 1995.

[14] CALM, *The Calm Handbook*, ISO TC204 and ETSI ERM TG37, Version 3, March 2006.

[15] http://www.serconline.org/payd/fact.html.

[16] http://www.norwichunion.com/pay-as-you-drive/index.htm.

[17] European Commission, "White Paper: European Transport Policy for 2010: Time to Decide," COM(2001) 370, September 2001.

[18] SEA, "D1: UOBU Requirements Discussion Document," SEA/05/TR/4887, November 2001.

[19] Europe News, "The Universal On Board Unit," December 2005, http://www.telematics update.com/SubPages.asp?News=47710.

[20] Estevan-Ubeda, N., *The Universal On-Board Unit Challenge*, Tolltrans, Traffic Technology International, 2005.

[21] Davies, P., and F. K. Sommerville, "Development of Heavy Vehicle Electronic Licence Plate Concept," *Transportation Research Record No. 1060*, 1986, pp. 121–127.

[22] Dodoo, N. A., and N. Thorpe, "A New Approach for Allocating Pavement Damage Between Heavy Goods Vehicles for Road-User Charging," *Journal of Transport Policy*, Vol. 12, No. 5, 2005, pp. 419–430.

[23] Dodoo, N. A., and N. Thorpe, "Towards Fair and Efficient Charging for Heavy Goods Vehicles," *Proc. 12th IEE Intl. Conference on Road Traffic Information and Control*, London, U.K., April 2004, pp. 231–236.

[24] Directive 2004/52/EC of the European Parliament and of the Council of April 29, 2004, on the Interoperability of Electronic Road Toll Systems in the Community.

[25] European Commission, "Draft Decision of the Commission on the Definition of the EETS," published as N1836 by CEN, April 12, 2006.

[26] European Commission, *Recommendations on Parameters to Be Stored in On-Board Equipment Designed for Use with the European Electronic Toll Service*, Report of Expert Group 2, 2005.

[27] Foresight IIS, "Intelligent Infrastructure Futures: Project Overview," U.K. DTI Publication 8153/2k/01/06/NP. URN 06/522, London, U.K., January 2006.

[28] http://www.foresight.gov.uk.

[29] Foresight IIS, "Intelligent Infrastructure Futures: Technology Forward Look," U.K. DTI Publication 8154/2k/01/06/NP. URN 06/520, London, U.K., January 2006.

[30] Blythe, P. T., "Intelligent Infrastructure," *Proc. IEE Intl. Conference on Automotive Electronics*, London, U.K., March 2006.

[31] Foresight IIS, "Intelligent Infrastructure Futures: The Scenarios Towards 2055," U.K. DTI Publication 8152/2k/01/06/NP. URN 06/521, London, U.K., January 2006.

[32] Blythe, P. T., "Intelligent Infrastructure and Smart Markets," *Proc. Conference on EU Prod User Charging*, Centaur Communications, London, U.K., January 2006.

[33] Schmalensee, R., et al., "An Interim Evaluation of Sulfur Dioxide Emissions Trading," *Journal of Economic Perspectives*, Vol. 12, No. 3, 1998, pp. 53–68.

[34] Joskow, P., R. Schmalensee, and E. Bailey, "The Market for Sulphur Dioxide Eimission," Vol. 88, No. 4, 1998, pp. 669–685.

[35] Mackie-Mason, J. K., and H. R. Varian, "Pricing Congestible Network Resources," *IEEE Journal of Selected Areas in Communications*, Vol. 13, No. 7, 1995, pp. 1141–1149.

[36] Buchanan, C., (ed.), *Traffic in Towns (The Buchanan Report)*, London, U.K.: Penguin & HMSO, 1963.

Glossary

2G	second generation mobile telephony
2.5G	an extension to 2G mobile telephony, such as GPRS (GSM)
3-DES	triple DES, a security encryption algorithm
3G	third generation mobile telephony, such as WCDMA
3GPP	Third Generation Partnership Project; the concept of a "Partnership Project" was pioneered by the European Telecommunications Standards Institute (ETSI) early in 1998 with the proposal to create a 3GPP focusing on global system for mobile (GSM) technology
4G	fourth generation mobile telephony
407 ETR	Electronic Toll Road 407 (Canada) (http://www.etr407.com)
A1	A project to assess the feasibility of interoperable full EFC systems and EFC applications and resulted in the publication of an interoperability specification with the same name (see http://www.cordis.lu/telematics/tap_transport/research/projects/a1.html)
A1+	an extension of the A1 specification to include on-board charging (maintenance of a balance held by an OBU or an ICC attached to the OBU)
A555	Köln-Bonn Motorway in Germany, where a trial of 10 different tolling systems was undertaken by the Federal Government in the mid-1990s, also known as the AGE trials
AAI	automatic account identification
AASHTO	American Association State Highway and Transportation Officials
ABE	agent-based economics
ACE	agent-based computational economics
ACM	automatic coin machine, for automatically processing and counting cash payments paid directly by road users on toll plazas
ADEPT	Automatic Debiting and Electronic Payment for Transport—an EU DRIVE II Research Project on advanced communications for road user charging (1991–1995); ADEPT II was a follow-on project

ADS	automatic debiting system
ADVICE	advanced vehicle classification and enforcement, a Fourth Framework European research project
AEI	automatic equipment identification (generic term although sometimes applies to assets such as shipping container)
AES 128	Advanced Encryption Standard, an encryption standard based on 128 bit keys
AFC	automatic fee collection (synonym for EFC) is the generic term for the procedure that allows data to pass between a device fitted to a vehicle moving at speed, and a fixed roadside charging station, as the vehicle passes, for the purpose of charging a toll. It is automatic in the sense that no action is necessary either by the driver or by the operator of the roadside equipment to achieve a transaction.
A-GPS	assisted global positioning system
AHS	automated highway systems
ALF	annual license fee
ALI	Autofarer Leitung und Informationsystem
ALI	automatic location identification
ALI-SCOUT	Autofarer Leitung und Informationsystem beacon-based dynamic route guidance system developed by Siemens and demonstrated in Berlin (see ALI)
Almanac	approximate position data for the satellite constellation
ALPR	automatic license plate recognition (usually for enforcement of electronic tolling systems)
ALS	Area Licensing Scheme—the Singapore paper licensing system used from 1975 to 1999
AN/LI	auto location number/location interface
ANI	automatic number identification
ANPR	automatic number plate recognition system (usually for enforcement of electronic tolling systems)
AOA	angle of arrival
AP	an Internet access point, an elemental part of a MANET
ASECAP	Association Européenne des Concessionnaires d'Autoroutes et d'Ouvrages à Péage (European Association of Companies with Concessions for Motorway, Bridge and Tunnel Tolls), http://www.asecap.com
ASETA	Spanish Association of Toll Road Operators
ASFA	L'Association des Sociétés Françaises d'Autoroutes et d'Ouvrages à Péage, the French toll road operators association
ASFINAG	Autobahnen und Schnellstrassen-Finanzierungs-Aktien Gesellschaft, the operator of the Austrian truck tolling scheme
ASIC	application-specific integrated circuit, a processing device within the OBU

ASTM	Formerly known as the American Society for Testing and Materials (ASTM), and currently known as ASTM International, it is a U.S.-based organization that develops and maintains technical standards for materials, products, systems, and services; and owns the E2213-03, part of the 5.9-GHz WAVE platform
ATD	absolute time difference
ATIS	advanced traveler information systems
ATMS	advanced traffic management systems; Singapore integrated management system, including the use of "floating vehicle" data
ATT	advanced transport telematics
Augmentation system	system that provides integrity and range correction data
Autograph	brand name of a PAYD system in Canada
AutoPASS	brand name for the Norwegian national road user charging system
Autopass	former name of an ETC services company in Hong Kong; since 1998, part of Autotoll Ltd.
AVC	automatic vehicle classification
AVI	automatic vehicle identification
AVL	automatic vehicle location
Bandwidth	width of the frequency band necessary for the transmission of information, or the bandwidth assigned to a channel
Bluetooth	short-range radio communications technology designed to replace wired local connections, usually using ISM frequencies around 2.4 GHz
BCD	binary coded decimal
BOOT	build, own, operate, and transfer
BOT	build, operate, and transfer
BST	beacon service table
BTS	base transmitter station, a term for a cellular mobile phone mast and its local controller
CALM	continuous air interface for long and medium range initiatives in DSRC
CAN	controller area network—a standard communications bus for cars and commercial vehicles
CAPTIVE	a European Fifth Framework project, dealing with cross-border enforcement and road traffic in Europe
CARDME	Concerted Action for Research on Demand Management in Europe (http://www.cardme.org)
CASH	Coordinated Action for the Standardisation of HADES, a Third Framework European project dealing with a European specification for tolling

CB Radio	citizen band radio, a two-way radio system often used by fleet and truck drivers
CBC	cell broadcast center
CBD	central business district
CC	congestion charging
CCTV	closed circuit television, a means of video-based surveillance
CCZ	congestion charging zone
CDMA	code division multiple access, a cellular communication standard used mainly in the United States
CEE	Central and Eastern Europe
CEN	European Committee for Standardization (Comité Européen de Normalisation, http://www.cenorm.be)
CENELEC	Comité Européen de Normalisation Electrotechnique
CEP	circular error probable
CEPT	Conférence Européene des Administrations des Postes et des Telecommunications—the European Committee of PTTs (in French)
CESARE	Common EFC System for ASECAP Road Tolling European System, the third part of which was launched in September 2003 as CESARE 3
CIH	Cross-Israel Highway
CISPR	The International Special Committee of the IEC on Radio Interference
CITIES	Cooperation for Integrated Traffic management and Information Exchange Systems
CJIB	Centraal Justitieel Incasso Bureau, the public prosecution services of the Netherlands, responsible for issuing, among other things, speeding fines
CL	current location
CLZ	Central London Zone
CN	cellular networks
COBI	common on-board interface
COE	Certificate of Entitlement (Singapore), a vehicle purchase quota system based on periodic auctions of the right to buy a vehicle
CONCERT	Cooperation for Novel city Electronic Regulation Tools
Congestion Charging	A road user charging scheme in which the toll fee varies depending on the level of congestion
Constellation	orbiting group of satellites
Contraflow	temporarily diverting one or more lanes of opposing traffic onto the opposite side of the road in segregated lanes
CP	charge point, applied to DSRC-based schemes, and includes charging and enforcement equipment unless otherwise stated

CPTC	California Private Transportation Company
CRM	customer relationship management
CS	commercial service; also central system—a generic term for the back office that is responsible for data gathering, customer relations, operations and maintenance, enforcement record processing, and reporting
CSC	customer service center
CTIA	Cellular Telecommunications & Internet Association (U.S.)
CVISN	Commercial Vehicle Information Systems and Networks refers to the ITS information system elements part of the U.S. commercial vehicle operations application.
CVO	commercial vehicle operations
DAB	digital audio broadcast
DBFO	design, build, finance, and operate—a public sector initiative to inject private capital into road-building
DCO	data clearing operator (OMISS definition)
Dead reckoning	measurement of distance traveled, usually by mechanical measurement of wheel, axle, or propshaft rotation
DES	Data Encryption Standard (a U.S. Government standard)
DETR	Department of the Environment, Transport and the Regions; transport-related duties now part of DfT
DfT	Department for Transport, a U.K. government agency, formerly known as DoT and DTLR (http://www.dft.gov.uk)
DG TREN	Directorate General for Transport and Energy, a directorate of the European Commission
DGPS	differential GPS
DIRECTS	Demonstration of Interoperable Road-user End-to-end Charging and Telematics Systems, a U.K. project sponsored by the Department for Transport to create an interoperability specification that could be used as the basis for a multivendor competitive EFC system procurements in the United Kingdom
DMV	Department of Motor Vehicles, local U.S. government agencies that register and record vehicle and ownership details
DOT	Department of Transport (generic), or U.S. Department of Transportation; U.K. Department of Transport, now renamed Department for Transport
DRC	Dartford River Crossing
DRIVE	dedicated road infrastructure for vehicle safety in Europe, a European research program on ITS (1987–1991)
DSRC	dedicated short-range communication, a standard for microwave- and infrared-based vehicle-to-roadside communications
DTC	Dartford Thurrock Crossing

DVLA	Driver and Vehicle Licensing Agency (United Kingdom)
E/AFLT	enhanced forward link trilateration (or advanced forward link trilateration), a method of location using measurements made by a terminal device (mobile phone or OBU) of fixed base station transmissions
EC	European community
E-CGI	enhanced cell global identity
E-CID	enhanced cell ID
EDI	electronic data interchange
EEA	European Economic Area
EETS	European Electronic Toll Service
EFC	electronic fee collection, a general term for a revenue collection scheme that uses a vehicle-to-roadside communication system for the secure transfer of account or user information to a fixed roadside system
EFTA	European Free Trade Area
EGNOS	European Geostationary Navigation Overlay System
EIRP	effective isotropic radiated power (i.e., radiated power relative to an isotropic source)
EMC	electromagnetic compatibility
EN	European norm (Standard)
ENCC	European Network of Certification Centers
ENP	electronic number plate
ENV	European prestandard
E-OTD	Enhanced Observed Time Difference (GSM), otherwise known as OTDOA (WCDMA/UMTS) or Enhanced Forward Link Trilateration (CDMA)
ERI	electronic registration identification (synonymous with EVI)
ERP	electronic road pricing
ERT	electronic registration tag (CEN ISO/TS 24534 Standard for Electronic Registration Identification)
ERTICO	Brussels-based organization with public and private members that collectively pursue the development and deployment of intelligent transport systems and services (ITS) in Europe
ESA	European Space Agency
ESSP	European Satellite Services Provider
ETC	electronic toll collection
ETSI	European Telecommunications Standards Institute (http://www.etsi.org), the home of all European telecoms standards
ETTM	electronic toll and traffic management
EU	European Union
EUREKA	European Research Coordination Agency

EUTELTACS	European Satellite Messaging and Positioning System
EVI	electronic vehicle identification (synonymous with ERI)
Externalities	impact of road use on others (including the environment), typically regarded as negative, including noise, accidents, delays due to congestion, pollution, and visual intrusion on the landscape
EZ-Pass	brand name of a multiagency ETC scheme in the northeastern United States (http://www.e-zpassiag.com)
FAIR lanes	Fast and Intertwined Regular lanes (U.S.); drivers using regular lanes during period of peak demand would be compensated with credits that could be used as payments for priced lanes
False positive	undetected error, such as (1) a high confidence ANPR record that contains an error, or (2) an erroneous bit stream from an OBU that contains an error
FAQ	frequently asked question
FCC	Federal Communications Commission (U.S.)
FDOT	Florida Department of Transport (http://www.dot.state.fl.us)
FHWA	Federal Highway Administration (U.S.)
FOV	field of view, the width of the capture zone of a camera
GDOP	geometric dilution of precision
GDP	gross domestic product
Geo-object	georeferenced data (i.e., an object such as a road, area, boundary or cordon that is described by its latitude and longitude)
GHz	gigahertz, a frequency of billion Hz
GIS	geographical information system
GJU	Galileo Joint Undertaking, a joint initiative between the European Commission (EC) and the European Space Agency (ESA), to provide Europe with its own independent civilian-controlled satellite navigation system
GLONASS	Global Orbiting Navigation Satellite System
GoL	Government office for London
GNSS	Global Navigation Satellite System, a generic term for any satellite-based location system
GNSS-1	first generation Global Navigation Satellite System (e.g., GPS + WAAS)
GNSS-2	second generation Global Navigation Satellite System
GPRS	general packet radio service (GSM-based)
GPS	global positioning system, a MEO constellation of 24 satellites that that circles the Earth every 12 hours, and constantly transmits location and time of day from on-board atomic clocks, allowing mobile receivers to determine their positions
Grade, at	intersection that links roads of the same type (grade) together

GSM	Groupe Spéciale Mobile, European cellular mobile phone standard for 2G and 2.5G systems and name of mass market personal wireless communication service developed in Europe
GSS	global specification for short-range communication, (http://www.kapsch.se/comweb/inter_o/pdf/GSS_30.pdf)
GST	general sales tax, equivalent to value added tax
HADES	High level Automatic Debiting European Specification, a Second Framework European research project that brought together the expertise of the ADEPT and VITA projects
HC-SDMA	High Capacity—Spatial Division Multiple Access, a communications access protocol
HCV	heavy commercial vehicle (Melbourne City Link definition)
HDOP	horizontal dilution of precision
HELP	Heavy Vehicle Electronic License Plate Program, a North American AVI systems for trucks and HGVs for interstate-border toll-payments
HGV	heavy goods vehicle
HMI	human-machine interface, synonymous with man-machine interface, the interface with the user of the equipment
HOSDB	Home Office Scientific Development Branch (U.K.), part of the Home Office that advises on scientific and technical policing issues
HOT	high occupancy and toll (lane)
HOV	high-occupancy vehicle lane, a traffic lane that can only be used by vehicles with a certain number of occupants (e.g., two passengers and driver)
HPMS	highway performance monitoring system
Hydraulic bollard	hydraulically operated post placed in the center of a travel lane to prevent unauthorized vehicle access
Hypothecation	A "pledge" in the context of RUC usually meant as ring-fencing net funds collected by a scheme to be applied specifically to support related operations or transport services
IAG	Interagency Group, U.S. tristate area, New York City, New York State, and Connecticut (http://www.e-zpassiag.com)
IAT	inertial aided technology, a means of determining change of direction based on measurement of acceleration (e.g., by using solid state gyroscopes)
IBTTA	International Bridge, Tunnel and Turnpike Association (http://www.ibtta.org)
ICC	integrated circuit(s) card, a term often used for a smart card
ICT	information and communications technologies
ID	identification
IEC	International Electrotechnical Commission

IEEE	Institute of Electrical and Electronics Engineers, Inc. (U.S.)
IIS	intelligent infrastructure systems
INS	inertial navigation systems
Intelligent client	OBU that is capable of estimating its position and matches this to on-board geodata of road segments
INTELSAT	International Telecommunication Satellite Organization
IP	Internet Protocol; also intellectual property
IPv6	Internet Protocol version 6, a derivative of IP specifically designed for the mobile environment
IR	infrared, a DSRC method
IrDA	Infrared Data Association
IRTE	Integrated Road Traffic/Transport Environment
ISM	industrial, scientific, and medical frequency band, a series of internationally recognized frequency bands that allow unlicensed communications services
ISO	International Organization for Standardization, the world's largest developers of standards (http://www.iso.ch)
ISO9001	internationally recognized standard for quality management of businesses, applicable to any product, process, or services delivery anywhere in the world
ISTEA	Intermodal Surface Transportation Efficiency Act
IT	information technology
ITIL	IT infrastructure library
ITS	intelligent transport systems
ITS AMERICA	Intelligent Transportation Society of America
ITU	International Telecommunication Union, the world body for telecommunications standards. The ITU now comprises a general secretariat, concerned with policy and strategic issues, and three sectors: Radiocommunication (ITU-R), Telecommunication Standardization (ITU-T), and Telecommunication Development (ITU-D)
IU	in-vehicle unit (Singapore)
IVE	in-vehicle equipment, an IVU that includes an integrated smart card reader and smart card
IVHS	Intelligent Vehicle-Highway System (now superseded by ITS)
IVR	interactive voice response
IVU	in-vehicle unit, and when integrated with an ICC reader and ICC, known as an IVE
JPO	ITS Joint Program Office of the FHWA, the coordinating body for U.S. ITS activities
KPI	key performance indicator
KSI	killed and serious injuries, a term for recording road accidents with human death or injury

LAN	local area network
LBS	location-based services
LCD	liquid crystal display
LCS	location services
LCV	light commercial vehicle
LED	light emitting diode
LES	location-enhanced mobile services
LKW Maut	LastKraftWagen Maut, literally "Truck Toll," usually meaning electronic toll collection system for heavy goods vehicles
LOBU	low-use on-board unit (U.K. LRUC scheme)
LPN	license plate number, the character set displayed on the license plate (otherwise known as number plate)
LRUC	lorry road user charging, an initiative by HM Revenue and Customs to charge heavy goods vehicles according to distance driven on U.K. roads
LSVA	Swiss Customs Authority
LTA	Land Transport Authority, executive agency in Singapore responsible for the ALS and road pricing schemes (http://www.lta.gov.sg)
MANET	working group on mobile ad hoc networking within the Internet Engineering Task Force (IETF), a collection of mobile computing devices that cooperate to form a dynamic multihop network; also means ad hoc network
MÅNS	achieving interoperability between the Nordic payment means for road user charges
Map-matching	Map-matching is used in satellite-based navigation and charging systems, to compensate for errors or uncertainties in the position as determined by GNSS; if a vehicle is known to be on a road, then the position as determined by the satellite system can be compared to a digital map, and adjusted to place the vehicle on the nearest road
MCLP	Melbourne City Link Project, Melbourne, Australia (http://www.citylink.com.au)
MEDIA	Management of EFC DSRC Interoperability in the Alpine area, a joint project of toll operators in France, Italy, Switzerland, Austria, and Slovenia
MEL	Midland Expressway Limited, the operator of the M6 Toll road, the United Kingdom's first privately funded motorway
MEO	medium Earth orbit, the orbital band of satellites between 1,600 and 15,000 miles above the Earth (e.g., GPS at 12,500 miles)
MERCOSUR	Mercado Común del Sur (Spanish) or Mercado Comum do Sul (Portugese), a trading zone comprising many of the South American nations

Metadata	"data about data," or information about another set of data, such as a vehicle's license plate number (metadata) extracted from an image (data) showing the license plate
MGV	medium goods vehicle
MGW	maximum gross weight, the maximum, fully loaded weight for which a vehicle is designed, often indicated on the vehicle or its registration documents
MHz	megahertz, frequency of million Hz
MIS	management information system
MISTER	Minimum Interoperability Specification for Tolling on European Roads
MLFF	multilane free-flow
MMI	man-machine interface, synonymous with human-machine interface, the interface with the user of the equipment
MNO	mobile network operator
MOPTT	Ministry of Public Works, Transport and Telecommunications, Santiago de Chile, Chile (http://www.moptt.cl)
Mote	literally "a speck of dust in the eye," used to describe a small device capable of establishing dynamic ad hoc connections with other similar devices for the purposes of message exchange
MOU	memorandum of understanding
MOVE-it	Motorway Operators Validate EFC for Interoperable Transport, a Fourth Framework European project
MPS	mobile positioning system
MPT	Ministry of Posts and Telecommunications
MSAS	Multifunctional Satellite Augmentation System
MTA	Metropolitan Transporation Authority
MTO	Ministry of Transportation
MTT	manual toll terminal, a keyboard/display used by a toll collection officer in manual lanes
NAFTA	North American Free Trade Agreement
NEXTEA	National Economic Crossroads Transportation Efficiency Act
NJTA	New Jersey Turnpike Authority (http://www.state.nj.us/turnpike)
NTCIP	National Transportation Communications for ITS Protocol (U.S.)
NY MTA	New York Metropolitan Transportation Authority (http://www.mta.nyc.ny.us)
NYSTA	New York State Thruway Authority (http://www.thruway.state.ny.us)
O&M	operations and maintenance (otherwise known as "OAM")
OAM	operations and maintenance (otherwise known as "O&M")
OBC	outline business case

OBE	on-board equipment, a general term for ITS subsystems located or integrated within the vehicle, which interact with roadside charging and enforcement functions
OBU	on-board unit, a monolithic DSRC communication device, or an OBE without an ICC
OCR	optical character recognition, a process within an ANPR system that automatically reads alphanumeric text from an image of a vehicle license plate, or finds the plate within the image
OECD	Organization for Economic Cooperation and Development
OEM	original equipment manufacturer (e.g., a car manufacturer)
OGC	Office of Government Commerce, a department of the U.K. Treasury (Finance Ministry)
OMIS	Open Minimum Interoperability Specification, an interoperability specification created and published by DfT, based substantially on OPMIS, defining the minimum requirements for end-to-end system interoperability
OMISS	Open Minimum Interoperability Specification Suite
OPMIS	Open Preliminary Minimum Interoperability Specification, a document from the U.K. DIRECTS project, defining the minimum requirements for end-to-end system interoperability
OPP	Ontario Provincial Police
ORSP	on-road service provider (OMISS definition)
ORT	open road tolling
OS	open service
PAMELA	Pricing and Monitoring Electronically of Automobiles, a Second Framework research project under the European DRIVE program delivering the fundamental research for DSRC
PAN	personal account number (U.K. DfT)
PANYNJ	Port Authority of New York and New Jersey (http://www.panynj.gov)
PATH	Partners for Advanced Transit and Highways, formerly Programs on Advanced Technology for Highways
PAYD	pay-as-you-drive, a vehicle insurance product
PBC	provisional business case
PCF	position calculation function
PCN	penalty charge notice, issued as part of an escalating penalty charge regime by Transport for London as part of the London Congestion Charging scheme, to vehicle owners recorded as nonpayers
PDA	personal digital assistant, a handheld computer running applications, including manual data logging
PFI	private finance initiative
PIN	personal identification number

PISTA	Pilot on Interoperable Systems for Tolling Applications
PKI	public key infrastructure
PLG	private and light goods
PLS	personal location systems
PNR	private nonresidential, refers to a class of parking spaces in the United Kingdom, on which local authorities can seek powers to levy a local tax
PPP	public private partnership
PREMID	Programmable Remote Identification, the brand name for a Swedish vehicle-to-roadside communications family of products
PROMETHEUS	Programme for a European Traffic with Highest Efficiency and Unprecedented Safety
Pseudolites	fixed transmitters that broadcast ranging information to mobile devices, such as OBUs
PSP	payment service provider (OMISS definition)
PTC	Pennsylvania Turnpike Commission (http://www.paturnpike.com)
QAP	quality assessment plan, or quality assurance plan
QMS	quality management system
QoS	quality of service
RCI	road charging interoperability, an ERTICO-led program to define and enable a framework for interoperability of RUC schemes in Europe
Red Team	Defined by DoD Directive 3600.3 DoD Information Operations Red Teaming as "an independent, threat-based, and simulated opposition force that uses passive, active, technical, and non-technical capabilities on a formal, time-bounded basis to expose and exploit information system vulnerabilities of friendly forces"; used here as an independent review panel that critically evaluates a draft bid against bid requirements and achievement of stated strategic objectives
RF	radio frequency
RFID	radio frequency identification
Rising curb	heavy duty platform in the shape of a solid wedge, hinged on one edge, that can be lowered to be flush with the surface of the road, permitting authorized vehicle access
ROCOL	Review of Charging Options for London, a report published in 2000 that considered the feasibility of alternative approaches to charging and enforcement in London
ROI	return on investment, the ratio of an amount gained or lost relative to a reference level, used to measure the performance of an investment or to compare alternative investment opportunities

RRLP	Radio Resource Location Services Protocol
RSE	roadside equipment, equivalent to RSS
RSS	roadside system, equivalent to RSE
RTD	real-time difference
RTK	real-time kinematics
RTT	road transport telematics
RUC	road user charging, otherwise known as road use charging
RZ	restricted zone
SANRA	South African National Road Agency Ltd., formerly the National Roads Agency (NRA) (http://www.nra.co.za)
SAW	surface acoustic wave
Section Control	speed enforcement regime practiced in the Netherlands that is based on the measurement of average speed over a measured distance
Shadow Tolling	An approach to funding road operations, which are payments by sponsoring public agencies to operating concessionaires based on measured traffic volumes and achieved service levels
SJTA	South Jersey Transportation Authority (http://www.sjta.com)
S/M	master-slave
SMLC	serving mobile location center
SMPRT	Smart Market Protocols for Road Transport
SMS	short message service, a GSM service typically used for transmission of messages between mobile terminals or between other GSM-capable devices
SNRA	Swedish National Road Administration (Vägverket) (http://www.vv.se)
SoL	Safety of Life, a specific service from Galileo that guarantees a minimum service level signal
SOV	single occupancy vehicle
SPV	special purpose vehicle, a company set up for a specific, limited task
SSL	secure socket layer
SUC	Santiago Urban Concessions
SUV	sports utility vehicle
TA	timing advance
TANFB	Taiwan Area National Freeway Bureau (http://www.freeway.gov.tw)
TCP/IP	Transmission Control Protocol/Internet Protocol
T-DES	Triple Data Encryption Standard, otherwise known as 3-DES
TDOA	time difference of arrival
TEN-T	Trans-European Transport Network
TERN	Trans-European Road Network

TfL	Transport for London
Thick client	See intelligent client
Thin client	An OBU that estimates its position, temporarily caches this information on-board, and whenever possible reports this information, with corresponding time stamps, to a central system for matching with a map database
TIS	Télépeage Inter-Sociétés (France), an initiative of the French toll road operators to have a national multivendor interoperable electronic tolling system
TOA	time of arrival
TORG	Transport Operations Research Group (University of Newcastle)
Transceiver	transmitter and receiver
Transit NZ	Transit New Zealand, the state agency responsible for all of New Zealand's state roads (10,900 km) (http://www.transit.govt.nz)
TRB	Transportation Research Board
TripSense	A PAYD trial in the United States
TSP	transport service provider
TTFF	time to first fix (GPS)
TTP	trusted third party, an organization responsible for the issue of keys or digital signatures
UHF	ultrahigh frequencies, in the region from 300 MHz to 3 GHz
UMTS	Universal Mobile Telecommunications Service, the ETSI third generation mobile radio standard
UNECE	United Nations Economic and Social Council
UOBU	universal on-board unit, a European Commission initiative to develop a specification for a generic OBU
USAP	Union des Societes d'Autoroutes a Péage
USDOT	United States Department of Transportation
UWB	ultrawideband, a spread spectrum communications technology, originally designed for military applications, and capable of precise TOA localization within the footprint of a multiple array of UWB receivers
VAS	value-added services
VAT	value-added tax, equivalent to GST
VDC	vehicle detection and classification
VDOP	vertical dilution of precision
VED	vehicle excise duty, an annual payment in the United Kingdom, for use of the public road network, otherwise known as annual car tax
VERA	Video Enforcement for Road Authorities
VERTIS	Vehicle, Road & Traffic Intelligence Society (Japan)

VES	video enforcement system
VICS	Vehicle Information and Communication System (Japan)
Video tolling	the use of digital imaging to trigger a charging event, which captures multiple images of a vehicle's passage, to reduce the potential false positive rate of automatic number plate recognition
VII	vehicle infrastructure integration
VIN	vehicle identification number
VISSUM	traffic microsimulation model from PTV
VITA	Vehicle Information and Transaction Aid, a Second Framework research project under the European DRIVE program, the first to try to bring together operator requirements for interoperable tolling
VMS	variable message sign
VMT	vehicle miles traveled
VP	value pricing
VPS	vehicle positioning system
VR	vehicle registration
VRM	vehicle registration mark, the character set displayed on the license plate or number plate, equivalent to license plate number
VST	vehicle service table
WAAS	Wide Area Augmentation System
Walled Garden	artificially restricting a user's choice of applications to those selected by the service provider, allowing the user to play anywhere, but only within the "walled garden"
WAN	wide area network
WAVE	Wireless Access for Vehicular Environments, a track 2 activity of the VII initiative in the United States; in parallel, the Omni-Air consortium is creating a U.S. standard (802.11p) for DSRC and a related equipment certification program "to enable the national (U.S.) deployment of effective, interoperable 5.9 GHz DSRC systems" (see http://www.omniair.org/mission.html)
WCDMA	wideband CDMA (otherwise known as UMTS)
WEZ	Western Extension, a geographic extension of the CLZ to the West in London
WIMP	weigh-in motion platform, a dynamic HGV weighing platform used in the HELP project in the United States and Canada
Windage	effect of wind on an object, such as fixed infrastructure to support enforcement or charging equipment
WLAN	wireless local area network
WLS	wireless location service
WRC	ITU World Radiocommunication Conference
ZigBee	wireless ad hoc communications standard operating within the 2.45-GHz ISM frequency band

About the Authors

Andrew T. W. Pickford is the principal of Transport Technology Consultants, located in Cambridge, United Kingdom. He is a chartered engineer and holds a B.Sc. in electrical engineering from the University of Bristol and an M.B.A. from Warwick Business School. Since managing the United Kingdom's first electronic toll collection scheme in 1988, he has been a management team member of three technology start-up companies in this field and has been instrumental in the development of charging and enforcement solutions for some of the most prominent multilane free-flow and congestion charging schemes in the world. His work includes development of future evidential strategies, drafting new legislation for road user charging, and evaluation of novel vehicle detection methods. He also chairs the ITS (United Kingdom) Interest Group on Road User Charging.

Philip T. Blythe is a professor of intelligent transport systems and the director of the Transport Operations Research Group, in the School of Civil Engineering and Geoscience at the University of Newcastle upon Tyne, United Kingdom. He is a chartered engineer and works mainly at the interface between ITS and charging technologies and policy. He has been involved in the research and development of road user charging schemes for 20 years, including the development of the world's first DSRC multilane free-flow system in 1991. He has advised governments and agencies around the world on pricing policies and technology options, and continues to lead innovative research on road user charging, currently using wireless ad hoc networks as an alternative technology. He also played a leading role in the recent U.K. government's groundbreaking future Foresight study on intelligent infrastructure systems and their development over the next 50 years.

Index

Recent Titles in the Artech House ITS Series

John Walker, Series Editor

Advanced Traveler Information Systems, Bob McQueen, Rick Schuman, and Kan Chen

Advances in Mobile Information Systems, John Walker, editor

Fundamentals of Intelligent Transportation Systems Planning, Mashrur A. Chowdhury and Adel Sadek

Incident Management in Intelligent Transportation Systems, Kaan Ozbay and Pushkin Kachroo

Intelligent Transportation Systems Architectures, Bob McQueen and Judy McQueen

Intelligent Vehicle Technology and Trends, Richard Bishop

Introduction to Transportation Systems, Joseph Sussman

ITS Handbook 2000: Recommendations from the World Road Association (PIARC), PIARC Committee on Intelligent Transport (edited by Kan Chen and John C. Miles)

Positioning Systems in Intelligent Transportation Systems, Chris Drane and Chris Rizos

Road User Charging and Electronic Toll Collection, Andrew T. W. Pickford and Philip T. Blythe

Sensor Technologies and Data Requirements for ITS, Lawrence A. Klein

Smart Highways, Smart Cars, Richard Whelan

Tomorrow's Transportation: Changing Cities, Economies, and Lives, William L. Garrison and Jerry D. Ward

Vehicle Location and Navigation Systems, Yilin Zhao

Wireless Communications for Intelligent Transportation Systems, Scott D. Elliott and Daniel J. Dailey